CONCRETE PORTABLE HANDBOOK

R. Dodge Woodson

AMSTERDAM • BOSTON • HEIDELBERG • LONDON
NEW YORK • OXFORD • PARIS • SAN DIEGO
SAN FRANCISCO • SINGAPORE • SYDNEY • TOKYO
Butterworth-Heinemann is an imprint of Elsevier

Butterworth-Heinemann is an imprint of Elsevier
225 Wyman Street, Waltham, MA 02451, USA
The Boulevard, Langford Lane, Kidlington, Oxford, OX5 1GB, UK

Library of Congress Cataloging-in-Publication Data
Woodson, R. Dodge (Roger Dodge), 1955-
Concrete portable handbook / R. Dodge Woodson.
p. cm.
Includes bibliographical references and index.
ISBN 978-0-12-382176-8 (alk. paper)
1. Concrete—Handbooks, manuals, etc. I. Title.
TA439.W584 2011
624.1'834—dc23
2011016242

British Library Cataloguing-in-Publication Data
A catalogue record for this book is available from the British Library.

Printed and bound by CPI Group (UK) Ltd, Croydon, CR0 4YY

Typeset by: diacriTech, India

This book is dedicated to my daughter, Afton, and her new husband, Giovanni. May Mr. and Mrs. Sinclair enjoy a healthy, happy, productive marriage for their future years.

Short Contents

Full Contents

Acknowledgments

I would like to thank the following sources for providing art and illustrations contained in this book: The United States Government; BigStock Photos.

About the Author

R. Dodge Woodson is an internationally known best-selling author who has written in excess of 100 nonfiction books. He is a licensed general contractor and a licensed master plumber who has built as many as 60 single-family homes a year. In addition to being a builder, he is a well-known remodeling contractor and plumbing contractor who has owned his own businesses since 1979. Additionally, Woodson is accredited as an expert witness and does consulting on construction and plumbing litigations. He lives in a remote part of Maine on 150 acres and enjoys being surrounded by nature and wildlife. Outdoor photography is his passion.

Introduction

Are you looking for a comprehensive guide to working with concrete? If so, you have found it. Take a moment to scan the table of contents. You will quickly see that this is a true handbook for concrete professionals.

R. Dodge Woodson is an internationally known expert in the field of construction. He has written dozens of professional reference books and worked for over 30 years in the field as a contractor and tradesman. Additionally, attorneys call upon Woodson to serve as an expert witness in cases involving construction matters. Without a doubt, Woodson is an authority in the field of construction.

Do you want to see case studies that provide proof of concrete performance? It is here. Are you involved in new construction? Check out the first half of this book. The second half will walk you through the repair and rehabbing techniques of concrete. Open the pages, it is all here.

CHAPTER 1

Basic Information

OUTLINE

Who controls the codes that you are to work with? Where are the minimum standards of good practice for concrete work found? Your local code enforcement office will mandate the code requirements in your personal jurisdiction. It is very likely that your local code requirements will be based on the recommendations and requirements set forth by the American Concrete Institute® (ACI).

I talked with people at the International Code Conference (ICC) and confirmed that the ICC codes are based around the ACI requirements for both structural concrete and reinforced concrete applications. Essentially, the ACI requirements are the foundation of most building codes.

There are some exceptions to the coverage of the ACI recommendations. For example, soil-supported slabs are often exempt from the rules of the ACI, but this is not always the case. If a slab transmits vertical loads or lateral forces from other portions of a structure to the soil, the slab requirements can be tied to the ACI regulations.

ACI regulations do not apply to the installation or design of structural concrete slabs cast on stay-in-place, composite steel form decks.

Areas subject to seismic risk levels are commonly governed by the local building code in conjunction with the ACI recommendations. Earthquake-resistance building principles are generally adopted from the ACI.

PAPERWORK

There are a host of requirements that must be met prior to the approval of a permit and these documents must bear the seal of a registered engineer or architect. Common elements required as part of an official submittal may include, but not be limited to, the following:

- Name and date of issue of the code requirements used to determine drawings and specifications
- Any supplement to the code used in the design and specifications
- Specified compressive strength of concrete at stated ages
- Each section of the structure has a compressive strength of concrete at stated stages of construction
- Specified strength or grade of reinforcement
- Size of all structural elements, reinforcement, and anchors
- Location or specified structural elements, reinforcement, and anchors
- Provision for dimensional changes that may result from creep, shrinkage, or temperature
- Magnitude of prestressing forces
- Location of prestressing forces
- Anchorage length of reinforcement
- Location and length of lap splices
- Types of mechanical and welded splices of reinforcement to be used
- Location of mechanical and welded splices of reinforcement to be used
- Details and location of all contraction or isolation joints specified for plain concrete
- Minimum compressive strength of concrete at the time of post tensioning
- Stressing sequences of post-tensioning tendons
- Statements pertaining to slabs on grade that are designed as structural diaphragms

Calculations that pertain to the design of concrete applications are filed with the drawings and specifications for a job when permit application is made. Computer-generated information is normally acceptable for calculations.

If model analysis is used, the process should be performed by an experienced engineer or architect.

INSPECTION

Code office inspections of concrete installations are required to assure that minimum standards are met. Workmanship is a key factor in the success of concrete construction. The best materials and designs will be inadequate if they are not used properly in an installation. On the rare occasions when a building code official is not available in a region, inspections may be done by qualified, registered design professionals or qualified inspectors.

Tip

Records of inspections are usually required to be kept by the inspecting engineer or architect for two years beyond the completion of construction.

When inspectors review a site location there are a number of components included in a proper inspection. The following list outlines common objectives reviewed during an official inspection:

- Quality of concrete materials
- Proportions of concrete materials
- Strength of concrete materials
- Construction methods and materials of forms
- Removal of forms and reshoring
- Placement of reinforcement and anchors
- Mixing concrete
- Placing concrete
- Curing of concrete
- Sequence of erection and connection of precast members
- Tendon tensioning
- Significant construction loadings on completed concrete structures
- General progress and workmanship

Tip

Ambient temperature changes below 40°F or above 95°F require that a record be kept of all concrete temperatures and any and all protection given to concrete during placement and curing.

SPECIAL CONSIDERATIONS

Sometimes a situation arises where a system designer wishes to stray from the traditional code requirements. When this is the case, design specifications can be brought to the attention of a local code officer. It may be necessary for special approvals to be requested from a board of examiners who are appointed by building officials. This board is typically made up of qualified engineers.

The board will consider testing procedures, load factors, deflection limits, and similar types of pertinent data.

CHAPTER

2

Concrete Materials

OUTLINE

Concrete materials are obviously a key element in concrete construction, and building officials have the power to require testing of any of these concrete materials. The purpose of this testing is to ensure that all of the materials meet the quality specified when applying for a permit. The results of these tests are generally required to be available for review for at least two years after construction is completed.

When it comes to discussing approved types of cement, specifications and standards will come from the American Standard Testing Methods (ASTM).

EXHIBIT: OVERVIEW OF PORTLAND CEMENT AND CONCRETE

Overview of Portland Cement and Concrete

Although the terms "cement" and "concrete" are often used interchangeably, cement is actually an ingredient of concrete. Cements are binding agents in concretes and mortars. Concrete is an artificial rock-like material, basically a mixture of coarse aggregate (gravel or crushed stone), fine aggregate (sand), cement, air, and water. Portland cement is a general term used to describe a variety of cements. Because they are hydraulic cements, they will set and harden by reacting chemically with water through hydration.

Current (2004) world total annual production of hydraulic cement is about 2 billion metric tons (Gt), with production spread unevenly among more than 150 countries. This quantity of cement is sufficient for about 14–18 Gt/year of concrete (including mortars), and makes concrete the most abundant of all manufactured solid materials. The current yearly output of hydraulic cement is sufficient to make about 2.5 metric tons per year (t/year) of concrete for every person worldwide (van Oss, 2005).

Cement and Cement Manufacturing

Hydraulic cements are the binding agents in concretes and most mortars and are thus common and critically important construction materials. Hydraulic cements are of two broad types: those that are inherently hydraulic (i.e., require only the addition of water to activate), and those that are pozzolanic. The term pozzolan (or pozzolanic) refers to any siliceous material that develops hydraulic cementitious properties in the presence of lime [$Ca(OH)_2$]. This includes true pozzolans and latent cements. The difference between these materials is that true pozzolans have no cementitious properties in the absence of lime, whereas latent cements already have some cementitious properties, but these properties are enhanced in the presence of lime. Pozzolanic additives or extenders can be collectively termed supplementary cementitious materials (SCM; van Oss, 2005).

Portland cement is the most commonly manufactured and used hydraulic cement in the United States (and the world). It is manufactured through the blending of mineral raw materials at high temperatures in cement rotary kilns. Rotary kilns produce an intermediate product called "clinker." Clinker is ground to produce cement. By modifying the raw material mix and, to some degree, the temperature of manufacture, slight compositional variations in the clinker can be achieved to produce Portland cements with varying properties.

Similar varieties of Portland cement are made in many parts of the world but go by different names. In the United States, the different varieties of straight Portland cement are denoted per the ASTM Standard C-150 as:

- Type I: general use Portland cement. In some countries, this type is known as ordinary Portland cement.
- Type II: General use Portland cement exhibiting moderate sulfate resistance and moderate heat of hydration.
- Type III: High-early-strength Portland cement.
- Type IV: Portland cement with low-heat hydration.
- Type V: Portland cement with high sulfate resistance.

For Types I, II, and III, the addition of the suffix A (e.g., Type IA) indicates the inclusion of an air-entraining agent. Air-entraining agents impart a myriad of tiny bubbles into the concrete containing the hydrated cement. This offers certain advantages to the concrete, such as improved resistance to freeze-thaw cracking. In practice, many companies market hybrid Portland cements; Type I/II is a common hybrid that meets the specifications of both Types I and II. Another common hybrid is Type II/V.

Blended Cements

Blended cements (called composite cements in some countries) are intimate mixes of a Portland cement base (generally Type I) with one or more SCM extenders. The SCMs make up about 5–30% by weight of the total blend, but can be higher.

In blended cements, the SCMs (or pozzolans) are activated by the high pH resulting from the hydroxide ions released during the hydration of Portland cement. The most commonly used SCMs are volcanic ashes called pozzolana, certain types of fly ash (from coal-fired power plants), ground-granulated, blast-furnace slag (GGBFS) — now increasingly being referred to as slag cement — burned clays, silica fume, and cement kiln dust (CKD). In general, incorporation of SCMs with

Portland cement improves the resistance of the concrete to chemical attack, reduces the concrete's porosity, reduces the heat of hydration of the cement (not always an advantage), potentially improves the flowability of concrete, and produces a concrete having about the same long-term strength as straight Portland cement-based concretes. However, SCMs generally reduce the early strength of the concrete, which may be detrimental to certain applications (van Oss, 2005).

Blended cements can be prepared at a cement plant for sale as a finished blended cement product or can be blended into a concrete mix. Most of the SCM consumption by U.S. concrete producers is material purchased directly for blending into the concrete mix. Concrete producers in the United States buy relatively little finished blended cement.

The designations for blended cements vary worldwide, but those currently in use in the United States meet ASTM Standard C-595, C 989, or C-1 157. ASTM Standard C-595 defines several types of blended cements. The main designations include (van Oss, 2005):

- Portland blast-furnace slag cement (IS): Contains 25–70% GGBFS.
- Portland-pozzolan cement (IP and P): Contains a base of Portland and/or IS cement and 15–40% pozzolans.
- Pozzolan-modified Portland cement (I(PM)): The base is Portland and/or Type IS cement with a pozzolan addition of less than 15%.
- Slag-modified Portland cement (I(SM)): Contains less than 25% GGBFS.
- Slag cement (S):[1] GGBFS content of 70% or more. Type S can be blended with Portland cement to make concrete or with lime for mortars; the latter combination would make the final cement a pozzolan-lime cement.

Chemical Composition of Portland Cement

Modern straight Portland cement is a very finely ground mix of Portland cement clinker and a small amount (typically 3–7%) of gypsum (calcium sulfate dihydrate) and/or anhydrite (calcium sulfate). Cement chemistry is generally denoted in simple stoichiometric shorthand terms for the major constituent oxides. Table 2.1 provides the shorthand notation for the major oxides in the cement literature. It also shows the typical chemical composition of modern Portland cement and its clinker. For clinker, the oxide compositions would generally not vary from the rough averages shown by more than 2–4%. The oxide composition of Portland cement would vary slightly depending on its actual gypsum fraction or whether any other additives are present.

Mineralogy of Portland Cement and Its Clinker

The major oxides in clinker are combined essentially into just four cement or clinker minerals, denoted in shorthand: tricalcium silicate or "alite" (C_3S), dicalcium silicate or "belite" (C_2S), tricalcium aluminate (C_3A), and tetracalcium aluminoferrite (C_4AF). These formulas represent averages, ignoring impurities commonly found in actual clinker. It is the ratios of these four minerals (and gypsum) that determine the varying properties of different types of Portland cements. Table 2.2 provides the chemical formulas and nomenclature for the major cement oxides as well as the function of each in cement mixtures.

As indicated in Table 2.2, some of the minerals in clinker serve different functions in the manufacturing process while others impart varying final properties to the cement. The proportion of C_3S, for example, determines the degree of early strength development of the cement. The "ferrite"

[1]True Type S cements are no longer commonly made in the United States. Instead, the name slag cement (but with no abbreviation) is now increasingly given to the unblended 100% GGBFS product (van Oss, 2005). ASTM C 989 now governs slag cement (GGBFS).

TABLE 2.1 Chemical Shorthand and Composition of Clinker and Portland Cement

Oxide Formula	Shorthand Notation	Percentage by Mass in Clinker	Percentage by Mass in Cement*
CaO	C	65	65.0
SiO_2	S	22	22.0
Al_2O_3	A	6	6.0
Fe_2O_3	F	3	3.0
MgO	M	2	2.0
$K_2O + Na_2O$	K + N	0.6	0.6
Other (including SO_3)	$\ldots(\ldots\bar{S})$	1.4	3.6
H_2O	H	"nil"	1.0

Source: van Oss, 2005.
Based on clinker shown plus 5% addition of gypsum ($CaSO_4 \cdot 2H_2O$).

TABLE 2.2 Typical Mineralogical Composition of Modern Portland Cement

Chemical Formula	Oxide Formula	Shorthand Notation	Description	Typical Percentage	Mineral Function
Ca_3SiO_5	$(CaO)_3SiO_2$	C_3S	Tricalcium silicate (alite)	50–70	Hydrates quickly and imparts early strength and set.
Ca_2SiO_4	$(CaO)_2SiO_2$	C_2A	Dicalcium silicate	10–30	Hydrates slowly and imports long-term (ages beyond 1 week) strength.
$Ca_3A1_2O_6$	$(CaO)_3A1_2O_3$	C_4AF	Tricalcium aluminate	3–13	Hydrates almost instantaneously and very exothermically. Contributes to early strength and set.
$Ca_4A1_2Fe_2O_{10}$	$(CaO)_4A1_2O_3Fe_2O_3$	C_4AF	Tetracalcium aluminoferrite	5–15	Hydrates quickly. Acts as a flux in clinker manufacture. Imparts gray color.
$CaSO_42H_2O$	$(CaO)(SO)_3(H_2O)_2$	$C\bar{S}H_2$	Calcium sulfate dehydrate (gypsum)	3–7	Interground with clinker to make Portland cement. Can substitute anhydrite ($C\bar{S}$). Controls early set.
$CaSO_4$	$(CaO)(SO_3)$	$C\bar{S}$	Anhydrous calcium sulfate	0.2–2	

Source: van Oss, 2005.

mineral's (C_4AF) primary purpose, on the other hand, is to lower the temperature required in the kiln to form the C_3S mineral, and really does not impart a specific property to the cement. Table 2.3 presents the common mineralogical compositions of Types I through IV cements and the unique properties of each type.

Physical Properties of Portland Cement

Portland cement consists of individual angular particles with a range of sizes, the result of pulverizing clinker in the grinding mill. Approximately 95% of cement particles are smaller than 45 μm, with the average particle around 15 μm. The fineness of cement affects the amount of heat released during hydration. Greater cement fineness (smaller particle size) increases the rate at which cement hydrates and thus accelerates strength development. Except for AASHTO M 85, most cement standards do not have a maximum limit on fineness, only a minimum. The fineness of Types I through V Portland cement are shown in Table 2.4 (Kosmatka, 2002). Values are expressed according to the Blaine air-permeability test (ASTM C 204 or AASHTO T 153), which indirectly measures the surface area of particles per unit mass.

TABLE 2.3 Typical Range in Mineral Composition in Portland Cements

ASTM C-150 Cement Type	Clinker Mineral Percent*				Properties of Cement
	C_3S	C_2S	C_3A	C_4AF	
I	50–65	10–30	6–14	7–10	General purpose
II	45–65	7–30	2–8	10–12	Moderate heat of hydration, moderate sulfate resistance
III	55–65	5–25	5–12	5–12	High early strength**
IV	35–45	28–35	3–4	11–18	Low heat of hydration
V	40–65	15–30	1–5	10–17	High sulfate resistance

Source: van Oss, 2005.
**Range of minerals is empirical and approximate rather than definitional.*
***High-early strength is typically achieved by finer grinding of Type I cement.*

TABLE 2.4 Fineness of Portland Cement

ASTM C-150 Cement Type	Fineness (cm^2/g, Blaine)	
	Range	Mean
I	3,000–4,210	3,690
II	3,180–4,800	3,770
III	3,900–6,440	5,480
IV	3,190–3,620	3,400
V	2,750–4,300	3,730

The specific gravity of Portland cement typically ranges from 3.10 to 3.25, with an average of 3.15. Bulk densities can vary significantly depending on how the cement is handled and stored. Reported bulk densities range from 830 to 1,650 kg/m^3 (Kosmatka, 2002).

The Clinker Manufacturing Process

Portland cement manufacturing is a two-step process beginning with the manufacture of clinker followed by the fine grinding of the clinker with gypsum and other additives to make the finished cement product. Grinding can occur on site or at offsite grinding plants.

The first step in clinker manufacture is the quarrying, crushing, and proportioning of raw materials. Due to the low unit value of these raw materials, they typically are mined within a few miles of the cement plant. The cost of transport renders long-distance transport of these low-cost raw materials uneconomical.

Once the raw mix, or raw meal, is prepared, it is fed into a cement kiln and converted into the clinker minerals through a thermochemical conversion, referred to as pyroprocessing because it involves direct flame interaction. Figure 2.1 provides a generalized flow diagram of the cement manufacturing process (van Oss, 2005).

The raw materials for clinker manufacture consist primarily of materials that supply four primary oxides: calcium oxide (CaO), silicon dioxide (SiO_2), aluminum oxide (Al_2O_3), and ferric oxide (Fe_2O_3). The composition of the raw mix typically includes about 80% calcium carbonate, about 10–15% silica, and small amounts of alumina and iron. Depending on the quality and quantity of these oxides available to the facility, other raw materials, referred to as accessory or sweetener materials, are added to correct for any deficiencies in the primary raw materials. Certain types of

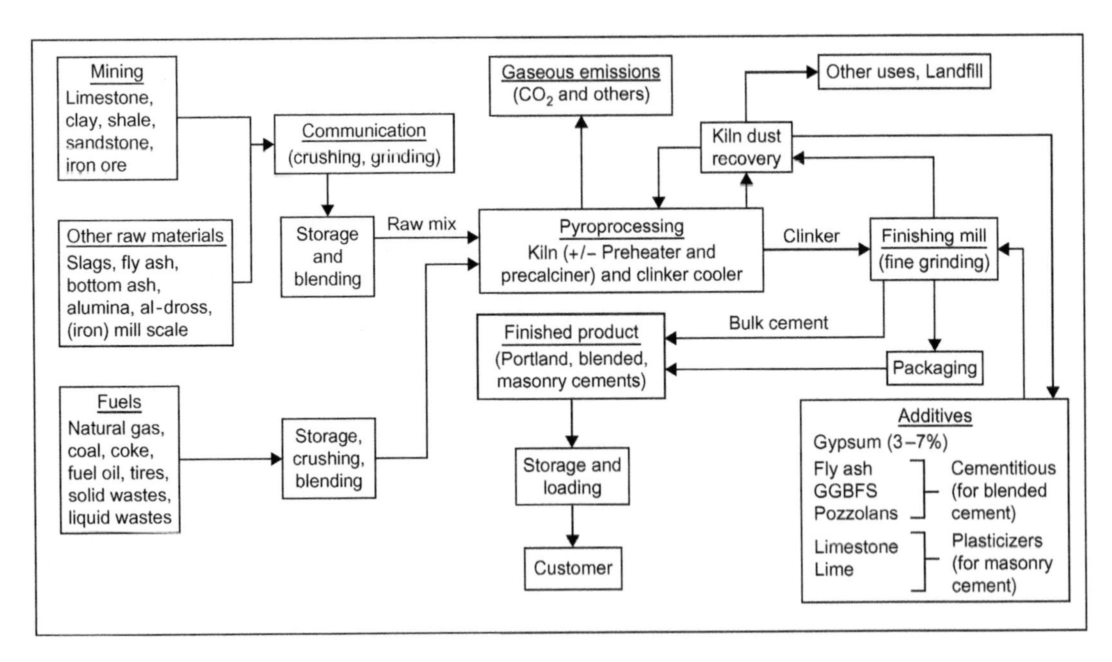

FIGURE 2.1 Cement manufacturing flow diagram.

fuel burned in the cement kiln can also contribute oxides (e.g., ash from coal combustion contributes silica oxides, steel belts in waste tires contribute iron oxide).

Calcium oxide (CaO or simply C in shorthand) is the primary ingredient in clinker, comprising about 65% of clinker by mass. A cement plant typically examines its source of C (typically limestone, marl, or chalk) and determines what other oxides need to be added to achieve the desired clinker composition. Clay, shale, slate, or sand provides the silica and alumina component, while iron, mill scale, or other ferrous materials provide the iron content. Preparing the raw mix for clinker production requires constant sampling, chemical testing, and adjusting of the inputs to maintain the desired clinker composition.

On average, it takes about 1.7 tons of non-fuel raw materials to produce 1 ton of clinker. Of the 1.7 tons of raw materials, approximately 1.5 tons is limestone or calcium oxide rich rock (van Oss, 2005). The lost mass takes the form of carbon dioxide (CO_2) driven off by the calcination of limestone and the generation of CKD. Nearly one ton of CO_2 is produced for every ton of clinker manufactured (van Oss, 2005). The CKD produced during clinker manufacture is carried "up the stack" and captured by emission control devices. A large portion of the CKD, although not all of it, is returned to the kiln as part of the feed stream.

Manufacture of Finished Cement from Clinker

After clinker has been cooled to about 100°C, it is ready to be ground into finished cement in a grinding mill, more commonly referred to as a finish mill. Generally, separate grinding and/or blending finish mill lines will be maintained at a plant for each of its major product classes (finished Portland cements, blended cements, masonry cements, ground slag). Additives that commonly require grinding at the mill include gypsum, limestone, granulated blast-furnace slag, and natural pozzolans. Additives that generally do not require significant grinding include coal fly ash, GGBFS, and silica fume, but the finish mill does provide intimate mixing of these with the Portland cement base.

Production

The U.S. Geological Survey (USGS) estimated that in 2005 approximately 97.5 million metric tons (Mt) of Portland plus masonry cement was produced at 113 plants in 37 states in the United States. The reported final production of masonry plus Portland cement was 99.3 Mt, with Portland cement alone accounting for 93.9 Mt of this total (van Oss, 2007). Figure 2.2 shows the locations of U.S. cement plants in 2005 based on information provided by the Portland Cement Association (PCA, 2006)…. The estimated value of cement production for 2005 was about $8 billion. The final reported actual value for Portland plus masonry cement production in 2005 was $11.6 billion. Of this total, $10.9 billion was for Portland cement alone (van Oss, 2007). Most of the cement was used to make ready mixed concrete (75%), while 14% went to concrete manufacturers, 6% to contractors, 3% to building materials dealers, and 2% to other users. Clinker production occurred at 107 plants, with a combined annual capacity of about 103 million tons. Actual U.S. cement imports in 2005 were reported at 30.4 Mt (excluding Puerto Rico), and clinker imports were 2.86 Mt (van Oss, 2007). Average mill prices for cement in 2005 were about $84 per ton. More than 172 million tons of raw materials were used to produce cement and clinker in the United States in 2004. Table 2.5 summarizes U.S. cement statistics for the years 2000 through 2005 (USGS, 2001; 2002; 2003; 2004; 2005; 2006). Table 2.6 summarizes raw materials used in the United States in 2003 and 2004 to produce cement and clinker (van Oss, 2004).

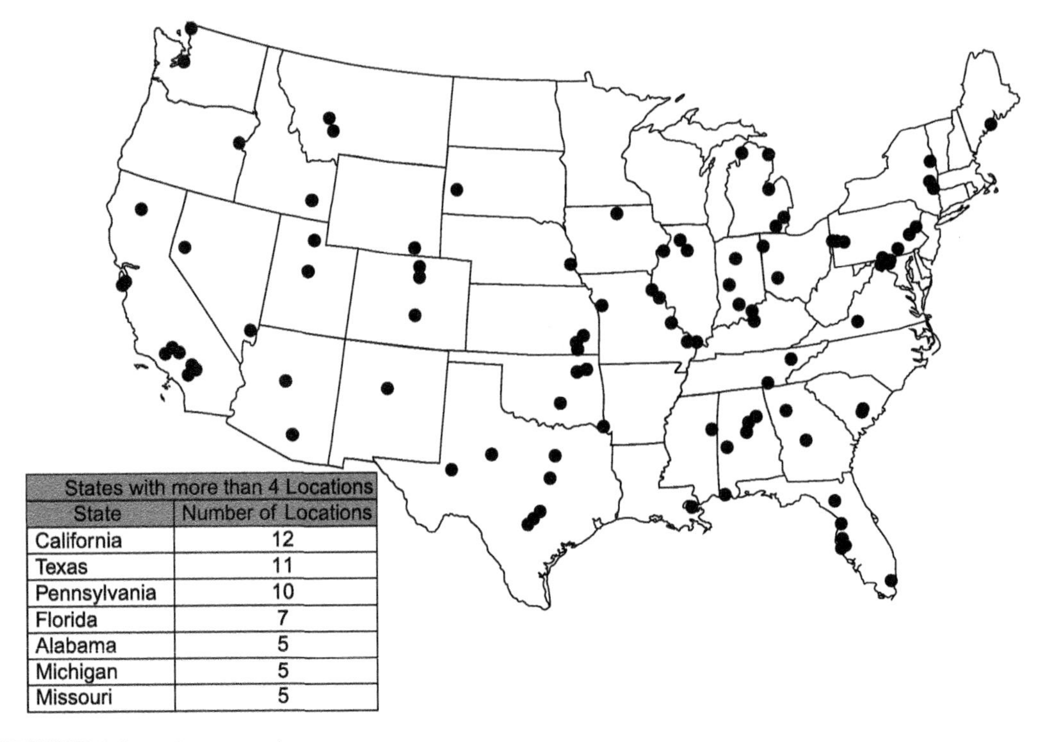

FIGURE 2.2 U.S. cement plants.

TABLE 2.5 U.S. Cement Statistics

Year	2000	2001	2002	2003	2004	2005
	Million Metric Tons					
Estimated cement production	87.8	88.9	89.7	92.8	97.4	97.5*
Clinker production	78.1	78.8	81.5	81.9	86.7	87.4
Imports of cement	24.6	23.7	22.2	21.0	25.4	29.0
Imports of clinker	3.7	1.8	1.6	1.8	1.6	2.8
Exports of cement and clinker	0.74	0.75	0.83	0.84	0.82	0.80
Average price, mill value, $/ton	78.56	76.50	76.00	75.00	79.50	84.00

* *Actual total masonry plus Portland cement final production for 2005 is reported at 99.3 Mt, of which 93.9 Mt was for Portland cement alone.*

Concrete

Concrete is basically a mixture of two components: aggregate and cement paste. The cement paste, comprised of a binder (usually Portland cement) and water, binds the aggregates (usually sand and gravel or crushed stone) into a rock-like mass as the paste hardens. The paste hardens because of a chemical reaction, called hydration, between the cement and water.

TABLE 2.6 Raw Materials Used in Producing Clinker and Cement in the United States

	2003		2004	
	Clinker	**Cement**	**Clinker**	**Cement**
Raw Materials	**Thousand Metric Tons**			
Limestone	109,000	1,530	125,000	1,810
Cement rock	12,700	44	12,700	2
Cement kiln dust	289	149	333	165
Lime	22	27	24	29
Other calcareous	235	32	23	19
Clay	3,950	—*	4,740	—
Shale	2,630	8	3,700	29
Other aluminous	618	—	661	—
Ferrous, iron ore, pyrites, mill scale, other	1,340	—	1,340	—
Sand and calcium silicate	2,860	2	3,150	—
Sandtone, quartzite soils, other	587	2	878	6
Coal fly ash	2,250	39	2,890	77
Other ash, including bottom ash	1,100	—	1,050	—
Granulated blast-furnace slag	17	333	104	345
Other blast-furnace slag	214	—	189	—
Steel slag	448	—	401	—
Other slags	113	—	53	—
Natural rock pozzolans	—	25	—	6
Other pozzolans	129	49	114	19
Gypsum and anhydrite	—	5,000	—	5,300
Other, not elsewhere classified	70	68	106	98
Clinker, imported	—	4,240	—	7,530
Total	**139,000**	**11,500**	**157,000**	**15,400**

* *Indicates none reported.*

The National Ready Mixed Concrete Association (NRMCA) estimates that ready mixed concrete production in the United States was approximately 349 million cubic meters in 2005. NRMCA estimates that there are approximately 6,000 ready mixed concrete plants in the United States, and that annual ready mixed concrete production is valued at more than $30 billion. Table 2.7 shows ready mixed concrete production by state in 2005 as reported by NRMCA. USGS estimates that total concrete production in the United States in 2005 was valued at more than $48 billion

TABLE 2.7 Ready Mixed Concrete Production by State (2005)

State	Production (Million Cubic Meters)	Percent of National Production
Alabama	4.9	1.4
Alaska	0.5	0.1
Arizona	13.1	3.8
Arkansas	3.4	1.0
California	43.4	12.4
Colorado	7.3	2.1
Connecticut	2.2	0.6
Delaware	0.6	0.2
District of Columbia	0.6	0.2
Florida	31.6	9.1
Georgia	12.4	3.6
Hawaii	1.2	0.3
Idaho	2.0	0.6
Illinois	11.6	3.3
Indiana	6.1	1.8
Iowa	5.4	1.6
Kansas	4.3	1.2
Kentucky	4.2	1.2
Louisiana	5.5	1.6
Maine	0.7	0.2
Maryland	4.4	1.3
Massachusetts	3.5	1.0
Michigan	8.2	2.4
Minnesota	5.7	1.6
Mississippi	3.0	0.9
Missouri	8.0	2.3
Montana	1.1	0.3
Nebraska	3.8	1.1
Nevada	7.3	2.1
New Hampshire	0.6	0.2

TABLE 2.7 (*Continued*)

State	Production (Million Cubic Meters)	Percent of National Production
New Jersey	5.5	1.6
New Mexico	2.5	0.7
New York	8.9	2.6
North Carolina	8.2	2.4
North Dakota	1.0	0.3
Ohio	10.9	3.1
Oklahoma	4.5	1.3
Oregon	3.5	1.0
Pennsylvania	9.3	2.7
Puerto Rico	5.2	1.5
Rhode Island	0.5	0.1
South Carolina	5.0	1.4
South Dakota	1.4	0.4
Tennessee	6.0	1.7
Texas	40.8	11.7
Utah	4.3	1.2
Vermont	0.4	0.1
Virginia	7.5	2.1
Washington	6.3	1.8
West Virginia	1.4	0.4
Wisconsin	6.6	1.9
Wyoming	1.3	0.4
Other	1.2	0.3
Total	**348.8**	**100.0**

(USGS, 2006). Although there are no data available on the amount of concrete placed annually in the United States, based on U.S. cement sales it can be estimated to be nearly one billion metric tons per year.

The character of concrete is determined by the quality of the cement paste (i.e., the cement and water mixture). The water–cement ratio — the weight of the mixing water divided by the weight of the cement — plus the quality and type of cement determines the strength of the paste, and hence the strength of the concrete. High-quality concrete is produced by lowering the water–cement ratio as much as possible without sacrificing the workability of fresh concrete. Generally, using less water produces a higher quality concrete provided the concrete is properly placed, consolidated, and cured.

For a typical concrete mix, 1 t of cement (powder) will yield about 3.4 to 3.8 m^3 of concrete weighing about 7 to 9 t (i.e., the density is typically in the range of about 2.2 to 2.4 t/m^3). Although aggregates make up the bulk of the mix, it is the hardened cement paste that binds the aggregates together and contributes virtually all of the strength of the concrete, with the aggregates serving largely as low-cost fillers. The strengths of the cement paste are determined by both the quality and type of the cement and the water–cement ratio (van Oss, 2005).

AGGREGATES

Aggregates are required to conform to ASTM standards. An exception to this is the use of aggregates that have been shown through approved special testing methods or actual service to produce concrete of adequate strength and durability. Any material used must be approved by the local building official.

Requirements on aggregate sizes for coarse aggregates come in three phases. For example, the largest aggregate used is to be no more than one-fifth the narrowest dimension between sides of forms. The aggregate cannot exceed a size greater than one-third the depth of slabs. Finally, aggregate maximum sizing is to be no more than three-quarters of the minimum clear spacing between individual reinforcing bars or wires, bundles of bars, individual tendons, bundles of tendons, or ducts.

The sizing rules for aggregates can be offset by a qualified engineer. To do this, an engineer must determine that variation in the aggregate requirements will not result in a substandard finished product and that no voids or honeycombs will occur in the concrete.

WATER

All water used to mix concrete must be clean. In most cases, potable (drinking) water is required for the mixing of concrete. There are always exceptions, and these will be reviewed shortly. Water used to mix concrete must also be free of injurious amounts of

- Oils
- Acids
- Alkalis
- Salt
- Organic materials
- All substances that can be deleterious to concrete or reinforcement used within the concrete

Non-Potable Water

Normally, non-potable water is not an acceptable mixing material for concrete. One exception is when the selection of concrete proportions is based on concrete mixes using water from the same source.

Test cubes for mortar can be made with non-potable mixing water. This requires both 7- and 28-day strengths equal to at least 90% of strengths of similar specimens that are made with potable water. All strength testing is to be done on identical samples with the only difference being the use of potable water with one sample and non-potable water with another sample.

STEEL REINFORCEMENT

Steel reinforcement used in concrete construction is usually deformed reinforcement. Plain reinforcement can be used for spirals or prestressing steel, reinforcement consisting of structural steel, steel pipe, or steel tubing. Deformed reinforcing bars must conform to one of several ASTM standards. Check your local code requirements to see which standards are approved in your region.

Tip

All welding or reinforcing bars must conform to "Structural Welding Code — Reinforcing Steel."

ADMIXTURES

Any admixtures used in concrete must be approved by a qualified engineer. Admixtures must be capable of maintaining essentially the same composition and performance throughout the work as the product used in establishing concrete proportions in an approved manner.

Prestressed concrete is not allowed to contain calcium chloride or admixtures containing chlorides, other than impurities from admixture ingredients. Chlorides are also not allowed in concrete containing embedded aluminum or in concrete cast against stay-in-place galvanized steel forms.

Air-entraining admixtures must conform to ASTM standards. The same is true of water-reducing admixtures, retarding admixtures, and accelerating admixtures. Fly ash and other pozzolans used as admixtures must also meet ASTM requirements. Blast-furnace slag is another example of an admixture with its own set of standards.

There are a great number of ASTM standards associated with concrete work. Examples of the standards that are commonly followed when dealing with admixtures are listed below:

- ASTM C 260
- ASTM C 494
- ASTM C 1017
- ASTM C 618
- ASTM C 989

As you can see, there are plenty of numbers involved in ASTM standards. The best way to avoid mistakes when working with concrete is to obtain a list of approved standards in your jurisdiction. A copy of ASTM standards can be obtained at the following address:

ASTM
100 Barr Harbor Drive
West Conshohocken, PA 19428

If you are, or plan to be, a concrete professional, you will find these standards very valuable when complying with local code requirements.

STORAGE OF MATERIALS

Cementitious materials and aggregates are required to be stored in a manner that will prevent deterioration or intrusion of foreign matter. Any material that has deteriorated or has been contaminated must not be used for concrete. This concludes our look at materials used in concrete construction.

References

1. Kosmatka, 2002.
2. PCA, 2006.
3. USGS, 2001.
4. USGS, 2002.
5. USGS, 2003.
6. USGS, 2004.
7. USGS, 2005.
8. USGS, 2006.
9. van Oss, 2005.

CHAPTER

3

Durability and Protection

OUTLINE

The durability of concrete must be assured through proper handling and protection. Common sense and experience play a role in this phase of concrete construction, but, code requirements mandate minimum standards and procedures, which will be discussed in this chapter.

To comply with your local code requirements, consult the tables provided in your local code book or the American Concrete Institute (ACI) guide to concrete installations and repairs. I could fill this chapter with sample tables, but they would only be examples. You need to work in the real world, so consult your local code materials to determine the finer points of maintaining the durability of concrete and protecting it.

Tip

Normalweight and lightweight concrete exposed to freezing and thawing or deicing chemicals must be air-entrained. Seek specifications from your local code for exact amounts of air content required.

SPECIAL EXPOSURE CONDITIONS

Special exposure conditions exist and are allowed for in code requirements. Four such special conditions include the following:

- Concrete meant to have low permeability when it is in contact with water
- Concrete that may freeze or thaw in moist conditions
- Concrete exposed to deicing chemicals
- Corrosion protection of reinforcement in concrete exposed to chlorides from deicing chemicals, salt, salt water, brackish water, seawater, or spray from any of these sources

When dealing with special conditions, use a code-provided table to determine what is needed; for example, if you need to know what the maximum water–cementitious material ratio by weight of normalweight concrete is used. You will also need to establish the pounds per square inch (PSI) allowable for these conditions.

FIGURE 3.1 Inspecting a fresh concrete pour.

Tip

Concrete that is to be exposed to sulfate-containing solutions must be made with a type of cement that provides sulfate resistance.

CORROSION PROTECTION

Corrosion protection is needed for the reinforcement materials used in concrete structures. This generally means the maximum water-soluble chloride ion concentrations in hardened concrete from 28 to 42 days old contributed from the ingredients including water, aggregates, cementitious materials, and admixtures that meet code requirements.

Concrete with reinforcement exposed to chlorides from deicing, chemicals, salt, salt water, brackish water, seawater, or spray from such sources must be protected.

CHAPTER

4

Mixing and Placing Concrete

OUTLINE

Concrete is required to be proportioned to provide an average compressive strength. This strength requirement is determined by your local code enforcement office. To test strength, it is common for core samples to be taken and tested. Splitting tensile strength tests must not be used as a basis for field acceptance of concrete.

The mixing of concrete is regulated in several ways:

- Workability and consistency must be such that the concrete used can be readily worked into forms and around reinforcements under the given circumstances and condition of the job description.
- Concrete is not allowed to have segregation or excessive bleeding.
- Any contact or special exposure risk must be taken into consideration.
- Separate testing is needed any time different materials are used for different portions of a job.
- All concrete proportions must be in compliance with the governing code used to establish minimum standards.
- The proportioning of concrete based on field experience and trial mixtures can be allowed if there is adequate, approved testing information available.

AVERAGE COMPRESSIVE STRENGTH

The average compressive strength of concrete is determined from mathematical tables provided by your local code book or recommended supplemental reading, such as the *Building Code Requirements for Structural Concrete and Commentary* by the American Concrete Institute (ACI).

This book is expensive, but it is a valuable tool that no concrete professional should be without.

Tip

When a job is small enough that a minimum of five test inspections would not be required based on volume, you must test a minimum of five batches of randomly selected concrete. If less than five batches are used, the testing must be done on the concrete batches that are used.

Sometimes there are changes in the materials planned for use. If this is the case, additional testing and documentation is needed. Standards for replacement materials will not be more stringent than those allowed for the planned materials. When establishing new testing and documentation, there are particular rules to follow.

Tip

When the total quantity of concrete to be placed is less than 50 cubic yards, strength testing of the batches is not required when there is evidence of satisfactory strength submitted to and approved by a building official.

Test records must consist of less than 30, but not less than 10, consecutive tests that are acceptable, and must encompass a period of time of not less than 45 days. The interpolation between strength and proportions of two or more of the test records that meet code requirements can be used to establish acceptable standards.

If records from field test results are not available, there are other options available to establish compliance with acceptable standards. Trial mixtures can be used, but they must comply with the following restrictions:

- Trial materials must be made using a minimum of three different water–cementitious materials ratios or cementitious materials with contents that produce a range of strengths.
- Trial mixtures are allowed a maximum slump within a plus or minus (±) range of 0.75 in.
- Trial mixtures can allow a maximum air content of ±0.5%.
- A minimum of three test cylinders are required for each trial-mixture test at each stage of testing.
- Trial mixtures must show ratings within approved limits to be acceptable.

Tip

Samples for strength tests are to be taken daily, or at least once for every 150 cubic yards of concrete placed and not less than once for every 5,000 square feet of surface area in walls or slabs.

Part of the strength of concrete is derived from suitable reinforcement. This can come in the form of welded wire or rebar. Figure 4.1 shows rebar that is properly installed and is being covered by fresh concrete.

FIGURE 4.1 Concrete being placed over rebar reinforcement.

SITE PREPARATION

You may not think much about it, but the code requirements extend to site preparation. Here is a checklist that you should keep in mind when preparing to begin a concrete job:

- Make sure all equipment that will come into contact with a concrete mix is clean.
- The placement location must be clear of any debris.
- In freezing weather, all ice from a pouring location must be removed.
- Check all concrete forms to make sure they are coated properly.
- Clean and drench any masonry filler units that will come into contact with fresh concrete.
- Inspect reinforcement materials to be sure that they are clean and free of ice.
- Under most conditions all water accumulated in a pour site must be removed.
- Any unsound material has to be removed prior to the placement of fresh concrete against hardened concrete.

Common sense goes a long way in prepping a site for a pour. Experienced concrete professionals know the drill. Just be aware that the preparation is governed by the building code, so don't underestimate the need for a pre-pour routine.

FIGURE 4.2 Pouring ready mixed concrete.

MIXING CONCRETE

When mixing concrete the goal is to achieve a mixture that offers a uniform distribution consistency. These materials must be used entirely before a mixer can be used to mix another batch of concrete. When using ready mixed concrete, as shown in Figure 4.2, you can assume that the mixture is in compliance with standards set forth by code requirements.

There are code requirements to be followed when mixing your own concrete on site. First, your mixer must be an approved type. Rotation speed of the mixer must be within the scale set forth by the manufacturer of the mixer.

How long do you have to mix a batch of concrete? The code requires mixing for a minimum of 90 seconds from the time the mixer is loaded and ready to mix. There can be exceptions to this if the material specifications are approved and call for a shorter mixing period.

When mixing on site, you must maintain records of your actions. Here is what you need to record:

- Number of batches mixed
- Proportions of materials that were used in a batch
- The approximate location of where the concrete was installed
- The time and the date of both mixing the concrete and placing the concrete

Depending upon the method used for concrete placement, you must ensure that the concrete is placed without any separation of ingredients. There can be no interruptions in placement that may result in plasticity between successive increments.

EXHIBIT: CONCRETE BATCHING

Process Description[1-5]

Concrete is composed essentially of water, cement, sand (fine aggregate), and coarse aggregate. Coarse aggregate may consist of gravel, crushed stone, or iron blast-furnace slag. Some specialty aggregate products can be heavyweight aggregate (of barite, magnetite, limonite, ilmenite, iron, or steel) or lightweight aggregate (with sintered clay, shale, slate, diatomaceous shale, perlite, vermiculite, slag pumice, cinders, or sintered fly ash). Supplementary cementitious materials, also called mineral admixtures or pozzolan minerals, may be added to make the concrete mixtures more economical, reduce permeability, increase strength, or influence other concrete properties. Typical examples are natural pozzolans, fly ash, ground granulated blast-furnace slag, and silica fume, which can be used individually with Portland or blended cement or in different combinations. Chemical admixtures are usually liquid ingredients that are added to concrete to entrain air, reduce the water required to reach a required slump, retard or accelerate the setting rate, to make the concrete more flowable, or perform other more specialized functions.

Approximately 75% of the U.S. concrete manufactured is produced at plants that store, convey, measure, and discharge these constituents into trucks for transport to a jobsite. At most of these plants, sand, aggregate, cement, and water are all gravity-fed from the weigh hopper into the mixer trucks. The concrete is mixed on the way to the site where the concrete is to be poured. At some of these plants, the concrete may also be manufactured in a central mix drum and transferred to a transport truck. Most of the remaining concrete manufactured includes products cast in a factory setting. Precast products range from concrete bricks and paving stones to bridge girders, structural components, and panels for cladding. Concrete masonry, another type of manufactured concrete, may be best known for its conventional $8 \times 8 \times 16$-inch block. In a few cases concrete is dry-batched or prepared at a building construction site. Figure 4.3 is a generalized process diagram for concrete batching.

The raw materials can be delivered to a plant by rail, truck, or barge. The cement is transferred to elevated storage silos pneumatically or by bucket elevator. The sand and coarse aggregate are transferred to elevated bins by front-end loader, clam shell crane, belt conveyor, or bucket elevator. From these elevated bins, the constituents are fed by gravity or screw conveyor to weigh hoppers, which combine the proper amounts of each material.

Emissions and Controls[6-8]

Particulate matter, consisting primarily of cement and pozzolan dust but including some aggregate and sand dust emissions, is the primary pollutant of concern. In addition, there are emissions of metals that are associated with this particulate matter. All but one of the emission points is fugitive in nature. The only point sources are the transfer of cement and pozzolan material to silos, and these are usually vented to a fabric filter or "sock." Fugitive sources include the transfer of sand and aggregate, truck loading, mixer loading, vehicle traffic, and wind erosion from sand

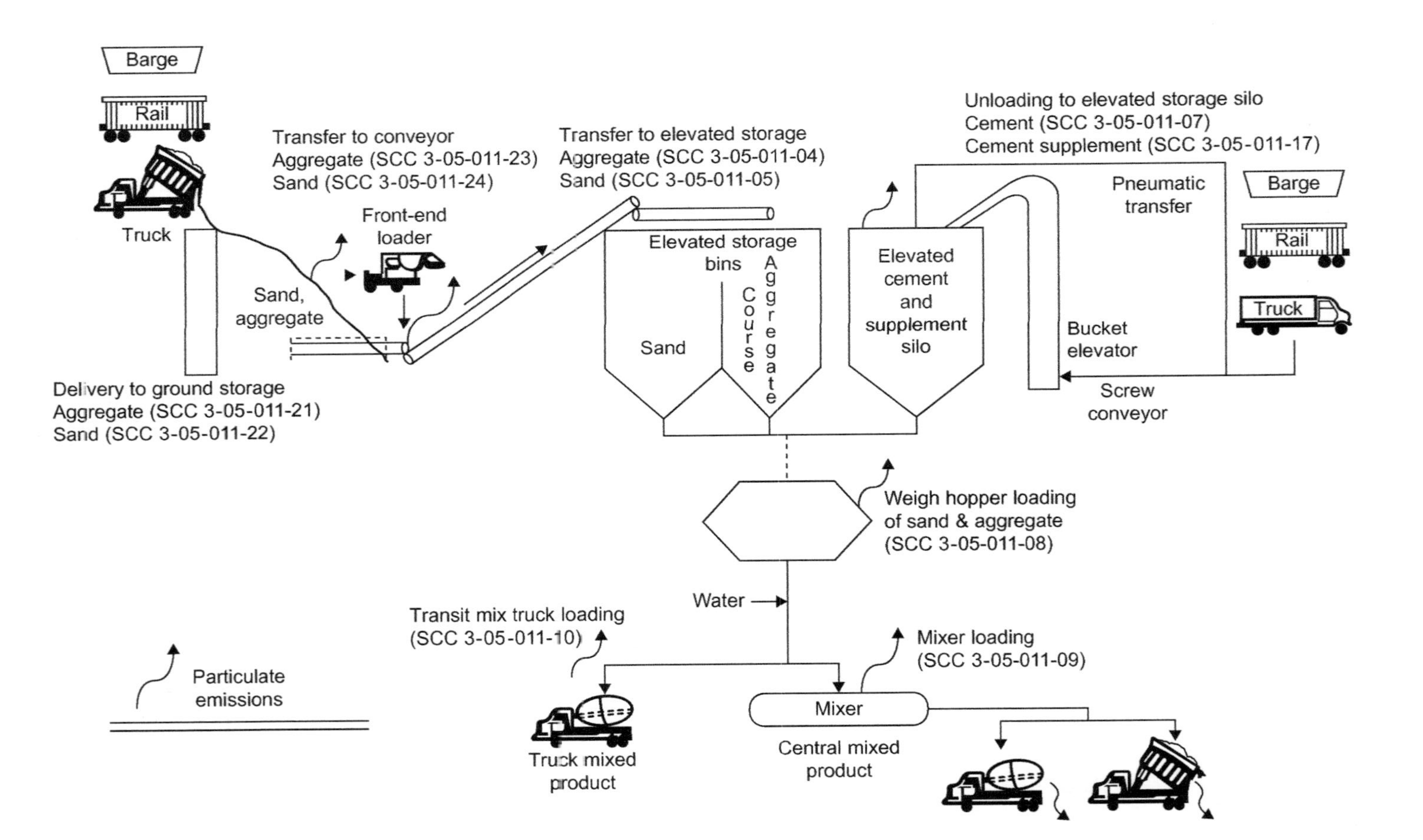

FIGURE 4.3 Typical concrete batching process.

and aggregate storage piles. The amount of fugitive emissions generated during the transfer of sand and aggregate depends primarily on the surface moisture content of these materials. The extent of fugitive emission control varies widely from plant to plant. Particulate emission factors for concrete batching are give in Tables 4.1 and 4.2.

Types of controls used may include water sprays, enclosures, hoods, curtains, shrouds, movable and telescoping chutes, central duct collection systems, and so forth. A major source of potential emissions, the movement of heavy trucks over unpaved or dusty surfaces in and around the plant, can be controlled by good maintenance and wetting of the road surface.

Whenever plant-specific data are available, they should be used with these predictive equations (e.g., Equations 4.1–4.3), in lieu of the general fugitive emission factors presented in Tables 4.1, 4.2, and 4.5–4.8, to adjust to site-specific conditions, such as moisture levels and localized wind speeds.

Updates since the 5th Edition

October 2001 — This major revision replaced emissions factors based upon engineering judgment and poorly documented and performed source test reports with emissions tests conducted at modern operating truck mix and central mix facilities. Emissions factors for both total PM and total PM_{10} were developed from these test data.

June 2006 — This revision supplemented the two source tests with several additional source tests of central mix and truck mix facilities. The measurement of the capture efficiency, local wind speed, and finer material moisture level was improved over the previous two source tests. In addition to quantifying total PM and PM_{10}, $PM_{2.5}$ emissions were quantified at all of the facilities. Single value emissions factors for truck mix and central mix operations were revised using all of these data. Additionally, parameterized emissions factor equations using local wind speed and fines material moisture content were developed from the newer data.

The particulate matter emissions from truck mix and central mix loading operations are calculated in accordance with the values in Tables 4.1 and 4.2 or by Eq. 4.1[14] when site-specific data are available.

$$E = k(0.0032)\left[\frac{U^a}{M^b}\right] + C \tag{4.1}$$

E = Emission factor in lb/ton of cement and cement supplement
k = Particle size multiplier (dimensionless)
U = Wind speed at the material drop point, miles per hour (mph)
M = Minimum moisture (% by weight) of cement and cement supplement
a, b = Exponents
c = Constant

The parameters for Eq. 4.1 are summarized in Tables 4.3 and 4.4.

To convert from units of lb/ton to units of kg/Mg, the emissions calculated by Eq. 4.1 should be divided by 2.0.

Particulate emission factors per yard of concrete for an average batch formulation at a typical facility are given in Tables 4.5 and 4.6. For truck mix loading and central mix loading, the emissions

TABLE 4.1 Emission Factors for Concrete Batching (Metric Units)[a]

	Uncontrolled				Controlled			
Source (SCC)	Total PM	Emission Factor Rating	Total PM_{10}	Emission Factor Rating	Total PM	Emission Factor Rating	Total PM_{10}	Emission Factor Rating
Aggregate transfer[b] (3-05-011-04,-21-23)	0.0035	D	0.0017	D	ND		ND	
Sand transfer[b] (3-05-011-05,22,24)	0.0011	D	0.00051	D	ND		ND	
Cement unloading to elevated storage silo (pneumatic)[c] (3-05-011-07)	0.36	E	0.23	E	0.00050	D	0.00017	
Cement supplement unloading to elevated storage silo (pneumatic)[d] (3-05-011-17)	1.57	E	0.65	E	0.0045	D	0.0024	
Weigh hopper loading[e] (3-05-011-08)	0.0026	D	0.0013	D	ND		ND	
Mixer loading (central mix)[f] 0(3-05-011-09)	0.272 or Eq. 4.1	B	0.067 or Eq. 4.1	B	0.0087 or Eq. 4.1	B	0.0024 or Eq. 4.1	
Truck loading (truck mix)[g] (3-05-011-10)	0.498	B	0.139	B	0.0280 or Eq. 4.1	B	0.0080 or Eq. 4.1	
Vehicle traffic (paved roads)	See AP-42 Section 13.2.1, Paved Roads							
Vehicle traffic (unpaved roads)	See AP-42 Section 13.2.2, Unpaved Roads							
Wind erosion from aggregate and sand storage piles	See AP-42 Section 13.2.5, Industrial Wind Erosion							

ND = No data

[a] *All emission factors are in kg of pollutant per Mg of material loaded unless noted otherwise. Loaded material includes course aggregate, sand, cement, cement supplement, and the surface moisture associated with these materials. The average material composition of concrete batches presented in References 9 and 10 was 846 kg course aggregate, 648 kg sand, 223 kg cement, and 33 kg cement supplement. Approximately 75 L of water was added to this solid material to produce 1,826 kg of concrete.*

[b] *References 9 and 10. Emission factors are based upon an equation from AP-42, Section 13.2.4, Aggregate Handling and Storage Piles: Equation 1 with $k_{PM\text{-}10} = 0.35$, $k_{PM} = 0.74$, $U = 10$ mph, $M_{aggregate} = 1.77\%$, and $M_{sand} = 4.17\%$. These moisture contents of the materials ($M_{aggregate}$ and M_{sand}) are the averages of the values obtained from References 9 and 10.*

[c] *The uncontrolled PM and PM_{10} emission factors were developed from Reference 9. The controlled emission factor for PM was developed from References 9–12. The controlled emission factor for PM_{10} was developed from References 9 and 10.*

[d] *The controlled PM emission factor was developed from Reference 10 and Reference 12, whereas the controlled PM_{10} emission factor was developed from only Reference 10.*

[e] *Emission factors were developed by using the AP-42 Section 13.2.4, Aggregate and Sand Transfer Emission Factors in conjunction with the ratio of aggregate and sand used in an average cubic yard of concrete. The unit for these emission factors is kg of pollutant per Mg of aggregate and sand.*

[f] *References 9, 10, and 14. The emission factor units are kg of pollutant per Mg of cement and cement supplement. The general factor is the arithmetic mean of all test data.*

[g] *References 9, 10, and 14. The emission factor units are kg of pollutant per Mg of cement and cement supplement. The general factor is the arithmetic mean of all test data.*

TABLE 4.2 Emission Factors for Concrete Batching (English Units)[a]

Source (SCC)	Uncontrolled				Controlled			
	Total PM	Emission Factor Rating	Total PM_{10}	Emission Factor Rating	Total PM	Emission Factor Rating	Total PM_{10}	Emission Factor Rating
Aggregate transfer[b] (3-05-011-04,-21,23)	0.0069	D	0.0033	D	ND		ND	
Sand transfer[b] (3-05-011-05,22,24)	0.0021	D	0.00099	D	ND		ND	
Cement unloading to elevated storage silo (pneumatic),[c] (3-05-011-07)	0.72	E	0.46	E	0.00099	D	0.00099	D
Cement supplement unloading to elevated storage silo (pneumatic)[d] (3-05-011-17)	3.14	E	1.10	E	0.0089	D	0.0049	E
Weigh hopper loading[e] (3-05-011-08)	0.0051	D	0.0024	D	ND		ND	
Mixer loading (central mix)[f] (3-05-011-09)	0.544 or Eq. 4.1	B	0.134 or Eq. 4.1	B	0.0173 or Eq. 4.1	B	0.0048 or Eq. 4.1	B
Truck loading (truck mix)[g] (3-05-001-10)	0.995	B	0.278	B	0.0568 or Eq. 4.1	B	0.0160 or Eq. 4.1	B
Vehicle traffic (paved roads)	See AP-42 Section 13.2.1, Paved Roads							
Vehicle traffic (unpaved roads)	See AP-42 Section 13.2.1, Unpaved Roads							
Wind erosion from aggregate and sand storage piles	See AP-42 Section 13.2.5, Industrial Wind Erosion							

ND = No data

[a] *All emission factors are in lb of pollutant per ton of material loaded unless noted otherwise. Loaded material includes course aggregate, sand, cement, cement supplement, and the surface moisture associated with these materials. The average material composition of concrete batches presented in References 9 and 10 was 1,865 lb course aggregate, 1,428 lb sand, 491 lb cement, and 73 lb cement supplement.*

Approximately 20 gal of water was added to this solid material to produce 4,024 lb (one cubic yard) of concrete.

[b] *References 9 and 10. Emission factors are based upon an equation from AP-42, Section 13.2.4, Aggregate Handling and Storage Piles: Equation 1 with* $k_{PM\text{-}10} = 0.35$, $k_{PM} = 0.74$, $U = 10$ *mph,* $M_{aggregate} = 1.77\%$, *and* $M_{sand} = 4.17\%$. *These moisture contents of the materials (*$M_{aggregate}$ *and* M_{sand}*) are the averages of the values obtained from References 9 and 10.*

[c] *The uncontrolled PM and* PM_{10} *emission factors were developed from Reference 9. The controlled emission factor for PM was developed from References 9–12. The controlled emission factor for* PM_{10} *was developed from References 9 and 10.*

[d] *The controlled PM emission factor was developed from References 10 and 12, whereas the controlled* PM_{10} *emission factor was developed from only Reference 10.*

[e] *Emission factors were developed by using the Aggregate and Sand Transfer Emission Factors in conjunction with the ratio of aggregate and sand used in an average cubic yard of concrete. The unit for these emission factors is lb of pollutant per ton of aggregate and sand.*

[f] *References 9, 10, and 14. The emission factor units are lb of pollutant per ton of cement and cement supplement. The general factor is the arithmetic mean of all test data.*

[g] *References 9, 10, and 14. The emission factor units are lb of pollutant per ton of cement and cement supplement. The general factor is the arithmetic mean of all test data.*

TABLE 4.3 Equation Parameters for Truck Mix Operations

Condition	Parameter Category	k	a	b	c
Controlled[1]	Total PM	0.19	0.8	0.3	0.013
	PM_{10}	0.13	0.32	0.3	0.0052
	$PM_{10-2.5}$	0.12	0.288	0.3	0.00468
	$PM_{2.5}$	0.3	0.48	0.3	0.00078
Uncontrolled[1]	Total PM		0.995		
	PM_{10}		0.278		
	$PM_{10-2.5}$		0.228		
	$PM_{2.5}$		0.050		

TABLE 4.4 Equation Parameters for Central Mix Operations

Condition	Parameter Category	k	a	b	c
Controlled[1]	Total PM	0.19	0.95	0.9	0.0010
	PM_{10}	0.13	0.45	0.9	0.0010
	$PM_{10-2.5}$	0.12	0.45	0.9	0.0009
	$PM_{2.5}$	0.3	0.45	0.9	0.0002
Uncontrolled[1]	Total PM	5.90	0.6	1.3	0.120
	PM_{10}	1.92	0.4	1.3	0.040
	$PM_{10-2.5}$	1.71	0.4	1.3	0.036
	$PM_{2.5}$	0.38	0.4	1.3	0

[1] *Emission factors expressed in lb/ton of cement and cement supplement.*

TABLE 4.5 Plant Wide Emission Factors per Yard of Truck Mix Concrete (English Units)[a]

	Uncontrolled		Controlled	
	PM(lb/yd^3)	PM_{10}(lb/yd^3)	PM(lb/yd^3)	PM_{10}(lb/yd^3)
Aggregate delivery to ground storage (3-05-011-21)	0.0064	0.0031	0.0064	0.0031
Sand delivery to ground storage (3-05-011-22)	0.0015	0.0007	0.0015	0.0007
Aggregate transfer to conveyor (3-05-011-23)	0.0064	0.0031	0.0064	0.0031
Sand transfer to conveyor (3-05-011-24)	0.0015	0.0007	0.0015	0.0007

TABLE 4.5 (*Continued*)

	Uncontrolled		Controlled	
	PM(lb/yd^3)	PM_{10}(lb/yd^3)	PM(lb/yd^3)	PM_{10}(lb/yd^3)
Aggregate transfer to elevated storage (3-05-011-04)	0.0064	0.0031	0.0064	0.0031
Sand transfer to elevated storage (3-05-011-05)	0.0015	0.0007	0.0015	0.0007
Cement delivery to silo (3-05-011-07 controlled)	0.0002	0.0001	0.0002	0.0001
Cement supplement delivery to silo (3-05-011-17 controlled)	0.0003	0.0002	0.0003	0.0002
Weigh hopper loading (3-05-011-08)	0.0079	0.0038	0.0079	0.0038
Truck mix loading (3-05-011-10)	See Eq. 4.2			

TABLE 4.6 Plant Wide Emission Factors per Yard of Central Mix Concrete (English Units)[a]

	Uncontrolled		Controlled	
	PM(lb/yd^3)	PM_{10}(lb/yd^3)	PM(lb/yd^3)	PM_{10}(lb/yd^3)
Aggregate delivery to ground storage (3-05-011-21)	0.0064	0.0031	0.0064	0.0031
Sand delivery to ground storage (3-05-011-22)	0.0015	0.0007	0.0015	0.0007
Aggregate transfer to conveyor (3-05-011-23)	0.0064	0.0031	0.0064	0.0031
Sand transfer to conveyor (3-05-011-24)	0.0015	0.0007	0.0015	0.0007
Aggregate transfer to elevated storage (3-05-011-04)	0.0064	0.0031	0.0064	0.0031
Sand transfer to elevated storage (3-05-011-05)	0.0015	0.0007	0.0015	0.0007
Cement delivery to silo (3-05-011-07 controlled)	0.0002	0.0001	0.0002	0.0001
Cement supplement delivery to silo (3-05-011-17 controlled)	0.0003	0.0002	0.0003	0.0002
Weigh hopper loading (3-05-011-08)	0.0079	0.0038	0.0079	0.0038
Central mix loading (3-05-011-09)	See Eq. 4.2			

[a] *Total facility emissions are the sum of the emissions calculated in Table 4.4 or 4.5. Total facility emissions do not include road dust and windblown dust. The emission factors in Tables 4.4 and 4.5 are based upon the following composition of one yard of concrete.*
Coarse Aggregate: 1,865 lb
Sand: 1,428 lb
Cement: 491 lb
Cement Supplement: 73 lb
Water: 20 gal (167 lb)

of PM, PM_{10}, $PM_{10-2.5}$, and $PM_{2.5}$ are calculated by multiplying the emission factor calculated using Eq. 4.2 by a factor of 0.140 to convert from emissions per ton of cement and cement supplement to emissions per yard of concrete. This equation is based on a typical concrete formulation of 564 lb of cement and cement supplement in a total of 4,024 lb of material (including aggregate, sand, and water). This calculation is summarized in Eq. 4.2.

$$\text{PM, PM}_{10}\text{, PM}_{10-2.5}\text{, PM}_{2.5}\text{ emissions}\left(\frac{\text{pounds}}{\text{yd}^3\text{ of concrete}}\right) = 0.140 \tag{4.2}$$

Metals emission factors for concrete batching are given in Tables 4.7 and 4.8. Alternatively, the metals emissions from ready mixed cement plants can be calculated based on (1) the weighted average concentration of the metal in the cement and the cement supplement (i.e., fly ash) and (2) on the total particulate matter emission factors calculated in accordance with Eq. 4.3. Emission factors calculated using Eq. 4.3 are rated D.

$$\text{Metal}_{EF} = \text{PM}_{EF}\left(\frac{aC + bS}{C + S}\right) \tag{4.3}$$

where:

$Metal_{EF}$ = Metal emissions, pounds as per ton of cement and cement supplement
PM_{EF} = Controlled particulate matter emission factor (PM, PM_{10}, or $PM_{2.5}$) pounds per ton of cement and cement supplement
a = ppm of metal in cement
C = quantity of cement used, pounds per hour
b = ppm of metal in cement supplement
S = quantity of cement supplement used, pounds per hour

This equation is based on the assumption that 100% of the particulate matter emissions are material-entrained from the cement and cement supplement streams. Equation 4.3 overestimates total metal emissions to the extent that sand and fines from aggregate contribute to the total particulate matter emissions.

DEPOSITING CONCRETE

The way in which concrete is deposited is controlled by the code. The first consideration is to avoid segregation due to re-handling or flowing. Make sure that concrete is poured when it is plastic and flows readily into spaces between reinforcements.

It is a code violation to use concrete that has hardened or that contains foreign materials. Any use of re-tempered concrete for construction requires the approval of an approved engineer.

Once a pour begins, it must be completed to the point that the structural element being created is completed. When vertically formed lifts are used the top surface of concrete should be fairly level.

It is an installer's responsibility to maintain concrete consistency and to work fresh concrete around all reinforcements and embedded fixtures and into corners of forms.

TABLE 4.7 Concrete Batch Plant Metal Emission Factors (Metric Units)[a]

	Arsenic	Beryllium	Cadmium	Total Chromium	Lead	Manganese	Nickel	Total Phosphorus	Selenium	Emission Factor Rating
Cement silo filling[b] (SCC 3-05-011-07) w/fabric filter	8.38e-07	8.97e-09	1.17e-07	1.26e-07	3.68e-07	1.01e-04	8.83e-05	5.88e-05	ND	E
	2.12e-09	2.43e-10	2.43e-10	1.45e-08	5.46e-09	5.87e-08	2.09e-08	ND	ND	E
Cement supplement silo filling[c] (SCC 3-05-011-17) w/fabric filter	ND	ND	ND	ND	ND	ND	ND	ND	ND	E
	5.02e-07	4.52e-08	9.92e-09	6.10e-07	2.60e-07	1.28e-07	1.14e-06	1.77e-06	3.62e-08	E
Central mix batching[d] (SCC 3-05-011-09) w/fabric filter	1.16e-07	ND	5.92e-09	7.11e-07	1.91e-07	3.06e-05	1.64e-06	1.01e-05	ND	E
	9.55e-09	ND	3.55e-10	6.34e-08	1.83e-08	1.89e-06	1.24e-07	6.04e-07	ND	E
Truck loading[e] (SCC 3-05-011-10) w/fabric filter	1.52e-06	1.22e-07	1.71e-08	5.71e-06	1.81e-06	3.06e-05	5.99e-06	1.92e-05	1.31e-06	E
	5.80e-07	5018e-08	4.53e-09	2.05e-06	7.67e-07	1.04e-05	2.39e-06	6.16e-06	5.64e-08	E

ND = No data

[a] *All emission factors are in kg of pollutant per Mg of material loaded unless noted otherwise. Loaded material includes course aggregate, sand, cement, cement supplement, and the surface moisture associated with these materials. The average material composition of concrete batches presented in References 9 and 10 was 846 kg course aggregate, 648 kg sand, 223 kg cement, and 33 kg cement supplement. Approximately 75 L of water was added to this solid material to produce 1,826 kg of concrete.*

[b] *The uncontrolled emission factors were developed from Reference 9. The controlled emission factors were developed form References 9 and 10. Although controlled emissions of phosphorous compounds were below detection, it is reasonable to assume that the effectiveness is comparable.*

[c] *Reference 10.*

[d] *Reference 9. The emission factor units are kg of pollutant per Mg of cement and cement supplement. Emission factors were developed from a typical central mix operation. The average estimate of the percent of emissions captured during each run is 94%.*

[e] *References 9 and 10. The emission factor units are kg of pollutant per Mg of cement and cement supplement. Emission factors were developed from two typical truck mix loading operations. Based upon visual observations of every loading operation during the two test programs, the average capture efficiency during the testing was 71%.*

TABLE 4.8 Concrete Batch Plant Metal Emission Factors (English Units)[a]

	Arsenic	Beryllium	Cadmium	Total Chromium	Lead	Manganese	Nickel	Total Phosphorus	Selenium	Emission Factor Rating
Cement silo filling[b] (SCC 3-05-011-07) w/fabric filter	1.68e-06 4.24e-09	1.79e-08 4.86e-10	2.34e-07 4.86e-10	2.52e-07 2.90e-04	7.36e-07 1.09e-08	2.02e-04 1.17e-07	1.76e-05 4.18e-08	1.18e-05 ND	ND ND	E E
Cement supplement silo filling[c] (SCC 3-05-011-17) w/fabric filter	ND 1.00e-06	ND 9.04e-08	ND 1.98e-10	ND 1.22e-06	ND 5.20e-07	ND 2.56e-07	ND 2.28e-06	ND 3.54e-06	ND 7.24e-08	E E
Central mix batching[d] (SCC 3-05-011-09) w/fabric filter	2.32e-07 1.87e-08	ND ND	1.18e-08 7.10e-10	1.42e-06 1.27e-07	3.82e-07 3.66e-08	6.12e-05 3.78e-06	3.28e-06 2.48e-07	2.02e-05 1.20-06	ND ND	E E
Truck loading[e] (SCC 3-05-011-10) w/fabric filter	3.04e-06 1.16e-06	2.44e-07 1.04e-07	3.42e-08 9.06e-09	1.14e-05 4.10e-06	3.62e-06 1.53e-06	6.12e-05 2.08e-05	1.19e-05 4.78e-06	3.84e-05 1.23e-05	2.62e-06 1.13e-07	E E

ND = No data

[a] *All emission factors are in lb of pollutant per ton of material loaded unless noted otherwise. Loaded material includes course aggregate, sand, cement, cement supplement, and the surface moisture associated with these materials. The average material composition of concrete batches presented in References 9 and 10 was 1,865 lb course aggregate, 1,428 lb sand, 491 lb cement, and 73 lb cement supplement. Approximately 20 gal of water was added to this solid material to produce 4,024 lb (one cubic yard) of concrete.*

[b] *The uncontrolled emission factors were developed from Reference 9. The controlled emission factors were developed form References 9 and 10. Although controlled emissions of phosphorous compounds were below detection, it is reasonable to assume that the effectiveness is comparable to the average effectiveness (98%) for the other metals.*

[c] *Reference 10.*

[d] *Reference 9. The emission factor units are lb of pollutant per ton of cement and cement supplement. Emission factors were developed from a typical central mix operation. The average estimate of the percent of emissions captured during each run is 94%.*

[e] *References 9 and 10. The emission factor units are lb of pollutant per ton of cement and cement supplement. Emission factors were developed from two typical truck mix loading operations. Based upon visual observations of every loading operation during the two test programs, the average capture efficiency during the testing was 71%.*

FIGURE 4.4 Floating fresh concrete for an even distribution.

CURING

With the exception of high-early-strength concrete, all concrete curing should be done at a temperature above 50°F. Conditions of the concrete must be moist for at least a week after placement. There are exceptions to this under the guidelines for accelerated curing.

High-early-strength concrete has to be cured at a temperature above 50°F. Moist conditions must be maintained for three days. Again, accelerated curing is an exception to this.

Accelerated Curing

It can be acceptable to quickly cure concrete with the use of high-pressure steam, steam at atmospheric pressure, heat, and moisture to accelerate the time it takes to achieve acceptable strength while reducing the time needed for curing. The strength of quick-cure concrete must be confirmed to be equal to the required design strength at load stage. Durability of concrete cured quickly must also be equivalent to conventionally cured concrete. Independent testing may be required to determine strength and durability if demanded by an engineer or architect.

FIGURE 4.5 Example of a swirl finish in concrete to avoid a slippery finished surface.

WEATHER CONDITIONS

Weather conditions can affect concrete placement and curing. If you are working in cold temperatures, you must provide equipment to protect concrete from freezing temperatures. This can mean covering the work area with tarps and supplying thermostatically controlled portable heaters.

If you are prepping a job, you must clear all contact areas of frost. This includes reinforcement materials, fillers, and the ground where the concrete will be deposited.

Working in hot conditions has a different set of requirements. If the temperature is excessively high you must protect concrete from curing in extreme temperatures. You must also control water evaporation that could impair the required strength or serviceability of a member or structure.

References

1. *Air Pollutant Emission Factors*, APTD-0923, U.S. Environmental Protection Agency, Research Triangle Park, NC, April 1970.
2. *Air Pollution Engineering Manual*, 2nd Edition, AP-40, U.S. Environmental Protection Agency, Research Triangle Park, NC, 1974. Out of Print.
3. Telephone and written communication between Edwin A. Pfetzing, PEDCo Environmental, Inc., Cincinnati, OH, and Richards Morris and Richard Meininger, National Ready Mix Concrete Association, Silver Spring, MD, May 1984.

4. *Development Document for Effluent Limitations Guidelines and Standards of Performance, The Concrete Products Industries, Draft*, U.S. Environmental Protection Agency, Washington, DC, August 1975.
5. Portland Cement Association. (2001). Concrete Basics. Retrieved August 27, 2001 from the World Wide Web: http://www.portcement.org/cb/.
6. *Technical Guidance for Control of Industrial Process Fugitive Particulate Emissions*, EPA-450/3-77-010, U.S. Environmental Protection Agency, Research Triangle Park, NC, March 1977.
7. *Fugitive Dust Assessment at Rock and Sand Facilities in the South Coast Air Basin*, Southern California Rock Products Association and Southern California Ready Mix Concrete Association, Santa Monica, CA, November 1979.
8. Telephone communication between T.R. Blackwood, Monsanto Research Corp., Dayton, OH, and John Zoller, PEDCo Environmental, Inc., Cincinnati, OH, October 18, 1976.
9. *Final Test Report for USEPA* [sic] *Test Program Conducted at Chaney Enterprises Cement Plant*, ETS, Inc., Roanoke, VA, April 1994.
10. *Final Test Report for USEPA* [sic] *Test Program Conducted at Concrete Ready Mixed Corporation*, ETS, Inc., Roanoke, VA, April 1994.
11. *Emission Test for Tiberi Engineering Company*, Alar Engineering Corporation, Burbank, IL, October 1972.
12. *Stack Test "Confidential"* (Test obtained from State of Tennessee), Environmental Consultants, Oklahoma City, OK, February 1976.
13. *Source Sampling Report, Particulate Emissions from Cement Silo Loading*, Specialty Alloys Corporation, Gallaway, TN, Reference number 24-00051-02, State of Tennessee, Department of Health and Environment, Division of Air Pollution Control, June 12, 1984.
14. Richards, J. and T. Brozell. *"Ready Mixed Concrete Emission Factors, Final Report"* Report to the Ready Mixed Concrete Research Foundation, Silver Spring, MD, August 2004.

CHAPTER

5

Concrete Formwork

OUTLINE

Concrete formwork regulations in the code generally are set at minimum standards. Ultimately, forms used in construction must produce a finished product that complies with approved design drawings and specifications. Any forms used have to be substantial and tight enough that mortar will not leak. In the construction process the appropriate securing and bracing of forms is required. If there is a previously placed structure in existence the use of new forms must not have a detrimental effect on the existing materials.

When you design formwork you must factor in the rate and method of placing concrete. All construction loads must be considered, including vertical, horizontal, and impact loads. Building shells, folded plates, domes, architectural concrete, or similar types of elements can require compliance with special regulations for formwork.

On occasions when prestressed concrete members are designed they must be formed with forms that can be removed without moving the structural members to a point where damage may occur (see Figure 5.1).

FORM REMOVAL

Form removal must be done in a manner that will not damage the supported concrete. The removal process must be done safely. Any concrete exposed after form removal must be strong enough to prevent damage during the removal process. A code officer may request a detailed plan of form removal prior to allowing forms to be removed.

FIGURE 5.1 Concrete forms and reinforcement in place.

EMBEDDED ITEMS

Embedded items that come into contact with concrete must be protected. This can be as simple as installing a sleeve of some type of material that is not compromised by being in contact with concrete.

Any aluminum piping or conduits installed within concrete must be coated or covered to prevent aluminum–concrete reaction or electrolytic action between the two materials.

When sleeves, conduits, or piping are installed and allowed to penetrate a slab, wall, or beam, the penetration must not degrade the strength of the concrete by any significant amount. When these items are placed in columns they must not displace more than 4% of the area of cross section on which concrete strength is determined.

Penetrations through concrete structures must not be larger than one-third of the overall thickness of the structure being penetrated.

The following list includes factors and regulations that must be observed when embedding reinforcement and fixtures in new concrete:

- Embedded items cannot be installed closer than three diameters or widths on center.
- Embedded items must not significantly impair the strength of construction.
- Embedded items cannot be installed in such a way that allows rusting or any other form of deterioration.
- Embedded items cannot be made of aluminum that is not coated for concrete contact.
- Galvanized steel pipe is not allowed to be embedded in concrete.
- Steel pipe placed in contact with concrete must not be thinner than Schedule 40 pipe.
- Embedded items must not have inside diameters in excess of 2 in.
- The spacing of embedded members must not be less than three diameter sizes when measured on center.
- All pipe, fittings, and other materials to be embedded in concrete must be designed for such a purpose.
- Piping that contains liquid, gas, or vapor, with the exception of water that does not exceed a temperature of 90°F or a pressure of 50 pounds per square inch (PSI), cannot be placed in piping embedded in concrete until the concrete has achieved its design strength.
- When piping is installed in a slab, it is to be placed between the top and bottom reinforcement. One exception to this is piping used for radiant heating or ice melting.
- The minimum concrete cover required over pipes, conduits, and fittings is 1½ in. when the concrete is exposed to earth or weather. Concrete protected from contact with ground and weather requires a minimum cover of ¾ in.
- Normal piping requires reinforcement that is not less than 0.002 times the area of the concrete section under review.
- The cutting, bending, or displacement of reinforcement from its proper location is not allowed.

CONSTRUCTION JOINTS

Construction joints must be clean and have all laitance removed. These joints must be wetted with standing water removed just prior to the placement of new concrete. The design strength of concrete must not be compromised by construction joints. The joints in floors are to be located within the middle third of spans of slabs, beams, and girders.

PRE-FORMED CONCRETE PRODUCTS

Pre-formed concrete products offer a great option for contractors doing routine work (see Figure 5.2). These products are required to meet code requirements when they are manufactured. This takes a lot of the guess work out of jobsite inspections. The burden of compliance is on the manufacturer of the products made.

These products can be used for everything from manhole construction to bridge construction to setting building walls in place in large sections (see Figure 5.3). When all cost factors are considered the use of precast products can be a very wise decision (see Figures 5.4 and 5.5).

FIGURE 5.2 Pre-formed concrete materials ready for use.

FIGURE 5.3 A large selection of pre-formed concrete products awaiting use.

FIGURE 5.4 Precast walls that have been set in place for building construction.

FIGURE 5.5 Notice the window sections in the precast products used in this photo for building construction.

Tip

Beams, girders, haunches, drop panels, and capitals are to be placed monolithically as part of a slab system, unless there is an exception approved in the drawings and specifications.

Construction joints in girders are to be offset a minimum distance of two times the width of intersecting beams. Any concrete columns or walls that will support beams, girders, slabs, or similar items must not be subjected to loads until the vertical columns are no longer plastic.

CHAPTER

6

Reinforcement

OUTLINE

Concrete is very dependent on proper reinforcement for use as a material for long-term construction. Without proper reinforcement, concrete is likely to crack, spall, and fall apart. Good reinforcement takes many forms. With this in mind, let's review some of the most common forms of concrete reinforcement.

Tip

It is a code violation to bend reinforcement partially embedded in concrete unless an engineer authorizes a technique and the work.

STANDARD HOOKS

When the code refers to standard hooks, it can mean one of two types of hooks. The first type of hook has a 180 degree bend. This must also include an extension that is not less than 2.5 in. at the free end of the bar. The other type of standard hook has a 90 degree bend with an extension at the free end of the bar.

Stirrup and Tie Hooks

Stirrup and tie hooks are limited to number 8, or smaller, bars. A 90 degree hook is limited to a number 5 or smaller bars. Welded wire reinforcement can be used for stirrups and ties.

Stirrup and tie hooks can use bars ranging from less than a number 5 to number 8. A number 5 or smaller bar has to have a 90 degree bend and an extension at the free end of the bar. A number 8 bar or something smaller is required to have a 135 degree bend and an extension at the end of the bar. On the one hand a number 5 bar requires a 90 degree bend, but a bar smaller than a number 8 bar requires a 135 degree bend, which seems a bit confusing. What establishes the difference?

The code book and the math formulas establish the difference. Equations and tables can be found in your local code book, which will dictate the final use of particular rods. There are some commonly accepted standards for rods used in the United States. For complete accuracy when calculating reinforcement rods refer to your local concrete code book.

Tip

All reinforcement is to be bent cold, unless an engineer states otherwise.

CONDITION OF REINFORCEMENT MATERIALS

All reinforcement materials must be free of coatings that may decrease bonding power, including mud, oil, and non-metallic coatings. With the exception of prestressing steel, reinforcement can be used if it is covered with rust or mill scale. Prestressed steel has to be clean of all oil, dirt, scale, pitting, and excessive rust. A light coating of rust is acceptable with prestressed steel.

All reinforcement, including tendons and post-tensioning ducts, are required to be placed so they provide adequate support (see Figure 6.1).

REINFORCEMENT SPACING

Reinforcement spacing is controlled by local code requirements. The spacing is not to be less than 1 in. If parallel reinforcement bars are placed over each other, they must align directly with each other and be separated by a minimum of 1 in. Spirally reinforced and tied reinforcements cannot be spaced closer than 1.5 in.

With the exception of concrete joists, primary flexural reinforcement in walls and slabs is not to be spaced more than 18 in. apart.

BUNDLING BARS

The process of bundling bars for concrete reinforcement must meet the following requirements:

- No more than four bars are allowed in a bundle.
- Bundled bars are to be enclosed within stirrups or ties.
- No bars larger than number 11 are to be bundled in beams.
- Other requirements may exist in your local code.

FIGURE 6.1 Example of dense, horizontal use of rebar reinforcement.

Tip

The bundling of post-tensioning ducts will be permitted if it is determined that concrete can be placed in a suitable manner without risk of steel breaking through ducts.

HOW MUCH COVERAGE IS NEEDED?

How much coverage is needed when pouring concrete over reinforcement material? It all depends on the type of installation. The following list contains some solid examples of recommended practices for providing adequate coverage of cast-in-place, non-prestressed concrete over reinforcement members.

- When concrete is cast against, and permanently exposed to, earth it requires a minimum coverage of 3 in. of concrete.
- Concrete exposed to earth or weather requires a minimum coverage of 2 in. of concrete for reinforcement bars ranging from number 6 to number 18 bars. When the bar size is number 5 and wire is W31 or D31 or smaller, the minimum required coverage is 1.5 in.
- Concrete that is not exposed to weather nor in contact with ground, such as slabs, walls, and joists, requires a minimum coverage of 1.5 in. of concrete for number 14 and number 18 bars. When bars are number 11 or smaller, the minimum coverage requirement is 0.75 in.
- Beams and columns used as primary reinforcement, ties, stirrups, and spirals call for a minimum coverage of 1.5 in. of concrete.
- Shells and folded plate members that are a number 6 or larger bar require 0.75 in. of coverage. Number 5 bars and W31 and D31 wire or smaller material require a minimum coverage of 0.5 in. of concrete.

You can find different coverage requirements for cast-in-place, prestressed concrete. There are also some differences in coverage requirements when dealing with precast concrete. Concrete must be protected from corrosive environments. To be safe, refer to your local code book prior to pouring any concrete.

Tips

Bundled bars require concrete coverage equal to the diameter of the bundle. Coverage is not required to exceed 2 in. There is one exception—concrete cast against and permanently exposed to earth requires a minimum coverage of 3 in.

COLUMN SUPPORT

There are a few rules to be followed when reinforcing concrete columns (see Figure 6.2). For example, the requirements for offset bent longitudinal bars include the following:

- The slope of the inclined portion of an offset bar with the axis of the column must not be more than 1 in 6.

- Any portion of a reinforcement bar above and below and offset is to be parallel to the axis of a column.
- Horizontal support at offset bends can be provided by lateral ties, spirals, or parts of the floor construction.
- Horizontal support must be able to resist 1½ times the horizontal component of computed force in an inclined portion of an offset bar.
- When lateral ties or spirals are used, they must not be more than 6 in. apart.
- The bending of offset bars is required to be done prior to placement in concrete forms.

FIGURE 6.2 Vertical rebar reinforcement installation.

SOME SPIRAL FACTS

The following list provides necessary facts about spirals.

- Spirals must consist of evenly spaced continuous bar or wire. The size of this material must be adequate to prevent distortion from designed dimensions when handled.
- Cast-in-place construction requires a minimum spiral size of 3/8 in.
- The spacing between spirals is to be not less than 1 in. and not more than 3 in.
- Spirals are required to have anchorage at the end of the spiral unit. Additional anchorage is required when there are 1½ extra turns in a spiral.
- Spiral reinforcement can be spliced.
- Spirals must extend from the top of a footing or slab in any story to the level of the lowest horizontal reinforcement in members supported above.
- Spirals must be held firmly in place and true to line.

This list is not comprehensive of all requirements that may be encountered in local code requirements, but it does give you a strong overview of the necessary requirements.

Tips

In columns with capitals, spirals must extend to a level at which the diameter or width of capital is two times that of the column.

TIES

Any non-prestressed reinforcement bar is required to be enclosed by lateral ties. Deformed wire or welded wire reinforcement is not allowed to be used as a lateral tie. The vertical spacing of ties must not exceed 16 longitudinal bar diameters, 48 tie bar or wire diameters, or least dimension of the compression member. Ties must be arranged in an approved manner. Confirm tie requirements with your local code enforcement office.

STRUCTURAL INTEGRITY

Structural integrity is a key factor in concrete design and implementation. What is required to assure structural integrity? The code will give you regional requirements. These can change considerably depending on where the work is done. Seismic requirements, for example, can be very different in California than in Maine.

All structural integrity is established with proper design procedures by qualified professionals and is the responsibility of contractors and installers. I could give you some benchmark data to work with, but this is far too serious for any guessing or mistakes. Take the personal responsibility to ensure that your work meets, or exceeds, minimum code requirements and design specifications for structural integrity.

CHAPTER

7

General Design Consideration

OUTLINE

General design considerations for concrete construction must deal with two types of loads — live and dead. Other types of loads can pertain to earthquake loads and wind loads. Adequate design for lateral loads is always an important consideration.

In addition to typical loads, there are other factors to consider when working with the design and analysis of concrete. Some of these factors are listed below.

- Prestressing
- Crane loads
- Creep
- Expansion
- Shrinkage-compensating concrete
- Unequal support settlement

ANALYSIS

There are different forms of analysis that can be used when working with concrete. Frame analysis is one type that is usually satisfactory for buildings of normal types of construction, spans, and story heights. It is not to be used with prestressed concrete.

Tip

The use of any set of reasonable assumptions is permitted for computing relative flexural and torsional stiffness of columns, walls, floors, and roof systems. The assumptions adopted must be consistent throughout analysis.

When dealing with continuous beams or one-way slabs, use approximate moments and shears provided by your local code regulations. This is an alternative to the frame-analysis method. To do this, you must know these facts:

- There is more than one span.
- Spans are to be approximately equal.
- The larger of two adjacent spans must not be greater than the shorter span by more than 20%.
- Loads must be uniformly distributed.
- Unfactored live loads must not exceed three times the unfactored dead load.
- Members must be prismatic.

Tips

Effect of haunches must be considered when determining moments and designing members.

LENGTH OF SPANS

It is acceptable for beams built integrally with supports to have their design based on the moments at faces of support. The span length of members that are not built integrally with supports must be considered as the clear span plus the depth of the member, but must not exceed the distance between centers of supports.

For analysis of frames or continuous construction for determination of moments, span length is the distance center-to-center of supports. You are allowed to analyze solid or ribbed slabs built integrally with supports, but clear spans must not be more than 10 feet as continuous slabs on knife-edge supports with spans equal to the clear spans of the slab and width of beams otherwise neglected.

COLUMNS

Axial forces must be considered when designing concrete columns. All columns created have to be able to resist axial forces of factored loads on all floors or roofs. Any unbalanced loads must be considered in the design of columns.

Gravity load moments in columns allow the assumption far ends of columns built integrally with the structure will be fixed. Resistance to moments at any floor or roof level must be provided by distributing the moment between columns immediately above and below the given floor in proportion to the relative column stiffnesses and conditions of restraint.

FIGURE 7.1 Multi-level concrete structure.

LIVE LOAD ASSUMPTIONS

Assumptions that can be made when designing the arrangement of live loads include:

- A live load is applied only to the floor being constructed.
- A live load is applied only to the roof being constructed.
- The far ends of columns built into a structure are to be fixed.
- The arrangement of live load is limited to combinations of factored dead load on all spans with full factored live load on two adjacent spans and factored dead load on all spans with full factored live load on alternative spans.

T-BEAM CONSTRUCTION

Flanges and webs in T-beam construction are required to be built integrally or otherwise effectively bonded together. The width of slab effective as a T-beam flange is not to exceed one-quarter of the span length of the beam, and the effective overhanging flange width on each side of the web must not exceed eight times the slab thickness and one-half the clear distance to the next web. This type of construction is common in multi-level concrete structures.

Tips

Transverse reinforcement must be spaced no farther apart than five times the slab thickness or farther apart than 18 in.

For beams with a slab on one side only, the effective overhanging flange width must not exceed one-twelfth the span length of the beam, six times the slab thickness, and one-half the clear distance to the next web.

Isolated beams that utilize a T-shape to provide a flange for additional compression area must have a flange no more than one-half the thickness of the width of the web. The effective flange width must not be more than four times the width of the web.

Tips

Transverse reinforcement must be designed to carry the factored load on the overhanging slab width assumed to act as a cantilever.

JOIST CONSTRUCTION

Joist construction is described as a monolithic combination of regularly spaced ribs and a top slab arranged to span in one direction or two orthogonal directions. Ribs must not be less than 4 in. wide. The depth of a rib is not allowed to be more than 3.5 times the minimum width of ribs.

Tips

The required maximum distance between ribs is 30 in.

The creation of a concrete slab over permanent fillers must be created to be no less than one-twelfth the clear distance between ribs and not less than 1.5 in.

The use of ribs in joist construction can trigger a variety of requirements. Many of these requirements depend on fillers, removable forms, slab thickness, shells, and conduits. Consult your local code requirements to assure compliance of your proposed work.

Tips

When joist construction does not meet required limitations, it must be designed as slabs and beams.

FINISHED FLOORING

Finished flooring is not included as a part of a structural member. One exception is if the flooring is placed monolithically with the floor slab. Finished floor coverings can be considered as a part of the required cover or total thickness for non-structural considerations.

CHAPTER

8

Requirements for Strength and Serviceability

OUTLINE

Requirements for strength and serviceability are based on load factors. A design must meet minimum requirements as set forth by the code. Several potential factors need to be addressed when designing for strength and serviceability. These factors include:

- Expansion of shrinkage-compensating concrete
- Temperature changes
- Creep
- Settlement
- Shrinkage

I have been in the construction trades for about 30 years. My career began in Virginia, and I have also worked in South Carolina, Colorado, New Hampshire, and Maine. The differences in geographical locations have affected day-to-day construction procedures and code requirements.

Tips

Any structures built in flood zones or where atmospheric ice loads are a consideration must be designed to withstand these potential forces.

Last week I was talking to a worker from Jamaica about this book and good concrete practices. He explained to me that concrete work done in Jamaica is done on a very different level. The man described a group of workers mixing batches of concrete in wheelbarrows and dumping, then starting another batch once the previous batch was mixed. In the example he was describing the construction for a home.

How much consistency do you think could be maintained by a variety of workers with varying skills mixing concrete in wheelbarrows? I doubt that the consistency would be very consistent at all.

Tip

A load factor of 1.2 is required to be applied to the maximum prestressing steel jacking force for post-tensioned anchorage zone designs.

Now think about how the batches were being poured. A wheelbarrow load here and there is not good procedure for a solid pour. Without a lot of reinforcement material, concrete made and poured in the manner described in Jamaica is not going to have significant strength or durability.

A couple of days ago I was talking to a friend of mine who is a very experienced builder in Maine. When I told him the Jamaican concrete story he scoffed. His definition of that type of work was that it was like stacking rocks on top of each other if there is not enough steel in it. We laughed, but he is right. A concrete pour needs to be consistent to meet suitable design criteria. This includes proper preparation, reinforcement, such as the use of welded wire and rebar, and an organized approach.

FIGURE 8.1 Preparing a site for a concrete pour.

DESIGN STRENGTH

Design strength takes a number of factors into consideration. For example, the strength provided by a member has to be computed. The same is true for connections between members. Cross sections have to be designed for strength. Then there is flexure, axial load, shear, and torsion to take into consideration. Here are some examples of the types of conditions that may be required to determine the viable strength of a concrete product.

- Strength reduction factor
- Tension-controlled sections
- Compression-controlled sections
- Members with spiral reinforcement
- Other reinforced members
- Tensile strain
- Shear
- Bearing on concrete
- Post-tensioned anchorage zones
- Strut-and-model ties
- Flexural sections in pre-tensioned members where strand embedment is less than the development length
- Where bonding of a strand does not extend to the end of a member
- Where strand embedment is assumed to begin at the end of the debonded length
- Special moment-resisting frames
- Special reinforced concrete structural walls to resist earthquake effects
- Joints
- Diagonally reinforced coupling beams

Tips

Reinforced concrete members subjected to flexure must be designed to have adequate stiffness to limit deflections or any deformations that adversely affect strength or serviceability of a structure.

COMPLICATED CALCULATIONS

The calculations used to define strength and serviceability of concrete can be fairly complicated for the average builder, contractor, or concrete worker. This is why design professionals normally create the plans and specifications for projects. In addition to what has already been covered in this chapter, there are many other factors to consider during the proper design of concrete structures:

- Prestressed concrete construction
- Composite construction
- Shored construction
- Unshored construction

FIGURE 8.2 Installing reinforcement for a commercial concrete pour.

LOCAL PRACTICES

Local practices for concrete construction vary. Weather is one factor that creates these changes. The risks from various types of loads are another factor in design criteria. For example, the design specifications for a column in Maine would likely be very different from the same column used in an area subject to extreme seismic action. The same could be true of a wall structure being built in a region where extremely high winds may be expected. It is essential to consult your local code requirements for specifics that apply to your region. Concrete workers must be diligent in placing adequate reinforcement materials in concrete walls and floors to overcome seismic action, frost heaves, and other conditions that may damage the finished concrete.

COMMON SENSE

Common sense is always a good resource to pull from. While experience and common sense are not always enough to rely on when working with concrete, it can go a long way in avoiding problems. The best designs can have flaws in them. I have been on numerous jobs where the plans and specifications looked great on paper, but they were not so fantastic on the construction site.

One example has nothing to do with concrete, but it is one of the war stories that I love to tell. My crews were installing a plumbing system in a new building. The architectural plans and specs were clear on where the main building sewer should be routed. The design paperwork had been approved by an architect and the local code enforcement office.

I happened to be on the job when the foreman discovered a glitch in the building plan. Dead in the middle of where the foreman was supposed to install a 6-in. diameter building drain was a huge metal HVAC duct trunk. Upon seeing this, I called the architect for instructions.

The architect proceeded to explain to me that the air duct could not be where I was saying it was. He went on to explain how the drawings would not have allowed for such a conflict. It made perfect sense on the phone and on paper, but the duct trunk was, in fact, there. So, what are you going to do? In this case, my crew had to find a way to reroute the piping. That made more sense than placing it inside of the heating-and-cooling duct.

Granted, my example has nothing to do with concrete, but it does define the type of problems that can arise in the real world. Here is another example.

I was building a subdivision in Virginia. I was using the plans and specs that detailed the footing requirements. When the site contractor dug the footing trench, he opened an underground stream. How do you pour a concrete footing into a trench that has become a stream? Simple, you don't.

It would seem like we might have been able to shift the foundation location. However, this subdivision had small lots and setback requirements, and there was very little room to navigate. Engineers were dispatched to come up with an answer for this one. It involved some diversion digging, some underground drain piping, and some temporary water pumping. Ultimately, the task was completed. I can only image what this must have cost. Fortunately, I did not have to pay the bill.

CHAPTER

9

Inspecting In-Place Concrete

OUTLINE

Proper evaluation of concrete requires several steps. The need for repairs or rehabilitation is generally established in one of three ways. In the worst case, there is a failure of structural integrity that prompts the need for an evaluation. Visual inspections often reveal a need for further evaluation and testing. If cracks, flaking, or other visual defects are encountered, a full investigation and evaluation is usually warranted. The third means of generating a need for an evaluation is periodic testing. This is the safest way to check the strength and dependability of concrete structures.

PLANS AND SPECS

Construction documents are typically filed before permits for construction are issued. For concrete construction, reviewing original documents and a paper trail over the history of the structure is one step in a thorough evaluation. Examples of existing records that should be available for consideration include:

- Design documents
- Plans
- Specifications
- History reports
- Inspection reports
- Site inspection results
- Laboratory test records
- Concrete records on the materials used in construction and the batch plant
- Instrumentation documents
- Operation reports
- Maintenance reports
- Monument survey data

JUDGING THE SITE

A site survey consists of a visual exam of all exposed concrete. Inspectors are looking for any implication of distress that may require repair or rehabilitation. Standard procedure is to create a map of potential defects that are encountered, including cracks, disintegration, spalling, or joint deterioration. The mapping process is typically done with fold-out sketches of the monolith surfaces. Mapping must include inspection and delineating of pipe and electrical galleries, filling, and emptying culverts, when possible. Core drilling is a common procedure when evaluating concrete. The material recovered from the drilling can be taken to a laboratory for scientific testing.

Code Consideration

It is not acceptable to use tests for splitting tensile strength to establish the acceptance of concrete in the field.

QUALITY OF INSTALLATION

A lack of quality installation can lead to significant problems in concrete structures. Site inspections can bring such problems to light, which often results in further testing. Some of the normal defects looked for in a visual inspection include:

- Cold joints
- Bug holes

FIGURE 9.1 Pre-formed concrete components awaiting inspection.

- Reinforcing steel that has become exposed
- Honeycombing
- Surface defects that can indicate serious problems

Workmanship should be monitored during the construction process. In a perfect world it would be, but in the real world, corners are cut and problems can result from the lack of professionalism. Since safety is a paramount concern, routine inspections are needed to confirm that the workmanship in a structure is as it should be (see Figure 9.1).

Code Consideration

Sometimes different materials are used on the same job for different purposes. When this is the case, each material has to be evaluated for each purpose it is used.

CRACKED CONCRETE

Cracking is a common problem in concrete construction (see Figure 9.2). Homeowners see it in basement floors, garage floors, and basement walls. Cracks occur in sidewalks, dams, bridges, and retaining walls. Any crack can be reason for concern and should result in a thorough inspection and investigation (see Table 9.1).

FIGURE 9.2 Cracked concrete.

TABLE 9.1 Terms Associated with Visual Inspection of Concrete

Construction Faults	Distortion or Movement
Bug holes	Buckling
Cold joints	Curling or warping
Exposed reinforcing steel	Faulting
Honeycombing	Settling
Irreqular surface	Tilting
Cracking	**Erosion**
Checking or crazing	Abrasion
D-cracking	Cavitation
Diagonal	**Joint-sealant failure**
Hairline	**Seepage**
Longitudinal	Corrosion
Map or pattern	Discoloration or staining

TABLE 9.1 (*Continued*)

Construction Faults	Distortion or Movement
Random	Exudation
Transverse	Efflorescence
Vertical	Incrustation
Horizontal	**Spalling**
Disintegration	Popouts
Blistering	Spall
Chalking	—
Delamination	—
Dusting	—
Peeling	—
Scaling	—
Weathering	—

Courtesy of United States Army Corps of Engineers.

Pattern Cracking

Pattern cracking is common. These cracks tend to be short in length and uniformly distributed throughout a concrete surface. There is usually one of two causes for pattern cracking: restraint of contraction on the surface layer by the backing or inner concrete or an increase of volume in the interior concrete.

You may hear pattern cracking referred to as map cracks, crazing, and checking. D-cracking is also a form or pattern cracking. This type of cracking is often found in the lower part of a concrete slab. It usually happens near a joint in the concrete. If you find moisture accumulation, you could find D-cracking.

Isolated Cracks

Isolation cracks appear as individual cracks. This type of cracking indicates tension on the concrete. The tension is usually perpendicular to the cracks. An individual crack can run in a diagonal, longitudinal, transverse, vertical, or horizontal direction.

Code Consideration

Code enforcement officers have the authority to require the results of strength tests of cylinders cured under field conditions.

Crack Depth

Crack depth is listed in four categories:

- Surface
- Shallow
- Deep
- Through

Crack Width

Crack width ranges from fine to medium to wide. Fine cracks are typically less than 0.04 in. wide. A medium crack would be from 0.04 to 0.08 in. Wide cracks exceed 0.08 in.

Code Consideration

When preparing to pour concrete, installers must ensure that the following steps are taken:

- Clean equipment
- No debris
- No ice
- Clean forms
- Any filler units in contact with concrete must be well drenched
- No deleterious coatings or ice on reinforcing materials
- No water in the path of concrete installation without consideration and approved methods acceptable to a code officer
- No unsound material allowed

CRACK ACTIVITY

Crack activity is the presence of a factor causing a crack. Determining crack activity is necessary to determine a mode of repair. Any crack that is currently moving or if a specific cause for a crack cannot be determined, the crack must be considered active.

Dormant cracks do not have current movement. Some cracks are considered dormant when any movement of the crack is minimal enough not to interfere with a repair plan.

Code Consideration

When concrete is mixed, it must be mixed to a uniform distribution of materials. All materials must be discharged completely before the mixer can be used again.

CRACK OCCURRENCE

Cracks can occur before or after concrete cures. The cracking can be structural or nonstructural. There may be trouble deciding if a crack is structural or nonstructural when only a

TABLE 9.2 Vibration Limits for Freshly Placed Concrete

Age of Concrete at Time of Vibration (hr)	Peak Particle Velocity of Ground Vibrations
Up to 3	102 mm/sec (4.0 in./sec)
3–11	38 mm/sec (1.5 in./sec)
11–24	51 mm/sec (2.0 in./sec)
24–48	102 mm/sec (4.0 in./sec)
Over 48	178 mm/sec (7.0 in./sec)

Hulshizer and Desci, 1984; Courtesy of United States Army Corps of Engineers.

visual inspection is done. A full analysis by a structural engineer is normally required to make a full determination of the type of cracking encountered.

Structural cracks tend to be wide. Their openings can increase as a result of continuous loading and creep of the concrete. As a rule of thumb, any crack that could be structural in nature should be treated as a structural defect and receive a full evaluation from appropriate experts (see Table 9.2).

DISINTEGRATION

Disintegration is the deterioration of any type of concrete. What is the difference between disintegration and spalling? The mass of particles being removed from the main body of concrete determines disintegration from spalling. When the loss is small, it is called disintegration; when the loss of intact concrete is large, it is called spalling.

Code Consideration

When concrete is poured, the final deposit method must not separate or lose materials in the process.

SCALING

Scaling is a form of disintegration. A common cause of scaling is freezing and thawing conditions. Localized flaking or peeling is normally a form of scaling. Scaling, which can also be referred to as spalling, is rated based on the depth of the defect. The rating system is as follows:

- Light spalling occurs when the loss of surface mortar does not expose any coarse aggregate.
- Medium spalling occurs when the damage has a depth of up to 0.4 in.

- Severe spalling is determined when the depth of the damage ranges from 0.2 to 0.4 in. There is also some loss of mortar surrounding aggregate particles with a depth range of 0.4 to 0.8 in.
- Very severe spalling is the loss of coarse aggregate particles as well as surface mortar and surrounding aggregate, generally to a depth greater than 0.8 in.

DUSTING

Dusting occurs when powdered material at the surface of hardened concrete develops. Horizontal concrete surfaces that receive heavy traffic are the most likely surfaces to exhibit dusting. Poor workmanship is generally at the root of the problem when dusting is found. Anyone in construction for any length of time has seen finishers spray water on a concrete surface during the finishing process. This practice is likely to create dusting.

Code Consideration

Poured concrete should be installed as closely as is practical to the final resting place of the mixture.

DISTORTION

Distortion of concrete is also known as movement. Put simply, this is a change in alignment of the components of a structure, which could be wall movement or support movement. When evaluating distortion, historic data can be helpful. Assuming that good records have been maintained over the years, research may turn up a history of continuing failure due to distortion.

EROSION

Erosion is a common concern in concrete construction. Experts generally break erosion down into one of two categories: abrasion and cavitation. Abrasion is recognized by the smooth surface that it leaves behind on concrete due to repeated rubbing and grinding of debris, equipment, gravel, or other items against concrete.

Repeated impact forces caused by a collapse of vapor bubbles in rapidly flowing water causes cavitation. Erosion caused by water does not generally leave a smooth surface on concrete, and a rough, pitted concrete surface is a sign of cavitation. In severe cases, cavitation can result in structural damage. Statistics indicate that cavitation generally requires a water velocity of at least 40 feet per second.

Code Consideration

Vertically formed lifts are required to be generally level.

SEAL FAILURE

Joint seals can fail. The seals are intended to repel water, but if a seal fails and water invades a concrete joint then buckling, cracking, erosion, and other problems may occur. Another purpose of joint seals is to prevent debris from entering an expansion joint. When debris embeds in a joint, it can result in expansion joint failure.

Seepage consists of water or other fluids moving through pores or interstices. When conducting a visual inspection, seepage can be checked by looking for the following:

- Water
- Dampness
- Moisture
- Corrosion
- Discoloration
- Staining
- Exudations
- Efflorescence
- Incrustations

Seepage is common around hydraulic structures, and any seepage found should be reported. These data are needed to maintain historical facts for future review.

Cracks associated with hardening concrete fall into two categories: those that happen while concrete is hardening and those that occur after concrete has hardened. Knowing when a crack occurs and what caused the cracking is always of interest to those who evaluate concrete construction.

Code Consideration

Concrete must remain plastic and flow readily while being installed.

SPECIAL CASES OF SPALLING

There are two types of special cases that involve spalling. The first type is known as a popout. These defects are shallow and are usually conical depressions in the surface of concrete. When concrete is poured in freezing conditions, especially if unsatisfactory aggregate particles exist, you may find popouts.

A popout is created when water finds its way into coarse aggregate and then freezing begins. The ice pushes off the top of the aggregate particle and superjacent layer of mortar. This leaves shallow pitting. Certain materials are more susceptible to popouts than others. If you are working with chert particles of low specific gravity, limestone that contains clay, or shaly material you may witness spalling.

The second special case of spalling involves the corrosion of reinforcement material. A visual inspection for this type of defect is fairly simple. It is usually identified by exposed reinforcement materials protruding through the concrete. Rust staining on the reinforcement material is also present in many cases.

Code Consideration

It is common practice to maintain a minimum temperature for most concrete installations at 50°F.

DELAMINATION

Delamination can occur when reinforcing steel is installed too close to the surface of a concrete structure. Chloride ions or air forming rust, or iron oxide, will corrode steel reinforcement material. A corresponding increase in volume up to eight times the original volume amount is possible. The increased volume can crack concrete.

There is a simple, inexpensive way to perform a preliminary test for delamination. All you need is a pair of safety glasses and a hammer. Use the hammer to tap the concrete being tested for defects. If the sound of the contact is a sharp ping, it is a good sign that delamination is not present. However, a hollow, echo sound should spur you on to perform further tests. This type of test is common when working with small surface areas.

Larger surface areas require extensive time to sound with a hammer. Horizontal surfaces can be tested by dragging a chain over the area. Listen for the same types of sound that you would when using a hammer. This is a simple way to cover more concrete in less time.

Infrared thermography is a more advanced method for inspecting concrete for delamination. The thermal gradients within concrete exposed to sunlight can be measured with thermography equipment, because delamination interrupts the heat transfer through concrete. Higher surface temperatures will be present if delamination is present. Infrared thermography is capable of identifying and recording large areas affected by delamination.

Code Consideration

Accelerated curing is an allowable practice.

CRACK SURVEYS

Crack surveys are needed for historical records and to expose potential problems. Obviously, cracks are not supposed to exist in concrete structures, but when they are found, the cracks must be identified. Typically, cracks are marked and researched. Once the type of crack and its cause is known, they are assessed and recorded for future review.

SIZING CRACKS

Sizing cracks is done with various methods. A simple card that contains lines of various widths can be used to estimate crack size. Crack monitors, which are small, hand-held microscopes, are used for more specific measurements. There are also transducers that can be used for crack measuring. Once the size of a crack is determined, it should be recorded in the historical data of the concrete structure. By doing this, the cracks can be monitored and measured to determine if they are growing in size.

Obtaining the depth of cracks is sometimes simple, but it can be extremely difficult to determine in some cases. A small measuring device, such as a feeler gauge, can be used to establish the depth of some cracks. It is not uncommon for simple methods to fail in determining the depth of a crack. When this is the case, inspectors must turn to more sophisticated methods, such as drilling or pulse-velocity measurements.

Code Consideration

At no time can forms, fillers, or the ground that concrete comes into contact with be covered with frost.

MAPPING

Surface mapping is important when establishing the history of a concrete surface over time. The types of defects that are sought in surface mapping may include:

- Cracking
- Spalling
- Scaling
- Popouts
- Honeycombing
- Exudation
- Distortion
- Unusual discoloration
- Erosion
- Cavitation
- Seepage
- Joint condition
- Joint materials
- Corrosion of reinforcement materials

Inspectors performing surface mapping by hand will need certain tools that may include:

- Structural drawings
- Historical data
- Documentation tools, such as a notebook computer or a note pad and pen
- Tape measure
- Ruler
- Feeler gauge
- Hand microscope
- Knife
- Hammer
- Fine wire
- String

- Flashlight
- Camera outfit
- Tape recorder

JOINT INSPECTIONS

Joint inspections can be done with a visual tour of all joints. Expansion, contraction, and construction joints should all be inspected. The condition of joints, both good and bad, should be noted for inclusion in the historical record. Potential defects to look for include:

- Spalling
- D-cracking
- Chemical attack
- Seepage
- Emission of solids

CORE DRILLING

Core drilling is the best method for testing concrete, but it is expensive. If the quality of concrete in a structure is suspected to be weakened after general inspections, core drilling may be the next step. Scaling, leaching, or pattern cracking can be a sign that core drilling is necessary.

How deep does core drilling go? It depends on the structure. For example, a massive structure may require core sampling to be done at a depth of up to two feet. The diameter of a core sample should be at least three times the nominal maximum size of aggregate. When there is little mortar bonding the concrete across the diameter of the core, you are likely to wind up with rubble, rather than a solid sample. Core samples must be properly labeled, oriented, and stored for future observation. Written records are also required to maintain consistency in the historical data.

There are times when a borehole camera is helpful. The use of such an instrument can reveal facts about the inner condition of a concrete structure. For example, if your core samples are coming up as rubble, the borehole camera may be your best alternative.

Code Consideration

All forms used for concrete installations are required to prevent leakage of mortar.

UNDERWATER CONCRETE

Underwater inspections are usually conducted by scuba divers. When a deep or long dive is required, a diver with surface-supplied air is a better option. Flexibility and speed is an advantage for the scuba diver. In clear water, a visual inspection can be done, but many types of structures are located in water that is not clear enough for this.

Fortunately, many types of test devices used above water have been adopted for use below water. Rebound hammers work underwater, and both direct and indirect ultrasonic pulse-velocity systems can be used below the water surface. These tools give a diver a good reading on the general condition of concrete that is surrounded by water.

Underwater Vehicles

Underwater vehicles are often used to inspect submerged concrete structures. These vehicles come in five different categories of manned units including:

- Untethered
- Tethered
- Diver lockout
- Observation or work bells
- Atmospheric diving suits

All of the these vehicles are operated by a person inside and have viewing ports, dry inside conditions, and some degree of mobility.

Unmanned vehicles are another option for underwater inspections. These units can include the following types:

- Tethered
- Free swimming
- Towed
- Towed midwater
- Bottom-reliant
- Bottom-crawling
- Structurally reliant
- Untethered

Unmanned vehicles are known as remotely operated vehicles (ROVs). Television cameras are mounted on the vehicles, and the vehicle is controlled from the water surface with some type of navigation system, such as a joystick. These vehicles can be fitted to perform inspections and maintenance.

ROVs can be operated at extreme depths, and can remain underwater for long periods of time. Repeated tasks can be completed accurately with ROVs. Another advantage is that ROVs can be operated in harsh conditions that would hamper general diving operations.

Code Consideration

All concrete forms must be installed and secured in a manner that ensures the forms will maintain the desired position and shape.

There are distinct advantages and disadvantages when using an ROV instead of a manned vehicle. Compared to ROVs manned vehicles are big, bulky, and expensive to operate. ROVs are small, flexible, and relatively inexpensive. An ROV provides a two-dimensional view while a manned unit can provide three-dimensional assessments. Both types of vehicles have their place in underwater inspections.

Photographic Tools

Photographic tools have come a long way over the years. An underwater inspection can involve the use of still cameras or video cameras, or both. Video systems can see through turbid water conditions; this is a big plus over the eyes of a diver.

HIGH-RESOLUTION ACOUSTIC MAPPING SYSTEM

High-resolution acoustic mapping systems can be used to check for erosion and faulting. These systems consist of three basic components: the positioning subsystem, the acoustic subsystem, and the compute-and-record subsystem.

An acoustic subsystem is made up of a boat-mounted transducer array and signal processing electronics. This type of system sends output back to a computer, and it calculates the elevation of the bottom surface from the information supplied.

A lateral positioning subsystem has a sonic transmitter on a boat and two or more transponders in the water at a known or surveyed location. The transponders receive a sonic pulse from the transmitter, and this information is then radioed to the survey vessel where a time and location are determined.

Compute-and-record subsystems provide computer-controlled operation of the system and for processing, display, and storage of data. Real-time mapping is done in a computerized manner.

While high-resolution systems are extremely accurate, they do have limitations. These systems typically work in depths ranging from 5 to 40 feet. Another drawback is that a high-resolution system works best in calm water. If there is wave activity that exceeds 5 degrees, this type of system shuts down.

Code Consideration

When removing concrete forms, workers must be sure to maintain all safety and serviceability of the concrete structure.

SIDE SCANNER

A side scanner sonar requires two transducers mounted in a waterproof housing. When a signal is sent from the scanner, it is called a sonograph. Darkened areas and shadows are used for evaluation. The width of shadows and the position of objects can be used to calculate height. Newer versions of scanners have far fewer limitations than early models. Side scanners have been proven useful in breakwaters, jetties, groins, port structures, and inland waterway facilities, such as locks and dams.

OTHER MEANS OF UNDERWATER TESTING

Other means of underwater testing include:

- Radar
- Ultrasonic pulse velocity
- Ultrasonic pulse-echo systems
- Sonic pulse-echo techniques for piles

All of these methods have their advantages and some are listed below:

- Advantages of radar systems
  - Electromagnetic signal emitted from radar travels very quickly.
  - Conductivity controls the loss of energy, and therefore, the penetration depth.
  - Dielectric constant determines the propagation velocity.

- Ultrasonic pulse-velocity advantages
  - Provides a nondestructive method for evaluating structures.
  - Measures the time of travel of acoustic pulses of energy through a material of known thickness.
  - Piezoelectric transducers are housed in metal casings and are excited by high-impulse voltages as they transmit and receive acoustic pulses.
  - An oscilloscope in the system measures time and displays acoustic waves.
  - Reliable in situ delineations of the extent and severity of cracks, areas of deterioration, and general assessments.
  - Capable of penetrating up to 300 feet of continuous concrete with the aid of amplifiers.
  - Can be transported easily.
  - Has a high data acquisition-to-cost ratio.
  - Can be converted for underwater use.

- Advantages of ultrasonic pulse-echo systems
  - Use piezoelectric crystals to generate and detect signals and the accurate time base of an oscilloscope to measure the time of arrival of a longitudinal ultrasonic pulse in concrete.
  - Can delineate sound concrete, concrete of questionable quality, deteriorated concrete, delaminations, voids, reinforcing steel, and other objects within concrete.
  - Can determine the thickness of concrete up to about 1.5 feet.
  - Can be adapted to water environments.

- Advantages of pulse-echo techniques for piles
  - Can determine the length of concrete piles, in tens of feet, in dry soil or under water.
  - Use a round-trip echo time in the pile to measure an accurate time base of an oscilloscope.
  - Can be used to calculate the reference between length and diameter ratios.

LABORATORY WORK

A great deal can be accomplished with site visits and visual inspections. However, it is often laboratory work that tells the tale of the tape. Petrographic exams use a branch of geology that deals with the descriptions and classifications of rocks. Hardened concrete is considered to be a synthetic sedimentary rock. What is usually checked with this type of examination? See the following list:

- Aggregate condition
- Pronounced cement–aggregate reactions
- Deterioration of aggregate particles in place
- Denseness of cement paste
- Homogeneity of concrete
- Settlement and bleeding of fresh concrete
- Depth and extent of carbonation
- Occurrence and distribution of fractures
- Characteristics and distribution of voids
- Presence of contaminating substances

CHEMICAL ANALYSIS

Chemical analysis of hardened concrete can be used to estimate the cement content, original water–cement ratio, and the presence and amount of chloride and other admixtures. This is another form of testing that is part of the larger puzzle in determining the qualities of concrete.

PHYSICAL ANALYSIS

Physical analysis is often done on core samples. This type of testing looks for these nine different elements:

- Density
- Compressive strength
- Modulus of elasticity
- Poisson's ratio
- Pulse velocity
- Direct shear strength of concrete bonded to foundation rock
- Friction sliding of concrete on foundation rock
- Resistance of concrete to deterioration caused by freezing and thawing
- Air content and parameters of the air-void system

NONDESTRUCTIVE TESTING

Nondestructive testing (NDT) is used to determine various relative properties of concrete: strength, modulus of elasticity, homogeneity, and integrity of concrete. There are many approaches to NDT and they require inspectors to have expertise in the given approach to arrive at accurate data.

Rebound Hammers

We talked about rebound hammers earlier. This is a form of NDT that is a fast and simple means of testing concrete. However, the test is imprecise and cannot accurately predict the strength of concrete. Some factors that can skew a test with a rebound hammer include:

- Smoothness of a concrete surface
- Moisture content
- Type of course aggregate
- Size, shape, and rigidity of specimen
- Carbonation of a concrete surface

Probes

Probes are another form of NDT. This device may use a powder cartridge to insert a high-strength steel probe into a section of concrete. The results of probe measurements can be converted to compressive strength values. There are reports, however, that the probes can be inaccurate.

Density is often what is tested with a probe, and a probe will embed deeper in concrete that is suffering from failure in density, subsurface hardness, and as the strength of concrete weakens. This type of testing is fine for on-site, general tests, but it is limited. Precise measurements are not available from probe testing, and the act of probing concrete will leave a hole in the concrete surface that should be repaired.

Ultrasonic Pulse-Velocity Testing

Ultrasonic pulse-velocity testing is probably the most frequently used NDT. The results of this testing can be calculated; high velocities indicate good concrete while low velocities reveal weak concrete. The system for this testing is portable and can penetrate about 35 linear feet of concrete. Testing of this type is fast; however, an inspector must have access to opposite sides of the section being tested, and this can sometimes be a problem.

Acoustic Mapping

Acoustic mapping provides comprehensive evaluation of the top surface wear of concrete in such structures as aprons, sills, lock chamber floors, and so forth. Fast, accurate evaluations of horizontal sections below water can be done with acoustic mapping. Dewatering is not needed. Accuracy falls off at depths greater than 30 feet.

Ultrasonic Pulse-Echo Testing

Ultrasonic pulse-echo testing is good for flat surfaces. It can detect steel and plastic pipe embedded in concrete. Resolution is good with this type of testing equipment, and improvements in this form of testing continue to move forward.

Radar

Radar is also an NDT. It does not require contact with concrete, and resolution and penetration is somewhat limited. Some opinions favor signal testing over radar, but the use of radar is growing in the field of concrete evaluation.

OTHER CONSIDERATIONS

There are other considerations when evaluating concrete. Additional levels are seen in the following list:

- Stability analysis
- Deformation monitoring
- Concrete service life
- Reliability analysis

CHAPTER

10

Concrete Failure

OUTLINE

The list of potential causes of concrete failures is a long one. A few examples include chemical reactions, shrinkage, weathering, and erosion. Many other potential causes exist, and we will explore them individually. Understanding the causes of concrete structure damage is an important element in the business of rehab and repair work.

UNINTENTIONAL LOADS

Unintentional loads are not common, which is why they are accidental. When an earthquake occurs and affects concrete structures, that action is considered to be an accidental loading. This type of damage is generally short in duration and few and far between in occurrences.

Visual inspection will likely find spalling or cracking when accidental loadings occur. How is this type of damage stopped? Generally speaking, the damage cannot be prevented, because the causes are unexpected and difficult to anticipate. For example, an engineer is not expecting a ship to hit a piling for a bridge, but it happens. The only defense is to build with as much caution and anticipation as possible.

CHEMICAL REACTIONS

Concrete damage can occur when chemical reactions are present. It is surprising how little it takes for a chemical attack on concrete to do serious structural damage. The following sections include examples of chemical reactions and how they affect concrete.

Acidic Reactions

Most people know that acid can have serious reactions with a number of materials, and concrete is no exception. When acid attacks concrete, it concentrates on its products of hydration. For example, calcium silicate hydrate can be adversely affected by exposure to acid. Sulfuric acid works to weaken concrete and if it is able to reach the steel reinforcing members, the steel can be compromised. All of this contributes to a failing concrete structure.

Visual inspections may reveal a loss of cement paste and aggregate from the matrix. Cracking, spalling, and discoloration can be expected when acid deteriorates steel reinforcements, and laboratory analysis may be needed to indentify the type of chemical causing the damage.

Code Consideration

It is a code violation to embed aluminum conduits and pipes in concrete, unless the aluminum is coated or covered to prevent aluminum–concrete reaction or electrolytic action between aluminum and steel.

How can you create a more defensive concrete where chemical reactions are anticipated? Portland cement concrete does not fare well when exposed to acid. When faced with this type of concrete, an approved coating or treatment is about the best that you can do. Using a dense concrete with a low water–cement ratio can provide acceptable protection against mild acid exposure.

Aggressive Water

Aggressive water is water with a low concentration of dissolved minerals. Soft water is considered aggressive water and it will leach calcium from cement paste or aggregates. This is not common in the United States. When this type of attack occurs, however, it is a slow process. The danger is greater in flowing waters, because a fresh supply of aggressive water continually comes into contact with the concrete.

If you conduct a visual inspection and find rough concrete where the paste has been leached away, it could be an aggressive-water defect. Water can be tested to determine if water quality is such that it may be responsible for damage. When testing indicates that water may create problems prior to construction, a non-Portland-cement-based coating can be applied to the exposed concrete structures.

Code Consideration

When conduits are installed in concrete, the diameter of the conduit must not be more than one-third of the overall thickness of the concrete slab where it is being installed.

There are many factors that can contribute to problems with a concrete installation. Table 10.1 outlines examples of such problems.

TABLE 10.1 Causes of Distress and Deterioration of Concrete

Cause	Type
Accidental loadings	
Chemical reactions	
	Acid attack
	Aggressive-water attack
	Alkali-carbonate rock reaction
	Alkali-silica reaction
	Miscellaneous chemical attack
	Sulfate attack
Construction errors	
Corrosion of embedded metals	
Design errors	
	Inadequate structural design
	Poor design details
Erosion	
	Abrasion
	Cavitation
Freezing and thawing	
Settlement and movement	
Shrinkage	
	Plastic
	Drying
Temperature changes	
	Internally generated
	Externally generated
	Fire
Weathering	

Courtesy of United States Army Corps of Engineers.

Alkali-Carbonate Rock Reaction

Alkali-carbonate rock reaction can result in damage to concrete, but it can also be beneficial. Our focus is on the destructive side of this action, which occurs when impure dolomitic aggregates exist. When this type of damage occurs, there is usually map or pattern cracking and the concrete can appear to be swelling.

Alkali-carbonate rock reaction differs from alkali-silica reaction because there is a lack of silica gel exudations at cracks. Petrographic examination can be used to confirm the presence of alkali-carbonate rock reaction. To prevent this type of problem, contractors should avoid using aggregates that are, or are suspected to be, reactive.

Code Consideration

Conduits embedded in concrete must be spaced a minimum distance that equals not less than three times the diameter of the conduit being installed.

Alkali-Silica Reaction

An alkali-silica reaction can occur when aggregates containing silica that is soluble in highly alkaline solutions may react to form a solid, nonexpansive, calcium-alkali-silica complex or an alkali-silica complex that can absorb considerable amounts of water and expand. This can be disruptive to concrete.

Concrete that shows map or pattern cracking and a general appearance of swelling could be a result of an alkali-silica reaction. This can be avoided by using concrete that contains less than 0.60% alkali.

Various Chemical Attacks

Concrete is fairly resistant to chemical attack. For a substantial chemical attack to have degrading effects of a measurable nature, a high concentration of chemical is required. Solid dry chemicals are rarely a risk to concrete. Chemicals that are circulated in contact with concrete do the most damage.

When concrete is subjected to aggressive solutions under positive differential pressure, the concrete is particularly vulnerable. The pressure can force aggressive solutions into the matrix. Any concentration of salt can create problems for concrete structures. Temperature plays a role in concrete destruction with some chemical attacks. Dense concrete that has a low water–cement ratio provides the greatest resistance. The application of an approved coating is another potential option for avoiding various chemical attacks.

Code Consideration

When concrete joints are created, they must be located in a manner that will not have an adverse effect on the strength of the concrete in which they are installed.

Sulfate Situations

A sulfate attack on concrete can occur from naturally occurring sulfates of sodium, potassium, calcium, or magnesium. These elements can be found in soil or in ground water. Sulfate ions in solution will attack concrete. Free calcium hydroxide reacts with sulfate to form calcium sulfate, also known as gypsum. When gypsum combines with hydrated calcium aluminate it forms calcium sulfoaluminate. Either reaction can result in an increase in volume. Additionally, a purely physical phenomenon occurs where a growth of crystals of sulfate salts disrupts the concrete. Map and pattern cracking are signs of a sulfate attack. General disintegration of concrete is also a signal of the occurrence.

Sulfate attacks can be prevented with the use of a dense, high-quality concrete that has a low water–cement ratio. A Type V or Type II cement is a good choice. If pozzolan is used, a laboratory evaluation should be done to establish the expected improvement in performance.

Code Consideration

Unless otherwise authorized, all bending of reinforcement material for concrete must be done while the material is cold.

Before you can correct problems with concrete installations, you must identify the problem. See Table 10.2 for reference material in completing this task.

TABLE 10.2 Relating Symptoms to Causes of Distress and Deterioration of Concrete

	Symptoms							
Causes	Construction Faults	Cracking	Disintegration	Distortion/ Movement	Erosion	Joint Failures	Seepage	Spalling
Accidental loadings		x						x
Chemical reactions		x	x				x	
Construction errors	x	x				x	x	x
Corrosion		x						x
Design errors		x				x	x	x
Erosion			x		x			
Freezing and thawing		x	x	x		x		x
Settlement and movement		x		x				x
Shrinkage	x	x						
Temperature changes		x				x		x

Courtesy of United States Army Corps of Engineers.

Poor Workmanship

Poor workmanship accounts for a number of concrete issues. It is simple enough to follow proper procedures, but there are always times when good practices are not employed. The solution to poor workmanship is to prevent it. This is much easier said than done. All sorts of problems can occur when quality workmanship is not assured and some of the key causes for these problems are as noted below:

- Adding too much water to concrete mixtures
- Poor alignment of formwork
- Improper consolidation
- Improper curing
- Improper location and installation of reinforcing steel members
- Movement of formwork
- Premature removal of shores or reshores
- Settling of concrete
- Settling of subgrade
- Vibration of freshly placed concrete
- Adding water to the surface of fresh concrete
- Miscalculating the timing for finishing concrete
- Adding a layer of concrete to an existing surface
- Use of a tamper
- Jointing

All concrete forms for the placement of concrete need to be inspected closely. This includes confirming that the forms are properly built and that all needed reinforcement is in place (see Figure 10.1).

FIGURE 10.1 Inspecting the formwork for a footing prior to concrete being poured.

CORROSION

Corrosion of steel reinforcing members is a common cause of damage to concrete. Rust staining will often be present during a visual inspection if corrosion is at work. Cracks in concrete can tell a story. If they are running in straight lines, as parallel lines at uniform intervals that correspond with the spacing of steel reinforcement materials, corrosion is probably at the root of the problem. In time, spalling will occur. Eventually, the reinforcing material will become exposed to a visual inspection.

Code Consideration

Unless otherwise authorized, it is a code violation to weld crossing bars that will reinforce concrete.

Techniques for stopping, or controlling, corrosion include the use of concrete with low permeability. In addition, good workmanship is needed. Some tips to follow include:

- Use as low a concrete slump as practical.
- Cure the concrete properly.
- Provide adequate concrete cover over reinforcing material.
- Provide suitable drainage.
- Limit chlorides in the concrete mixture.
- Pay special attention to any protrusions, such as bolts and anchors.

Designer Errors

Designer errors are divided into two categories: those that are a result of inadequate structural design and those that are a result of a lack of attention to relatively minor design details. In the case of structural design errors, the result can be anticipated and will generally result in a structural failure.

Identifying structural design mistakes concentrates on two types of symptoms, spalling and cracking. Spalling indicates excessively high compressive stress. Cracking and spalling can also indicate high torsion or shear stresses. High-tensile stresses will cause cracks. Petrographic analysis and strength testing of concrete is required if any of the concrete elements are to be reused after such failures. The best prevention requires careful attention to detail. Design calculations should be checked thoroughly. Flaws in design details account for most of these types of problems. See the following list for examples of design factors to consider:

- Poor design details
- Abrupt changes in section
- Insufficient reinforcement at reentrant corners and openings
- Inadequate provision for deflection
- Inadequate provision for drainage
- Insufficient travel in expansion joints

- Incompatibility of materials
- Neglect of creep effect
- Rigid joints between precast units
- Unanticipated shear stresses in piers, columns, or abutments
- Inadequate joint spacing in slabs

Abrasion

Abrasion damage can occur from waterborne debris, which typically rolls and grinds against concrete when it is in the water and in contact with concrete structures. Spillway aprons, stilling basin slabs, and lock culverts and laterals are the most likely types of structures to be affected by abrasion. The reason for this is often a result of poor hydraulic design. Another cause for abrasion can be a boat hull hitting a concrete structure.

Code Consideration

When groups of reinforcing bars are bundled together as a reinforcement device for concrete, the bundle must not contain more than four bars.

When abrasion occurs, concrete structures tend to wind up with a smooth surface. Long, shallow grooves in a concrete surface and spalling along monolith joints indicate abrasion.

The three major factors in avoiding abrasion damage are design, operation, and materials. For prevention strategies, review the list of tips below:

- Use hydraulic model studies to test designs.
- A 45 degree fillet installed on the upstream side of the end sill results in a self-cleaning stilling basin.
- Recessing monolith joints in lock walls and guide walls will minimize stilling basin spalling caused by barge impact and abrasion.
- Balanced flows should be maintained into basins by using all gates to avoid discharge conditions where eddy action is prevalent.
- Periodic inspections are needed to locate the presence of debris.
- Basins should be cleaned periodically.
- All materials used must be tested and evaluated.
- Install abrasion-resistant concrete.
- Fiber-reinforced concrete should not be used for repairing stilling basins or other hydraulic structures subject to abrasion.
- Coatings that produce good results against abrasion include polyurethanes, epoxy-resin mortar, furan-resin mortar, acrylic mortar, and iron aggregate toppings.

Cavitation

Cavitation erosion is a result of complex flow characteristics of water over concrete surfaces. For damage to occur, the rate of water flow normally needs to exceed 40 feet per second. Fast water and irregular surface areas of concrete can result in cavitation. Surface irregularity and water speed creates bubbles. The bubbles are carried downstream and have a lowered vapor pressure. Once the bubbles reach a stretch of water that has normal pressure, the bubbles collapse. The collapse is an implosion that creates a shock wave. Once the shock wave reaches a concrete surface, the wave causes a very high stress over a small area. When this process is repeated, pitting can occur. This type of cavitation has affected concrete spillways and outlet works of many high dams. Prevention has to do with design, materials, and construction practices. The following list highlights some of the key considerations:

- Include aeration in a hydraulic design.
- Use concrete designed with a low water–cement ratio.
- Use hard, dense aggregate particles.
- Steel-fiber concrete and polymer concrete can aid in the fight against cavitation.
- Neoprene and polyurethane coatings can assist in the fight against cavitation; however, coatings are rarely used as they might prevent the best adhesion to concrete. Any rip or tear in the coating can cause a complete stripping of the coating over time.
- Maintain approved construction practices.

Code Consideration

Bundle reinforcing bars must be enclosed within either stirrups or ties.

FREEZING AND THAWING

Freezing and thawing during the curing of concrete is a serious concern. Each time the concrete freezes, it expands. Hydraulic structures are especially vulnerable to this type of damage. Fluctuating water levels and under-spraying conditions increase the risk. Using deicing chemicals can accelerate damage to concrete with resultant pitting and scaling. Core samples will probably be needed to assess damage.

Prevention is the best cure. Provide adequate drainage, where possible, and work with low water–cement ratio concrete. Use adequate entrained air to provide suitable air-void systems in the concrete. Select aggregates best suited for the application, and make sure that concrete cures properly.

SETTLEMENT AND MOVEMENT

Settlement and movement can be the result of differential movement or subsidence. Concrete is rigid and cannot stand much differential movement. When it occurs, stress cracks and spall are likely to occur. Subsidence causes entire structures or single elements of entire structures to move. If subsidence is occurring, the concern is not cracking or spalling; the big risk is stability against overturning or sliding.

A failure via subsidence is generally related to a faulty foundation. Long-term consolidations, new loading conditions, and related faults are contributors to subsidence. Geotechnical investigations are often needed when subsidence is evident.

Cracking, spalling, misaligned members, and water leakage are all evidence of structure movement. Specialists are normally needed for these types of investigations.

Code Consideration

Spiral reinforcement for cast-in-place concrete must not be less than 3/8 in. in diameter.

SHRINKAGE

Shrinkage is caused when concrete has a deficient moisture content. It can occur while the concrete is setting or after it is set. When this condition happens during setting, it is called plastic shrinkage; drying shrinkage happens after concrete has set.

Plastic shrinkage is associated with bleeding, which is the appearance of moisture on the surface of concrete. This is usually caused by the settling of heavier components in a mixture. Bleed water typically evaporates slowly from the surface of concrete. When evaporation occurs faster than water is supplied to the surface by bleeding, high-tensile stresses can develop. This stress can lead to cracks on the concrete surface.

Cracks caused by plastic shrinkage usually occur within a few hours of concrete placement. These cracks are normally isolated and tend to be wide and shallow. Pattern cracks are not generally caused by plastic shrinkage.

Code Consideration

Spacing requirements for shrinkage and temperature reinforcement must be spaced no farther apart than five times a slab's thickness and no farther apart than 18 in.

Weather conditions contribute to plastic shrinkage. If the conditions are expected to be conducive to plastic shrinkage, protect the pour site with windbreaks, tarps, and similar arrangements to prevent excessive evaporation. In the event that early cracks are discovered, revibration and refinishing can solve the immediate problem.

Drying shrinkage is a long-term change in volume of concrete caused by the loss of moisture. A combination of this shrinkage and restraints will cause tensile stresses and lead to cracking. The cracks will be fine and the absence of any indication of movement will exist. The cracks are typically shallow and only a few inches apart. Look for a blocky pattern to the cracks. They can be confused with thermally induced deep cracking, which occurs when dimensional change is restrained in newly placed concrete by rigid foundations or by old lifts of concrete.

To reduce drying shrinkage, try the following precautions:

- Use less water in concrete.
- Use larger aggregate to minimize paste content.
- Use a low temperature to cure concrete.

- Dampen the subgrade and the concrete forms.
- Dampen aggregate if it is dry and absorbent.
- Provide adequate reinforcement.
- Provide adequate contraction joints.

FLUCTUATIONS IN TEMPERATURE

Fluctuations in temperature can affect shrinkage. The heat of hydration of cement in large placements can present problems. Climatic conditions involving heat also affect concrete; for example, fire damage, while rare, can also contribute to problems associated with excessive heat.

Code Consideration

The code allows one to assume that the ends of columns built integrally with a structure will remain fixed. This assumption is used while computing gravity load moments on columns.

Hydration of concrete can raise the temperature of freshly placed concrete by up to 100 degrees. Rarely is the temperature increase consistent throughout the concrete, which can generate problems such as shallow and isolated cracks. How can this be avoided? See the tips below:

- Use low-heat cement.
- Pour concrete at the lowest reasonable temperature.
- Select aggregates with low moduli of elasticity and low coefficients of thermal expansion.

External temperature changes can result in cracking that will appear as regularly spaced cracks. There may be spalling at expansion joints. Using contraction and expansion joints can help prevent this damage.

There are many potential causes for concrete failure. Extended education, experience, and scientific testing is often required to clearly identify these causes. There is always more to learn. Keeping an open mind and immersing yourself in the components of concrete is the best route to success.

CHAPTER

11

Concrete Repair Preparation

OUTLINE

Concrete requires repair from time to time. A successful repair relies on numerous factors, such as the repair strategy, material, and procedure. The best attempt at a repair is likely to fail if the wrong material is used.

When planning a concrete repair, various options must be considered and the best type of material to implement in the repair or rehabilitation of concrete structures must be evaluated. Your choice for repair may be controversial in some venues, so appropriate experts may need to be consulted for assured success. There are, however, many facts that are proven and can be trusted, and these will be reviewed in this chapter.

COMPRESSIVE STRENGTH

How much compressive strength is needed in material for concrete repairs? If it is determined that the existing concrete structure is of adequate compressive strength, then the repair material should be of a similar compressive strength. There are few instances where beefing up the compressive strength in a repair is beneficial. An exception is the repair of concrete that is damaged by erosion. When erosion is at fault for a defect, using a higher compressive strength is a valid decision.

Code Consideration

T-beam construction requires that the flange and web must be built integrally or otherwise effectively bonded together.

MODULUS OF ELASTICITY

What is modulus of elasticity? It is a measure of stiffness with higher modulus materials exhibiting less deformation under load compared to low modulus materials. When making a repair, it should be similar to that of the concrete substrate. This allows for uniform load transfer across a repaired section. Materials with a lower modulus of elasticity will exhibit lower internal stresses. This reduces the potential for cracking and delamination of a repair.

Code Consideration

Clear spacing between ribs in joist construction cannot be more than 30 in.

THERMAL EXPANSION

Thermal expansion is going to happen with concrete. If a polymer is used as a repair material, the result will often be cracking, spalling, or delamination of the repair. The coefficient of thermal expansion has to be considered for suitable repair materials. When you compare a polymer to concrete, the coefficient of thermal expansion for a polymer is likely to be up to 14 times greater than that of concrete. Large repairs and overlays are especially vulnerable to cracking due to thermal expansion.

Code Consideration

Slab thickness in joist construction must not be less than 2 in. in depth.

BONDING

The bonding between repair material and concrete is a key element in a successful repair. For concrete to accept a good bond with a repair, the concrete should be properly prepared. Polymer adhesives provide a better bond of plastic concrete to hardened concrete than can be obtained with a cement slurry or the plastic concrete alone. But this method is controversial, and many experienced people feel that polymer bonds are less than 25% better than properly prepared concrete surfaces without adhesives.

DRYING SHRINKAGE

Drying shrinkage is a concern in concrete repairs. Existing concrete that is in need of repair is unlikely to shrink. However, patches and repairs that are made fresh are subject to shrinking, and this can compromise the repair. To avoid shrinkage, the material used for repairing old concrete should be made with a low water–cement ratio. The goal is to use a material that will provide minimum shrinkage.

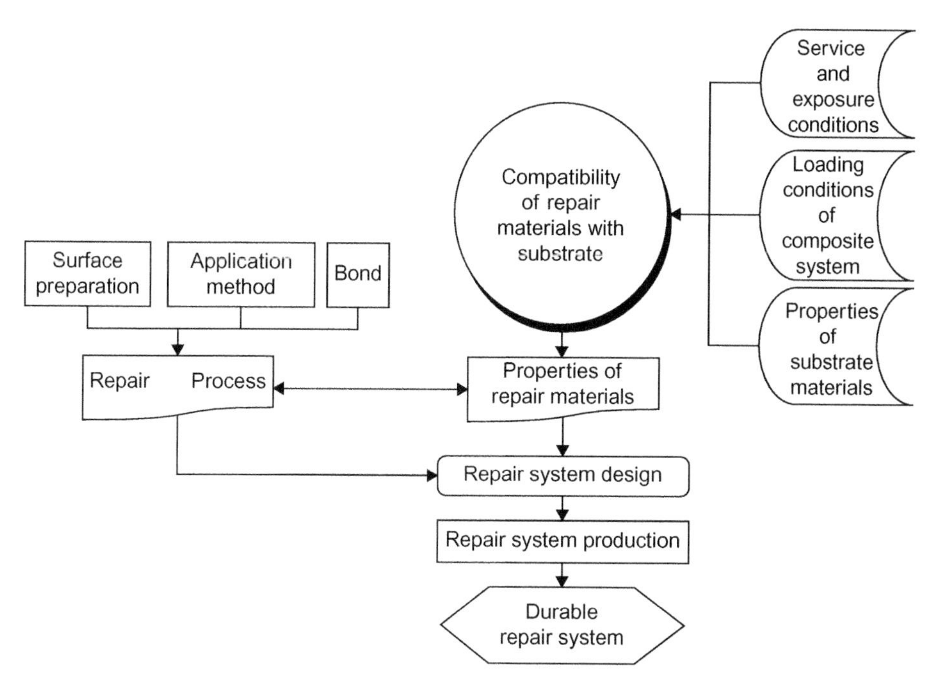

FIGURE 11.1 Factors affecting the durability of concrete repair systems. *Source: From Emmons and Vaysburd (1995); courtesy of United States Army Corps of Engineers.*

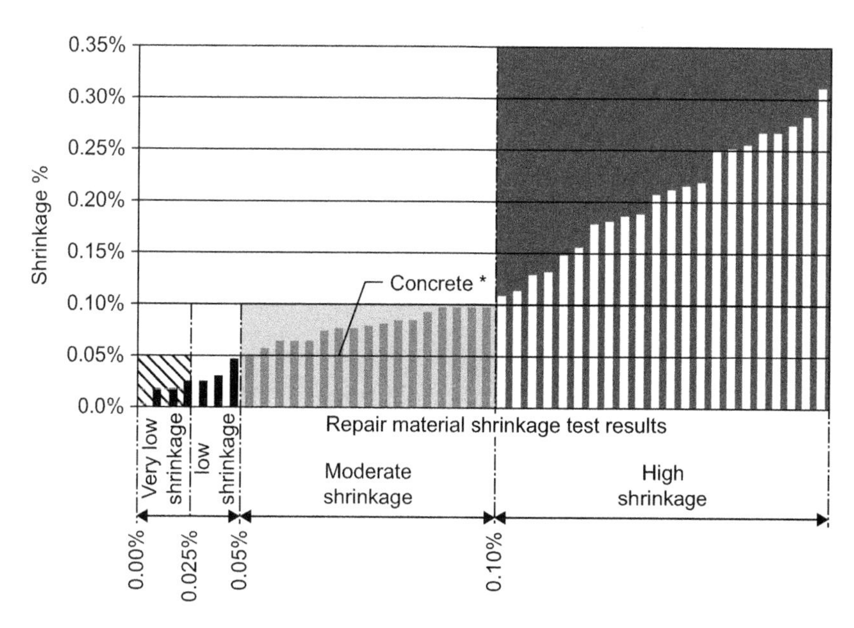

FIGURE 11.2 Classification of repair materials based on drying shrinkage. *Source: Emmons and Vaysburd (1995); courtesy of United States Army Corps of Engineers.*

Code Consideration

When applying concrete at the minimum thickness required to cover material, the floor finish may be counted for nonstructural considerations.

CREEP

Repair materials should have a creep factor consistent with the material being repaired. Stress relaxation through tensile creep reduces the potential for cracking. Refer to manufacturer reports when selecting an appropriate creep rate for various repairs.

PERMEABILITY

Good concrete is relatively impermeable to liquids; however, moisture evaporates at a surface and replacement liquid is pulled to the evaporating surface by diffusion. This has to be considered when making a repair. Any large patch or overlay made with an impermeable material can trap moisture between the existing concrete surface and the seal made by the repair. If this happens, the repair is likely to fail. In this situation use a repair material that has low water absorption and high water vapor transmission characteristics.

Code Consideration

The minimum number of longitudinal bars in compression members is four bars within rectangular or circular ties.

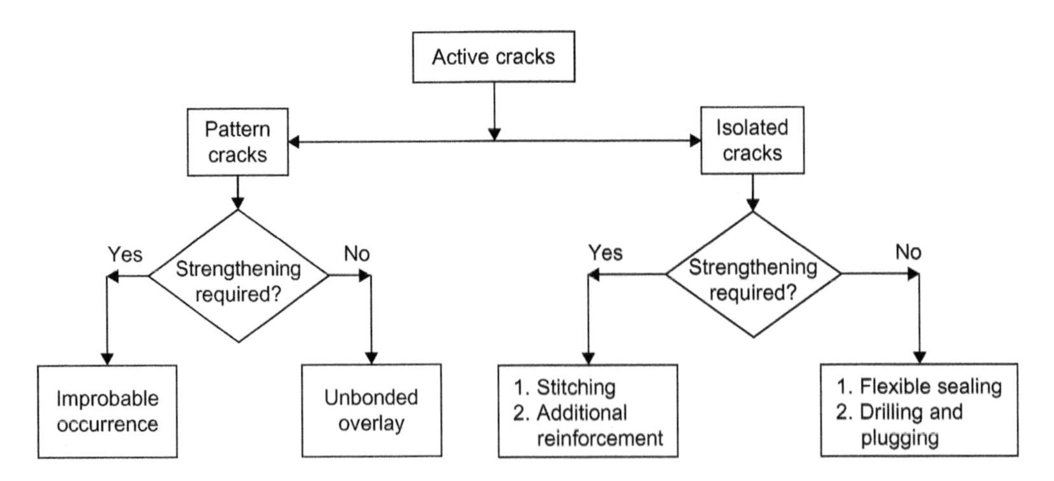

FIGURE 11.3 Selection of repair method for active cracks. *Source: After Johnson (1965); courtesy of United States Army Corps of Engineers.*

PLANNING A REPAIR

When planning a concrete repair there are many factors to consider:

- Application conditions
- Geometry
- Temperature
- Moisture
- Location
- Service conditions
- Downtime
- Traffic
- Chemical attack
- Appearance
- Service life

Depth and orientation of a concrete repair is an important consideration. At the time of curing, some repair materials generate heat, and thermal stress can reach an unacceptable level. Shrinkage is another concern. Thin layers of concrete used as a repair are subject to spalling. An advantage to polymer materials is that they can be used in thin layers. Aggregate size is determined by the thickness of a repair. Repairs made overhead must be made in such a way so that they will not sag. If you are dealing with dormant cracks, you can use the methods noted in Figure 11.4 for guidance.

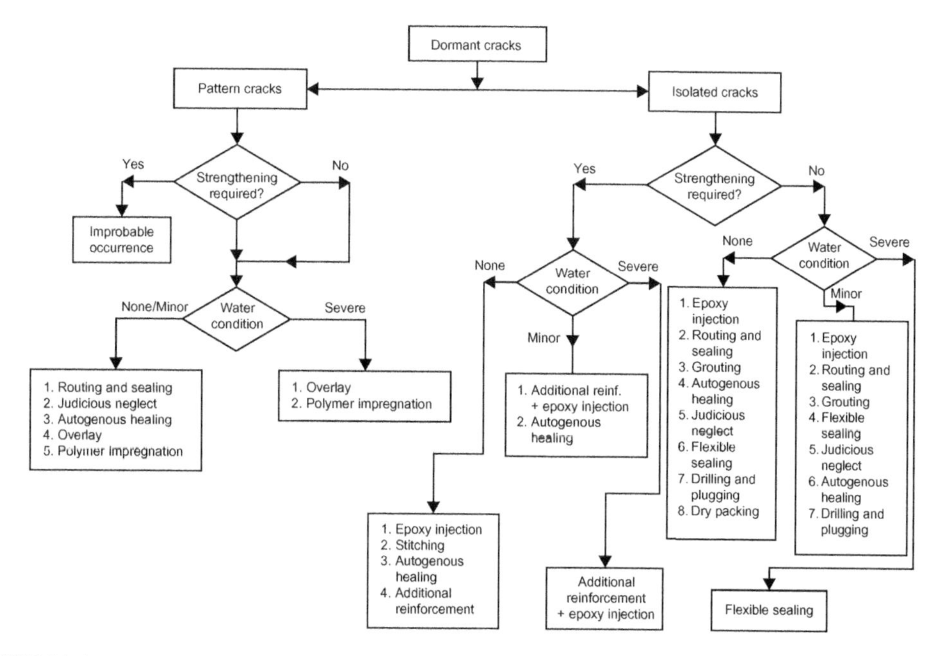

FIGURE 11.4 Selection of repair method for dormant cracks. *Source: After Johnson (1965); courtesy of United States Army Corps of Engineers.*

Code Consideration

The minimum number of longitudinal bars in compression members is three bars within triangular ties.

Portland cement hydration stops at, or near, freezing. Latex emulsions do not coalesce to form films at temperatures below about 45°F. Materials that can be used in colder temperatures generally require longer setting times. In contrast, high temperatures may make a repair material set faster resulting in a decrease in the working life of the material.

Having water come into contact with fresh concrete is not acceptable in most repairs. Grouting, external waterproofing, or diversion systems are commonly used to prevent interference from moving water while working with concrete. If you are working with polymers, some of them will not adhere in moist conditions, and some are not affected by moisture. See Table 11.1 for causes and suggested repair efforts when concrete is spalling or disintegrating. Table 11.2 describes additional repair options.

TABLE 11.1 Causes and Repair Approaches for Spalling and Disintegration

Cause	Deterioration Likely to Continue		Repair Approach
	Yes	No	
1. Erosion (abrasion, cavitation)	X		Partial replacement Surface coatings
2. Accidental loading (impact, earthquake)		X	Partial replacement
3. Chemical reactions Internal	X		No action Total replacement
External	X	X	Partial replacement Surface coatings
4. Construction errors (compaction, curing, finishing)	X		Partial replacement Surface coatings No action
5. Corrosion	X		Partial replacement
6. Design errors	X	X	Partial or total replacement based on future activity
7. Temperature changes (excessive expansion caused by elevated temperature and inadequate expansion joints)	X		Redesign to include adequate joints and partial replacement
8. Freezing and thawing	X		Partial replacement no action

NOTE: This table is intended to serve as a general guide only. It should be recognized that there are probably exceptions to all of the items listed.

TABLE 11.2 Repair Methods for Spalling and Disintegration

Repair Approach	Repair Method
1. No action	Judicious neglect
2. Partial replacement (replacement of only damaged concrete)	Conventional concrete placement
	Drypacking
	Jacketing
	Preplaced-aggregate concrete
	Polymer impregnation
	Overlay
	Shotcrete
	Underwater placement
	High-strength concrete
3. Surface coating	Coatings Overlays
4. Total replacement of structure	Remove and replace

Code Consideration

The minimum number of longitudinal bars in compression members is six bars enclosed by spirals.

The location of a needed repair can have an impact on materials and procedures. Some locations limit the type of equipment that can be used. Some repair materials are odorous, toxic, or combustible. All of these factors must be considered when planning a repair.

Sometimes materials that set quickly are needed to reduce downtime. In the case of heavy vehicular traffic, repair materials need to have a high-strength rating and good abrasion and skid resistance. Concrete deterioration must also be considered.

High-service temperatures can affect the performance of some polymers. Whereas polymers can be sensitive to solvents, most polymers resist most acids and sulfates. Soft water can damage Portland cement products. Matching patches and repairs can be very difficult. If appearance is a factor, you will have to do your research to come up with a close, visual match. How long will the repair be expected to last? This can influence the type of repair you choose.

MANUFACTURER'S DATA

Most manufacturers offer information on the qualities of their products such as:

- Compressive strength
- Tensile strength
- Slant-shear bond
- Modulus of elasticity

Even with this information, there may be other information not found in the manufacturer-supplied data such as:

- Drying shrinkage
- Tensile bond strength
- Creep
- Absorption
- Water vapor transmission

If the information you need for your planning phase is not provided, request it from the manufacturers.

The planning phase is critical to a successful outcome. You may have to spend a few hours on research, but the end result should be a better repair. Don't guess. Make your decisions based on facts.

CHAPTER

12

Removal and Repair

OUTLINE

It is common for concrete to be removed prior to making a repair to a concrete structure. Many jobs require existing concrete to be prepared properly to receive a repair. When the damage done to concrete does not threaten the structural integrity of a structure, the concrete will probably not have to be removed.

Environmental impact from concrete removal is often an issue. This is especially true when the work is done near a waterway. In some cases, concrete debris can be allowed to enter a waterway, because some of the aggregate used in concrete is natural river gravel. Therefore, allowing it to fall into the water is essentially returning the gravel to its place of origin. Another possibility, when allowed, is to dispose of large pieces of concrete in water so that it can create an artificial reef for fish. When this is an acceptable means of disposal, it reduces the cost of deconstruction considerably.

It is common for multiple methods to be used on a single structure to remove concrete. For example, a pre-splitting method might be used to weaken concrete so that an impacting method can be used to complete the concrete removal. Another example is cutting a section of concrete to delineate an area in which an impacting method will be used for concrete removal. Once the concrete is removed and sound concrete revealed, repairs can begin.

When demolishing concrete, you have to be aware of any utilities or other embedded items in the existing concrete. A review of as-built drawings can render assistance in projecting obstacle locations.

REMOVAL METHODS

There are a number of methods available to contractors for removal of damaged concrete:

- Blasting
- Crushing
- Cutting
- Impacting
- Milling
- Pre-splitting

BLASTING

Blasting is done with boreholes that employ rapidly expanding gas to create controlled fracture and concrete removal. The use of explosive blasting is usually not considered for jobs that require the removal of less than 10 in. of concrete from the face of a structure.

Blasting is a fast, and often cost-effective, method of removing concrete, but there is a downside. There is always a risk of damaging sound concrete that is not meant to be disturbed. A solution to this is to use a method known as smooth blasting.

Code Consideration

Whenever there is a possibility of lateral forces causing a transfer of moment on columns, the shear resulting from this action must be prohibited by the use of reinforcement.

Smooth blasting uses detonating cord to distribute blast energy throughout a borehole. This can avoid energy concentrations that might damage surrounding structural elements. An extended version of smooth blasting is a little used method called cushion blasting. This is essentially the same as smooth blasting except that cushion blasting requires the filling of boreholes with wet sand.

Cutting is often used in conjunction with blasting. Selected cuts are made at prime locations to control the breakaway of concrete when blasting is employed. Sequential blasting allows for more delays to be used with each firing. This optimizes the amount of explosive detonated with each firing. Air-blast pressures are maintained at acceptable levels along with ground vibrations and fly rock.

CRUSHING

Crushing is done with hydraulically powered jaws. Boom-mounted mechanical crushers and portable mechanical crushers are both used. When total demolition is desired, a boom-mounted crusher is normally used. This type of equipment is most effective when not used for partial removal. Portable crushers work best when the shearing plane depth is 12 in. or less.

CUTTING

Cutting is used to remove full sections of concrete. The methods used for cutting include:

- Abrasive water jets
- Diamond saws
- Diamond-wire
- Stitch drilling
- Thermal tools

Cutting with abrasive water jets can be used to make cutouts through slabs, walls, and other concrete members. This type of cutting is capable of cutting through steel reinforcing and hard aggregates. A drawback to abrasive water cutting is that it tends to be slow and costly. Other concerns are controlling the waste water and maintaining personal safety when working with very noisy, high-pressure systems.

Diamond blades are used when the cutting depth is less than 2 feet. Blade selection is dependent on the type of aggregate and reinforcing that must be cut. The time required for blade cutting depends on the hardness of the material being cut, but the process can be slow.

Diamond-wire cutting is used when the cutting depth is greater than that which can be achieved with a diamond blade. This type of cutting is well suited for cuts where access is limited or difficult. Wire cutting is specialized and tends to be expensive, but sometimes it is the best alternative available.

Code Consideration

Shearhead arms must not be interrupted within a column section.

Stitch Cutting

Stitch cutting is used when the cut depth is greater than what can be reached with a diamond saw blade. This is most often the case when only one face of the concrete is accessible. If two faces are accessible, wire cutting is more likely to be used.

Thermal Cutting

Thermal cutting is used for cutting through heavily reinforced concrete structures where site conditions will allow efficient flow of molten concrete from cuts. Flame tools are favored for this type of cutting up to depths of 2 feet. Cutting lances are used when the cutting depth is greater. Thermal cutting is expensive and is limited in the places it can be used. The dangers associated with thermal cutting include working with compressed and flammable gases, hot flying rock, and high temperatures.

Code Consideration

Negative moment reinforcement in a continuous member must be anchored in or through the supporting member by embedment length, hooks, or mechanical anchorage.

IMPACTING METHODS

Impacting methods involve the use of some sort of mass to hit concrete repeatedly. A simple example of this would be banging concrete with a sledgehammer. The impact is meant to fracture and spall concrete. Reinforcements must be cut out when impacting methods are used for concrete removal.

BOOM-MOUNTED CONCRETE BREAKERS

Boom-mounted concrete breakers can be found attached to backhoes and may be operated by compressed air or hydraulic pressure. This type of breaker is fast and cost-effective. Saw cuts are typically used to outline the breakout area, which reduces feathered edges. Concrete that remains after this type of breaking can suffer from microcracking. When it is found, a high-pressure water jet could be used to remove microfractured concrete. The water pressure should be set at a minimum of 20,000 PSI.

Code Consideration

Continuous reinforcement is required at interior supports of deep flexural members to provide negative moment tension.

SPRING-ACTION HAMMERS

Spring-action hammers are normally used on thin concrete. This method is most appropriate for total demolition. Cutting concrete along a boundary for a cutout is recommended when spring-action hammers are used. Microcracking along with exposed reinforcing steel is likely to occur.

HAND-HELD IMPACT BREAKERS

Hand-held impact breakers (which are basically jackhammers) work well for limited concrete removal. The breaker may be powered by compressed air, hydraulic pressure, self-contained gasoline engines, or self-contained electric motors. Small, shallow sections of concrete can be demolished with great mobility when a hand-held breaker is used.

Code Consideration

Both mechanical and welded splices are allowed by code.

HYDROMILLING

Hydromilling is sometimes called hydrodemolition or water-jet blasting. This type of concrete removal is effective on concrete with a depth of up to 6 in. Sound concrete and reinforcements are not normally damaged when hydromilling is used; however, this method is expensive and can be slow when sound concrete is encountered. Another problem with this method is the likelihood of blowouts (holes through a concrete member). During hydromilling a great deal of potable water has to be available with some units using up to 1,000 gallons of water per hour. Flying rock is also common.

ROTARY-HEAD MILLING

Rotary-head milling is used to remove deteriorated concrete from mass structures. If the compressive strength of concrete is 8,000 PSI or greater, rotary-head milling is not practical. The remaining concrete can suffer from microcracking.

Code Consideration

When feasible, splices should be located away from any points of maximum tensile stress.

PRE-SPLITTING

Pre-splitting depends on wedging forces in a designed pattern of boreholes to produce a controlled cracking of concrete to facilitate removal of concrete by other means. The extent of pre-splitting planes is affected by the pattern, spacing, and depth of boreholes. Chemical-expansive agents and hydraulic splitters are used in pre-splitting. Reinforcing steel significantly decreases the pre-splitting plane. A loss of control can occur if boreholes are too far apart or are drilled in severely deteriorated concrete.

CHEMICAL AGENTS

Chemical agents can be used for pre-splitting concrete if the depth of the boreholes is 10 times their diameter or more. This method is expensive; however, it works very well when pre-splitting vertical planes of significant depth. In the early hours of use, chemical agents can blow out boreholes and cause personal injury. It is common for rotary-head milling or mechanical-impacting methods to be used in conjunction with chemical pre-splitting methods to obtain a completed job.

Code Consideration

When dealing with spirally reinforced compression members, lap length in a splice must not exceed 12 in.

PISTON-JACK SPLITTERS

If you are pre-splitting a concrete face that is 10 in. or thicker, a piston-jack splitter is a good way to go. Boreholes for this type of work must have a minimum diameter of 3.5 in. The cost of using piston-jack splitters can be prohibitive.

PLUG-FEATHER SPLITTER

What is a plug-feather splitter? Sounds like a strange device, doesn't it? Plug-feather splitters are used on concrete surfaces that have a depth of no more than 4 feet. The direction of pre-splitting can be controlled with this type of splitter. Because the body of this type of splitter is wider than the borehole it is used on, it cannot be reinserted into boreholes to continue pre-splitting after a section has been pre-split.

PREP WORK

The prep work done prior to a repair is one of the most important steps in making successful concrete repairs. Preparation work varies and depends on the type of repair being made. In general, the concrete being prepared should be sound, clean, rough-textured, and dry. There are exceptions, but this is the normal rule of thumb.

Normally, all deteriorated concrete should be removed prior to a repair. This is often done with impact tools, either hand-held or boom-mounted. After secondary removal, wet or dry sandblasting or water-jet blasting can be used to clean the sound concrete. The options for cleaning concrete for a repair include:

- Chemical cleaning
- Mechanical cleaning
- Shot blasting
- Blast cleaning

- Acid etching
- Bonding agents

Chemical Cleaning

Chemical cleaning is needed when concrete is contaminated with oil, grease, or dirt. Detergents and other concrete cleaners can be used to rid the concrete of contaminants. However, the cleaners must also be removed before a repair is made. Avoid the use of solvents. When they are used, contaminants can be pushed deeper into the concrete. While muriatic acid is used for etching concrete, it is not very effective for removing grease or oil.

Code Consideration

A middle strip is bounded by two column strips.

Mechanical Cleaners

Mechanical cleaners include scabblers, scarifiers, and impact tools. Different heads on the tools allow for different types of abrasive material. Secondary cleaning will be necessary after using mechanical cleaners, and this is done with wet or dry sandblasting or water jetting.

Shot Blasting

Thin overlay repairs cry out for shot blasting. This is the use of steel shot being blasted against the concrete to create a uniform surface. Once the blasting is done, the steel shot is gathered by a vacuum and saved for later use. The end result is a dry surface that is ready to accept a repair.

Blast Cleaning

Blast cleaning is done with water jetting and both wet and dry sandblasting. Sandblasting requires the use of an effective oil trap to prevent contamination of concrete surfaces during the cleaning operation. Water-jetting equipment with operating pressures of 6,000 to 10,000 PSI is commercially available for cleaning concrete.

Acid Etching

Acid etching is used to remove laitance and normal amounts of dirt. Cement paste is removed by the acid, which provides a rough surface. General opinion is that acid etching should only be used when other choices are not suitable.

Code Consideration

Every middle strip must be designed and installed to proportion the resistance of the sum of moments assigned to its two half middle strips.

Bonding Agents

Bonding agents should be used on repairs that are less than 2 in. thick. Thicker repairs can be made without bonding agents when the receiving surface is properly prepared. Many types of bonding agents are available. Always refer to the manufacturer's recommendations for use of their product.

REINFORCING STEEL

Corrosion is the enemy of reinforcing steel. When the steel reinforcing material has to be replaced, concrete must be removed from around the existing reinforcement. A jackhammer is the tool of choice in such cases. All weak, damaged, and easily removable concrete should be removed. Circumstances may allow existing reinforcing steel to be cleaned and left in place. To do this, at least one-half of the existing steel must not be corroded.

If an air compressor is used to clean corrosion, make sure that the compressor is either an oil-free or has an oil trap; otherwise, the concrete may be tainted by blowing oil. Dry sandblasting is the best method for cleaning reinforcing steel. Wet sandblasting and water jetting can also be effective.

Code Consideration

When walls or columns are built integrally with a slab system, they must be made to resist moments caused by factored loads on the slab system.

ANCHORS

Anchors require the drilling of holes. This is usually done with a rotary carbide-tipped or diamond-studded drill bit. Using a jackhammer to create anchor holes is not recommended because of potential damage to in-place concrete. Anchor holes should be cleaned out and protected from debris entering the holes.

Bonded anchors can be headed or headless bolts, threaded rods, or deformed reinforcing bars and may be either grouted or chemical. Grouted anchors are embedded in pre-drilled holes with neat Portland cement, Portland cement and sand, or other commercially available premixed grout. An expansive grout additive and accelerator are commonly used with cementitious grouts.

Chemical anchors are embedded in pre-drilled holes with two-component polyesters, vinylesters, or epoxies. These anchors are available in the four forms noted in the following list:

- Glass capsules
- Plastic cartridges
- Tubes
- Bulk

Once inserted into a pre-drilled hole, the chemical anchor casing is broken or opened, and the two components mix. Epoxy products are used in bulk systems and mixed in a pot or pumped through a mixer and injected into a hole.

Expansion anchors are placed in pre-drilled holes and expanded by tightening a nut, hammering the anchor, or expanding into an undercut in the concrete. Expansion anchors that rely on side point contact to create frictional resistance should not be used where anchors are subjected to vibratory loads. Certain wedge-type anchors don't perform well when subjected to impact loads. Undercut anchors are suitable for dynamic and impact loads.

After any anchors are placed and ready for use, the anchors' strength should be tested. Don't overlook this step. It is also desirable to specify a maximum displacement in addition to the minimum load capacity.

CHAPTER

13

Rehabilitation Work

OUTLINE

Some types of concrete failures can be repaired by adding additional reinforcement. For example, a failing bridge girder may be able to be saved by post tensioning. Cracks might be able to be overcome with drillholes at 90 degree angles to the cracks. There are many ways to work with damaged concrete. Our goal here is to review both the materials and methods available for repairing and rehabilitating concrete. I already mentioned the use of drillholes at 90 degree angles to strengthen cracked concrete. To give you an example of what to expect in this chapter, let's explore this procedure more closely.

The goal is sealing any cracks and adding reinforcement bars. Holes are drilled with a ¾-in. diameter at 90 degree angles to the crack plane. Holes must be cleaned of all loose dust. The drilled holes and the crack plane are filled with an adhesive that is usually an epoxy. This is done by pumping the adhesive into the holes and crack plane with low pressure that typically ranges from 50 to 80 PSI. Reinforcing bars, using either number 4 or number 5 bars, are placed in the drilled holes. The rods should extend at least 18 in. on each side of the crack.

When the components are all in place, the adhesive bonds the bar to the walls of the hole, fills the crack plane, and bonds the cracked concrete surfaces together in one monolithic form. This reinforces the compromised concrete section.

Temporary elastic crack sealant is required for a successful repair. Gel-type epoxy crack sealants work well within their elastic limits. When cold weather is a factor, silicone or elastomeric sealants work well. Sealant should be applied in a uniform layer approximately 1/16 of an inch to 3/32 of an inch. This layer should span the crack by a minimum of 3/4 of an inch on each side.

Reinforcing bar placement is subject to individual repair requirements. The spacing and pattern of the bars will be determined by a repair design that is relevant to the fault at hand. Bars can be used in a number of ways to reinforce damaged concrete. Let's look at another example.

Code Consideration

All concrete walls must be designed to accept eccentric loads and any lateral or other loads that may have an impact on them.

Reinforcement bars can be used externally, and in conjunction with other means, to place new, reinforcing concrete. Longitudinal reinforcing bars and stirrups or ties around concrete members can be encased with shotcrete or cast-in-place concrete. Girders and slabs have been reinforced by the addition of external tendons, rods, or bolts that are prestressed.

PRESTRESSING STEEL

You can create compressive force by using prestressing steel strands or bars. When this is done, there must be adequate anchorage, and you have to be sure that the compressive force will not adversely affect other portions of a concrete member.

Cracks found in slabs on grade can sometimes be repaired with steel plates. To do this, concrete is cut across the cracks. The cuts are usually 2–3 in. deep and tend to span the crack so that there is 6–12 in. of the cut on each side of the cracks. These cuts are filled with epoxy, and steel plates are forced into the epoxy-filled cracks. The result is new strength in the cracked concrete. Figure 13.1 shows one method of crack repair that is used with conventional reinforcement. Figure 13.2 illustrates the use of compressive force for crack repair.

AUTOGENOUS HEALING

Autogenous healing is a natural process of crack repair that can occur in the presence of moisture and the absence of tensile stress. This type of healing does not occur with dormant cracks. Moisture is necessary for the process to occur, but running water tends to dissolve and wash away the lime deposits. Instead, damp moisture that evaporates is used for autogenous healing.

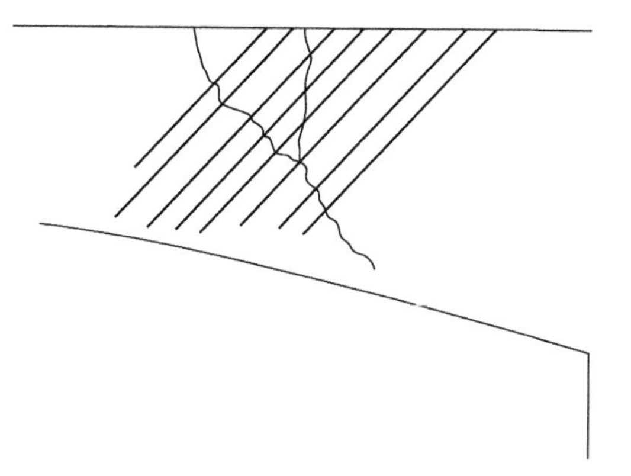

FIGURE 13.1 Crack repair using conventional reinforcement with drill holes 90 degrees to the crack plane. *Courtesy of the United States Army Corps of Engineers.*

How does the healing happen? It is the result of carbonation of calcium hydroxide in cement paste by carbon dioxide, which is present in surrounding air and water. You will find that calcium carbonate and calcium hydroxide crystals precipitate, accumulate, and grow within the cracks of concrete. This is a type of chemical bonding that restores some of the strength of cracked concrete. To develop substantial strength, water saturation of the crack and adjacent concrete is needed. Continuous saturation speeds up the healing process. Even a single cycle of drying and re-immersion can result in a drastic reduction in the degree of healing that is accomplished.

Code Consideration

Generally, the thickness of nonbearing walls must not be less than 4 in. in width.

CONVENTIONAL PLACEMENT

Conventional placement of concrete uses new concrete to repair damaged concrete. The repair concrete must be able to make an integral bond with the base concrete. A low water–concrete ratio and a high percentage of coarse aggregate are needed in the repair concrete to minimize shrinkage cracking.

When is concrete replacement used? It should be used when defects extend through a wall or beyond the reinforcement structure within the concrete. It is also a desirable solution when there are large sections of honeycombing in the concrete. Replacement concrete should not be used when there is an active threat of deterioration that caused existing concrete to fail.

Concrete repair with new concrete requires the removal of existing, damaged concrete. The goal is to get down to solid concrete to give the new concrete something to bond with. Generally, the depth desired is about 6 in. A light hammer is normally used to find sound concrete. This is done to clean the surface of the good concrete.

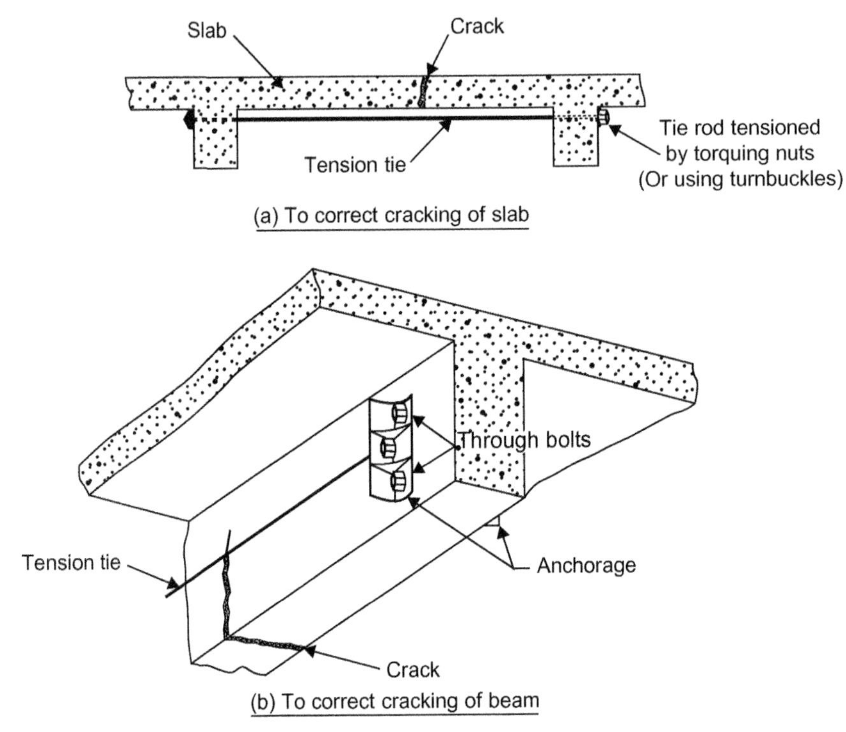

FIGURE 13.2 Crack repair with use of external prestressing strands or bars to apply a compressive force. *Courtesy of the United States Army Corps of Engineers.*

Repair of vertical sections of concrete must meet certain specifications. The cavity should have the following:

- A minimum of spalling or feather edging at the periphery of the repair area.
- Vertical sides and horizontal top at the surface of the member.
- Inside faces are generally normal to the formed surface, except that the top should slope up toward the front at about a 1:3 slope.
- Keying as necessary to lock the repair into the structure.
- Sufficient depth to reach at least one-quarter of an inch plus the dimension of the maximum size aggregate behind any reinforcement.
- All interior corners rounded with a radius of about 1 in.

All sound concrete has to be clean before repair concrete is applied. This is best accomplished with sandblasting, shot blasting, or an equally acceptable process. Only the surface receiving new concrete should be sandblasted. Final cleaning should be done with compressed

Code Consideration

When calculating the moments and shears permitted on footings on piles, assume that the reaction from any pile will be concentrated on the center of the pile.

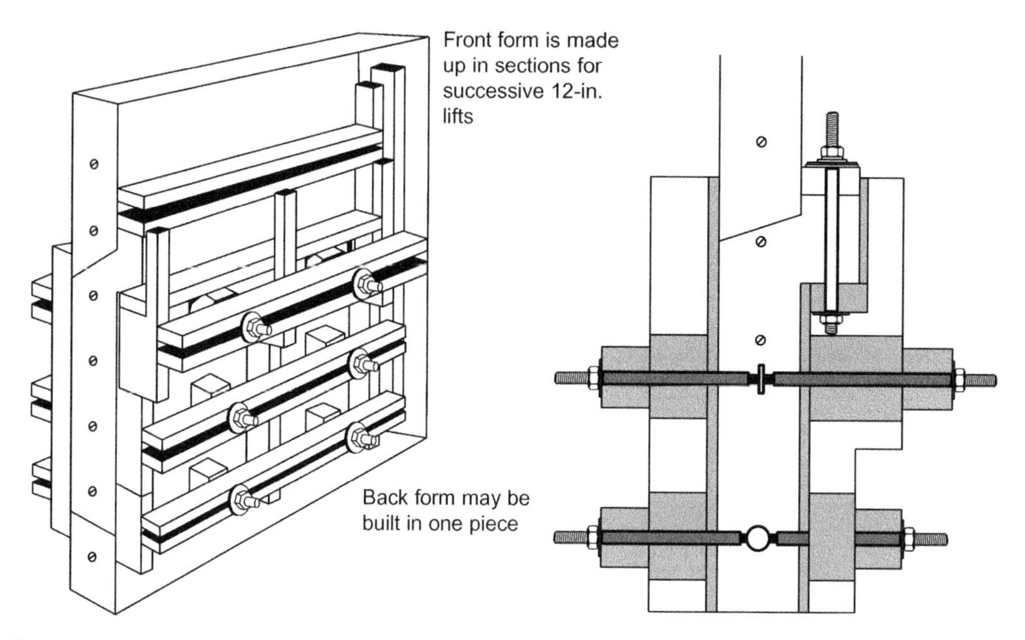

FIGURE 13.3 Detail of form for concrete replacement in walls after removal of all unsound concrete. *Courtesy of the United States Army Corps of Engineers.*

air or water. It is common for dowels and other reinforcements to be installed to make a concrete patch self-sustaining and to anchor it to the underlying concrete to provide an added safety factor. This is shown in Figure 13.3.

When repairing large vertical sections with new concrete, forming will be necessary. The forms have to be strong and mortar tight. Front panels of a form should be constructed as placing progresses so that concrete can be conveniently placed in lifts. If a back panel is needed for a form, it can be a single section.

Concrete prepared to receive repair concrete should be dry. When a thin layer of repair concrete is to be applied, for example, around 2 in. thick or less, a bonding agent should be used. Repairs of greater thickness usually do not require a bonding agent.

It is best when repair concrete is similar in content to existing concrete. This helps to avoid strains caused by temperature, moisture change, shrinkage, and so forth. Every lift of concrete should be vibrated thoroughly. Internal vibration is the preferred method.

If external vibration is required, the cavity should have a pressure cap placed inside the chimney immediately after filling the cavity. Pressure should be maintained during the vibration. This type of vibration should be repeated at 30-minute intervals until the concrete hardens and no longer responds to vibration. The projection left by the chimney is normally removed on the second day after the pour, and proper curing is essential.

Code Consideration

Pre-cast members must be designed to stand the forces and deformations that occur in and adjacent to connections.

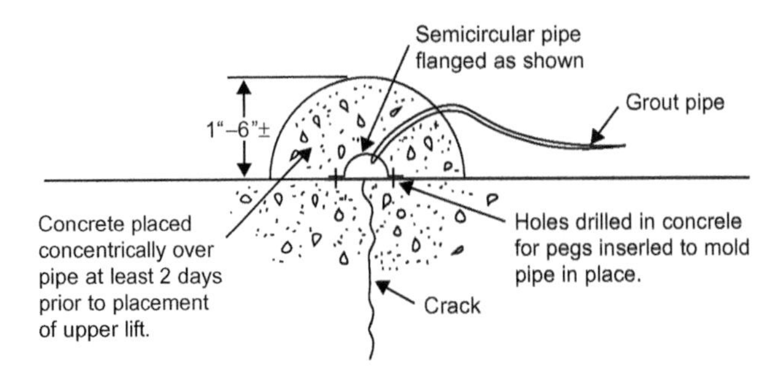

FIGURE 13.4 The use of a semicircular pipe in the crack arrest method of concrete repair. *Courtesy of United States Army Corps of Engineers.*

CRACK ARREST TECHNIQUES

Crack arrest techniques are used to stop cracking that can be caused by restrained volume change of concrete installations. The techniques are not suitable for cracks created by excessive loading. These techniques are typically used during the construction of massive concrete structures.

There is one simple technique that can be used. This consists of installing a grid of reinforcing steel over a cracked area. The reinforcing steel is then surrounded by conventional concrete rather than the mass concrete used in the structure.

Another, somewhat more complex, method involves the use of a piece of semicircular pipe. A 16-gauge pipe with an 8-in. diameter is cut in half. Then it is bent into a semicircular shape with about a 3-in. flange on each side. The area surrounding cracked concrete should be clean and the pipe section should be centered on the crack. Sections of the pipe are then welded together. Holes are cut into the pipe to receive grout pipes. Then the pipe section is covered with concrete placed concentrically by hand methods. Grout pipes can be used for grouting at a later date to attempt to restore the structural integrity of the cracked section. See Figure 13.4 for an example of this method.

Code Consideration

When spacing transverse ties perpendicular to floor or roof slabs, the spacing must not exceed the distance between the spacing of bearing walls.

DRILLING AND PLUGGING

Drilling and plugging a crack consists of drilling down the length of the crack and grouting it to form a key. This procedure is normally used on cracks that run basically in a straight line and that are accessible at one end. Vertical cracks in walls are the most likely cracks to be corrected with drilling and plugging.

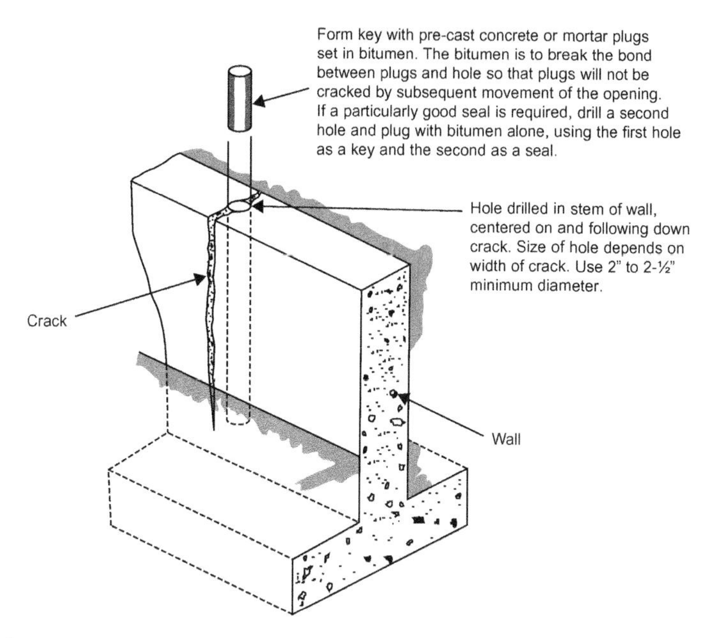

FIGURE 13.5 Repair of crack by drilling and plugging. *Courtesy of United States Army Corps of Engineers.*

How is drilling and plugging done? (Figure 13.5 shows crack repair with a drilling and plugging method.) Basically, a hole with a diameter of 2–3 in. is drilled in the center of a crack following the direction of the crack. The diameter has to be large enough to intersect the crack along its full length and to provide enough repair material to structurally take the loads exerted on the key. Drilled holes have to be cleaned and then filled with grout. Once the grout key is in place, it prevents transverse movement of sections of concrete that are adjacent to a crack. Another function of the grout key is to reduce heavy leakage through a crack and to reduce the loss of soil from behind a leaking wall.

On occasions when watertightness is essential and structural load transfer is not, the drilled hole should be filled with a resilient material of low modulus, such as asphalt or polyurethane form in lieu of Portland cement grout. If you have to deal with watertightness and the keying effect, you can use resilient material in a second hole and grout the first hole.

DRYPACKING

When you ram or tamp a low water–cement ratio mortar into a confined area, you are drypacking. There is minimal shrinkage with this type of repair. Patches are normally tight and of good quality — durable, strong, and watertight. No special equipment is required for drypacking, which makes it a preferred method when it can be used.

Circumstances that call for drypacking include patching rock pockets, formation of tie holes, and small holes with a relatively high ratio of depth to area. Shallow defects and active cracks should not be treated with drypacking. See Figures 13.6 and 13.7 for crack repair with bond breakers and flexible seals.

Code Consideration

A nominal strength in tension of not less than 16,000 lb is required when ties are around the perimeter of a floor or roof.

How is drypacking done? Start by undercutting the concrete to be repaired so that the base width is slightly greater than the surface width. Dormant cracks should be expanded to a point where the surface area is about 1 in. wide and 1 in. deep. A power-driven sawtooth bit is the best tool for this procedure. Undercut the slot slightly. Clean and dry the affected area. Apply a bond coat and then apply drypack mortar immediately.

Drypack mortar usually consists of one part cement and two and a half parts sand. Sometimes there may be three parts sand. The sand should be able to pass through a number 16 sieve. Use only enough water to allow the mortar to stick together when it is made into a ball by slight pressure with your hands. The mortar ball should leave your hands dry.

Latex-modified mortar is another option. Preshrunk mortar is also available. The preshrunk mortar is a low water–cement ratio mortar that has been mixed and allowed to stand idle for 30–90 minutes. This time depends on the air temperature. Remixing is required after the waiting period has expired.

When drypacking is used, the layers should be about 3/8 of an inch in thickness and should be tamped into place. Portland cement can be used as a part of the mixture if the desire is to match colors with existing concrete.

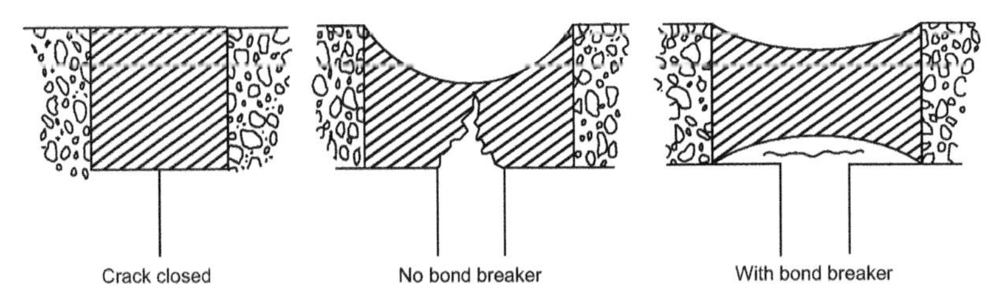

FIGURE 13.6 Effect of bond breaker involving a field-molded flexible sealant. *Courtesy of United States Army Corps of Engineers.*

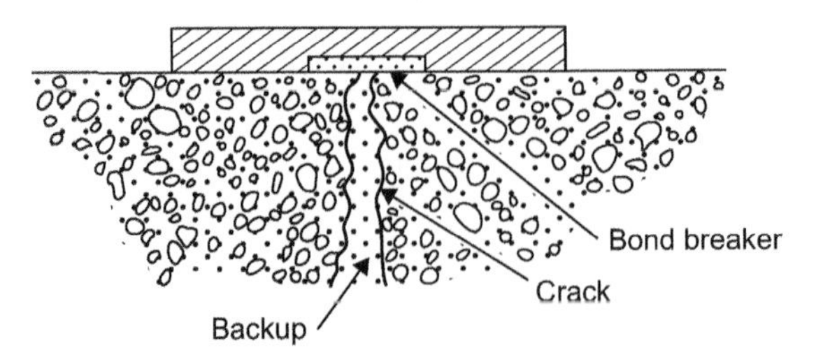

FIGURE 13.7 Repair of a narrow crack with flexible surface seal. *Courtesy of United States Army Corps of Engineers.*

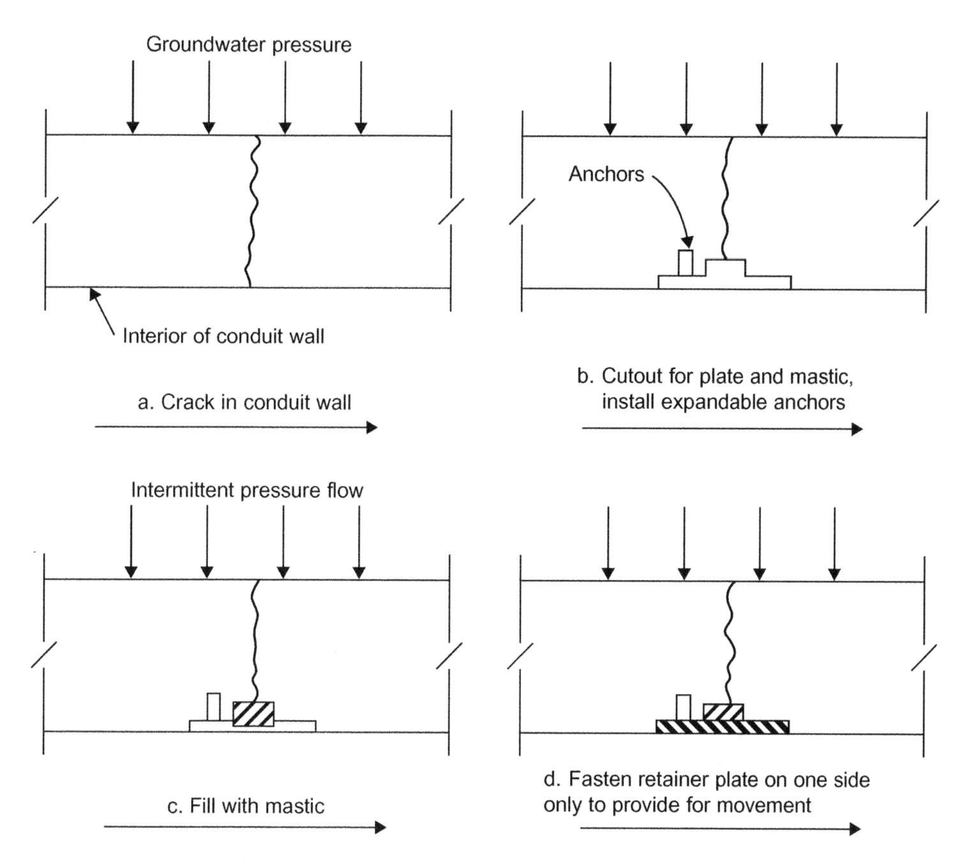

FIGURE 13.8 Repair of crack by using a retainer plate to hold mastic in place against external pressure. *Courtesy of United States Army Corps of Engineers.*

FIBER-REINFORCED CONCRETE

Portland cement that contains discontinuous discrete fibers is known as fiber-reinforced concrete. Fibers made from steel, plastic, glass, and other natural materials are added to the concrete in the mixer. Fiber-reinforced concrete is often used for the repair of pavement. It is not unusual for increased vibration to be needed for this type of concrete. The prep work for fiber-reinforced concrete is essentially the same as it would be for regular concrete, and it can be pumped with up to 1.5% fibers by volume in it. See Figure 13.8 for examples of repairing concrete cracks with retainer plates that hold in mastic against external pressure.

Code Consideration

All pre-cast members must be designed and built to withstand curing, stripping, storage, transportation, and erection.

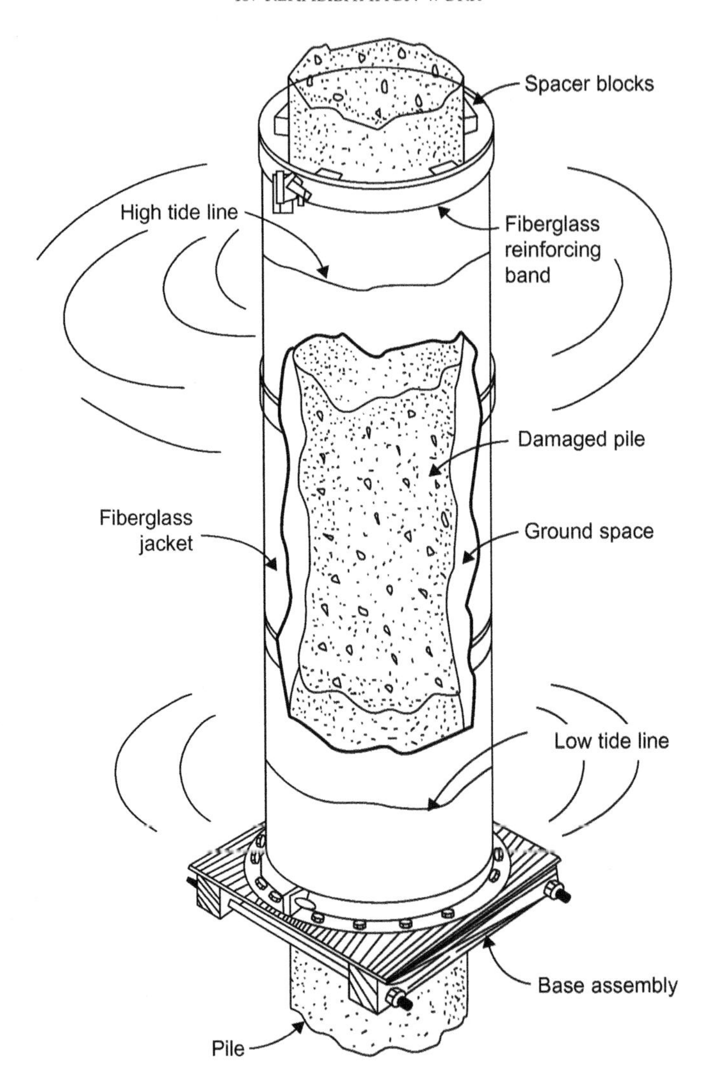

FIGURE 13.9 Typical pre-formed fiberglass jacket used to repair a concrete pile. *Courtesy of United States Army Corps of Engineers.*

FLEXIBLE SEALING

How is flexible sealing done? It involves routing and cleaning a crack and filling it with a suitable field-molded flexible sealant. This is not the same as routing and sealing, because an actual joint is made. It is not just a crack being filled. This method can be used to fill active cracks, but don't be fooled. The process is not likely to increase the structural capacity of a cracked section of concrete. See Figure 13.9 as an illustration of a pre-formed fiberglass jacket being used to repair a concrete pile.

Bond breakers are used at the bottom of a crack slot to prevent a concentration of stress on the bottom of a structure. This can be a polyethylene strip, a pressure-sensitive tape, or some other material that will not bond to the sealant before or during the curing process.

GRAVITY SOAK

High molecular weight methacrylate is poured or sprayed onto a horizontal concrete surface and spread by either a broom or squeegee to make a gravity soak. This method penetrates very small cracks, and is done by gravity and capillary action. Its purpose is to prevent access to reinforcing steel.

The types of surfaces where gravity soak is used often include:

- Horizontal concrete surfaces
- Bridge decks
- Parking decks
- Industrial floors
- Pavement

If you plan to use this process, make sure that the concrete surface is cured and air-dried. It may be necessary with older concrete to clean off any oil, grease, tar, or other contaminants. Sandblasting is the preferred method for this type of cleaning.

Code Consideration

Entire composite members, or portions thereof, that are used to resist shear are allowed.

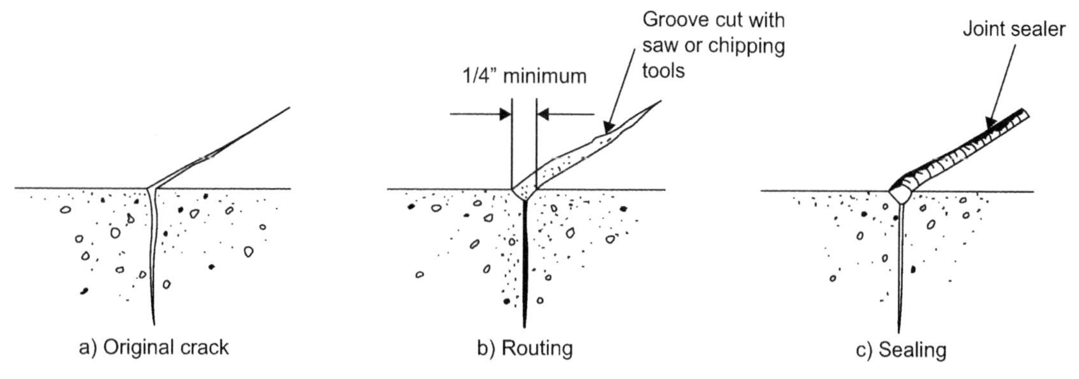

FIGURE 13.10 Repair of crack by routing and sealing. *Courtesy of United States Army Corps of Engineers.*

CHEMICAL GROUTING

Chemical grouting consists of two or more chemicals that combine to create a gel or solid precipitate as opposed to cement grouts that consist of suspensions of solid particles in a fluid. This process decreases fluidity and a tendency to solidify and fill voids in the material into which the grouting is injected.

There are pros and cons to grouting. The pros include the following benefits:

- Can be used in moist environments
- Wide limits of control of gel time
- Can be used to fill very fine cracks

The cons to grouting are

- A high degree of skill is needed to perform professional grouting.
- Can be a lack of strength in the existing concrete.
- Some grouts dry out too early.
- Some grouts are highly inflammable and cannot be used in enclosed spaces.

See Figure 13.10 to see how cracks are fixed with the use or routing and sealing.

HYDRAULIC-CEMENT GROUTING

Hydraulic-cement grouting is a common method of repairing cracks in concrete that is most often used in dormant cracks. This type of grouting tends to be less expensive than chemical grouts, and is well suited for large volume applications. However, the grouting is subject to pressure and may not fill a crack entirely. This is normally the type of grout preferred for sealing cracks in foundations.

To use hydraulic-cement grouting, first clean the cracks that are to be sealed. Install built-up seats at intervals astride the crack that will provide a pressure-tight contact with the injection tool. Seal the crack between the seats.

HIGH-STRENGTH CONCRETE

Concrete with a 28-day design compressive strength over 6,000 PSI is considered high-strength concrete. Admixtures can be used with this type of concrete. It resists chemical attacks, provides less abrasion, and has improved resistance to freezing and thawing. It is an expensive concrete, but worth it in various circumstances. Curing is more critical with

Code Consideration

Allowable ties to transfer horizontal shear can include the following:

- Single bars
- Wire
- Multiple leg stirrups
- Vertical legs of welded wire reinforcement

high-strength concrete than it is with normal concrete, and water curing is the preferred method with this type of concrete.

JACKETING

Jacketing is a method of encasing an existing structural member in new concrete. The member that is being encased can be made of concrete, steel, or wood. Pilings are one of the most common types of structures to benefit from jacketing, which can provide additional strength and protection. This type of repair is especially helpful when a construction member is submerged in water. A form is required for jacketing, and any type of concrete form can work. Typical forms for jacketing are made from either steel, fiberglass, or fabric.

JUDICIOUS NEGLECT

Judicious neglect is a decision not to take any action in the repair of a concrete member. A complete investigation is required before a decision to ignore a problem can be made. Sometimes no action is the best action. This may seem a bit strange, but if existing damage is not likely to escalate and is not causing structural damage, the defect may be monitored instead of being repaired.

POLYMER OVERLAYS

The type of polymer overlay used in a repair is generally decided based on the thickness of the overlay. Thin overlays (about 1 in. thick), are normally done with epoxy mortar or concrete. Thicker overlays, those from 1–2 in. thick, use a latex-modified concrete. Conventional Portland cement is used when an overlay will be more than 2 in. thick. Epoxy mortars work well for repairs when existing concrete is being aggressively attacked by acid or some other material.

It is not advisable to use an overlay with a vapor barrier on slab-on-grade applications or concrete walls that are backfilled in freezing climates. Bridge decks are often resurfaced with latex-modified concrete overlays. Epoxy-modified concrete is also used for this type of repair.

The aggregates used for conventional concrete can normally be used with epoxy-resin mixtures. All aggregates should be clean and dry when repairs are made. Maximum size for aggregates used in an epoxy-based concrete is 1 in. in diameter. The proportion of aggregate to the mixed resin in epoxy concrete can go as high as a 12:1 ratio.

The mixing of epoxy-modified concrete can be done by hand or machine. Large commercial mixers have proved to be very effective in mixing this type of concrete. When mixing, fine aggregate should be introduced into the mix first followed by coarse aggregates.

Code Consideration

You must neglect the tensile strength of concrete when computing the requirements for reinforcement.

A primer coat of epoxy should be applied to the surface being repaired with a brush or a trowel, or some other viable method. When the primer is tacky to the touch, it is time to pour the new concrete. Overlays deeper than 2 in. should be made in layers to minimize heat dissipation. Applications made in layers of 2 in. at a time are preferred. A time delay between layers is needed to allow for heat removal. However, epoxy-modified concrete does require prompt action as it doesn't last long in a mixer. A hand tamper can be used to compress layers and speed the application process.

Latex

Latex-modified overlays are often made with styrene-butadiene. This type of admixture is dealt with in much the same manner as conventional concrete. Temperature, however, can be a problem in hot climates. The latex-concrete should be maintained at a temperature range between 45 and 85°F. When air temperatures are higher than 85°F, a night placement of the concrete may be needed to maintain a working temperature range. Another risk of hot weather is rapid drying that results in shrinkage cracks.

A bond coat is typically made by eliminating coarse aggregate from the mix. Then the mixture is broomed onto the concrete surface. Placing of the concrete is done with standard methods. Finish work can be done with vibratory or oscillating screeds, but a rotating cylindrical drum is preferred.

A moistened material is used to cover new concrete. This covering is left in place for one or two days, and then air drying is allowed for about 72 hours. By following these guidelines, the risk of shrinkage cracks is reduced.

Code Consideration

Unbonded tendons require watertight and continuous sheathing.

Portland Cement

Portland cement overlays range in depth from 4 to 24 in.; however, thinner layers can be used. This type of overlay repairs spalling and surface cracks, and is appropriate for repairing concrete damaged by abrasion-erosion and deteriorated pavements. Bridge decks often benefit from a concrete overlay.

If concrete is damaged by acid or some other aggressive means, Portland cement should not be used for repairs. Active cracking is another defect that should not be repaired with conventional concrete.

POLYMER COATINGS

Polymer coatings aid in the protection of concrete from abrasion, chemical attacks, and freezing-thawing damage. Epoxy is favored because it allows the coating to be impermeable to water and to resist chemical attacks. There are types of epoxy resins that can adhere to damp surfaces and surfaces immersed in water.

The preferred temperature range for applying epoxy resins is from 60 to 89°F. When working outside of this range, special precautions are needed. When foot traffic is expected on a finished surface, installers should broadcast a sharp sand to the fresh surface of new concrete.

POLYMER CONCRETE

Polymer concrete (PC) is a composite material. Aggregates are bound together in a dense matrix with a polymer binder. This type of concrete sets up quickly, bonds well, and has strong chemical resistance. Additionally, polymer concrete is high tensile, flexural, and compressive in its strength.

When a polymer is added to Portland cement the result can be a stronger, more adhesive mixture. Other benefits of mixing polymers with conventional cement include:

- Resistance to freezing and thawing
- High degree of permeability
- Improved resistance to chemical attacks
- Better resistance against abrasion and impact

POLYMER IMPREGNATION

Polymer impregnated concrete (PIC) is a Portland cement concrete that is subsequently polymerized. A technique known as a monomer system is used to create this concrete. Water does not mix well with this type of concrete, and there is a varying degree of volatility, toxicity, and flammability related to these chemicals. When this admixture is heated, it becomes a tough, strong, durable plastic. Placed in concrete it adds considerably to a number of qualities and is very good for filling cracks.

Code Consideration

Any post-tensioning ducts installed in areas subject to freezing temperatures must be protected against water accumulation in the ducts.

POLYMER INJECTION

Polymer injections can be rigid or flexible systems. Epoxies are a common type of rigid system and they are often used for repairing cracks. A polyurethane system is much more flexible and is used on active cracks and to stop water infiltration. A high-pressure injection is preferred.

Rigid repairs have been used for cracks on bridges, in buildings, dams, and other types of concrete structures. Normally, a rigid repair is made only on dormant cracks. There are some exceptions, but a rigid repair is not normally recommended for active cracks. This type of repair can be used for delaminations in bridge decks.

TP-21763-05 (9/2003)

Minnesota Department of Transportation

Concrete Aggregate Worksheet

S.P.	Plant:	Date:	Agg. Source(s) #: FA – CA -
Engineer:	Tester:	Time:	CA – CA -

Sieve Analysis of Coarse Aggregate

Agg. Fract.	CA - ____ Mix Prop. ____ %				CA - ____ Mix Prop. ____ %				CA - ____ Mix Prop. ____ %			
	Test No. Sample Wt.		Quality Sample Submitted. By ____ Date		Test No. Sample Wt.		Quality Sample Submitted. By ____ Date		Test No. Sample Wt.		Quality Sample Submitted. By ____ Date	
Sieve Sizes	Weights		%	Grad.	Weights		%	Grad.	Weights		%	Grad.
Pass – Ret.	Ind.	Cum.	Pass	Req.	Ind.	Cum.	Pass	Req.	Ind.	Cum.	Pass	Req.
2" – 1 1/2"												
1 1/2" – 1 1/4"												
1 1/4" – 1"												
1" – 3/4"												
3/4" – 5/8"												
5/8" – 1/2"												
1/2" – 3/8"												
3/8" - #4												
#4 – Btm												
Check Total		± 0.3% or 0.2 lb of Sample Wt.				± 0.3% or 0.2 lb of Sample Wt.				± 0.3% or 0.2 lb of Sample Wt.		

Coarse Aggregate Percent Passing #200 Sieve Test

	(CA -)		(CA -)		(CA -)
(A) Dry weight of original sample					
(B) Dry weight of washed sample					
(C) Loss by washing (A – B)					
(D) % Passing #200 (C ÷ A) × 100					

Composite Gradation for (CA -)

Agg. Fract.	CA -	CA -	CA -	Composite	Grad.
Proportions	____ %	____ %	____ %	100%	Req.
2"					
1 1/2"					
1 1/4"					
1"					
3/4"					
3/8"					
#4					

Washing Data for Sieve Analysis of Fine Aggregate

(A) Dry sample and record weight	
(B) Wash and dry sample, record weight	
(C) Loss by washing (A – B)	
Enter (C) to the right, for fine sieve analysis	

Sieve Analysis of Fine Aggregate

Quality Sample Submitted. By: ____ Date :				
Test No. ____ Sample Wt.				
Sieve Size Pass Ret.	Weights Ind.	Cum.	% Pass	Grad. Req.
3/8" - #4				100
#4 - #6				95 – 100
*#6 - #8				**
#8 - #16				80 – 100
#16 - #30				55 – 85
#30 - #50				30 – 60
#50 - #100				5 – 30
#100 - #200				0 – 10
*#200 - Btm				0 – 2.5
Loss by washing				
Check Total		± 0.3% of Sample Wt.		
Fineness Modulus	Within ± 0.20			

* #6 and #200 not included in Fineness Modulus
** #6 is recommended as filler sieve

FIGURE 13.11 Concrete aggregate worksheet. *Courtesy of the Minnesota Department of Transportation.*

Active cracks should be filled with flexible repairs. The grout used will make the crack into a moveable joint. There are water-activated polyurethane grouts that are hydrophobic or hydrophilic. High-pressure injections are typically done at pressures of 50 PSI or higher. The steps for injection are listed below:

- Clean cracks.
- Seal surfaces.
- Drill holes to insert injection fittings.
- Bond cracks that are not V-grooved with a flush fitting.
- Mix the filler.
- Inject the filler.
- Remove any surface seals.

PRE-CAST CONCRETE

Pre-cast concrete can offer several benefits in the repair of damaged concrete structures. Some of these benefits include:

- Ease of construction
- Rapid construction
- High quality
- Durability
- Economy

What types of repairs use pre-cast concrete? They can be employed in a variety of repairs from the following list:

- Navigation locks
- Dams
- Channels
- Flood walls
- Levees
- Coastal structures
- Marine structures
- Bridges
- Culverts
- Tunnels
- Retaining walls
- Noise barriers
- Highway pavement

PREPLACED-AGGREGATE CONCRETE

Preplaced-aggregate concrete is used to fill voids in concrete. It is injected with a Portland cement sand grout and can be pumped into a form to fill voids. Water in the voids is displaced and the concrete forms a solid mass. This is normally used on large repairs. There is a slow shrinkage rate when this method is used. Figure 13.11 shows a worksheet that can be used when calculating the needs for aggregates in concrete.

RAPID-HARDENING CEMENT

Rapid-hardening cement has a minimum compressive strength of 3,000 PSI that is developed in approximately 8 hours or less. Some cements can obtain strength in as little as one hour. Magnesium-phosphate cement is one such cement. You have to work fast with this product, but it reduces downtime on a jobsite.

High alumina cement can have as much strength in 24 hours as conventional concrete will have after 28 days of setting. High humidity and air temperatures of 68°F or greater can reduce the strength of this mixture.

Other types of fast-setting cements are available:

- Regulated-set Portland cement
- Gypsum cement
- Special blended cements
- Packaged patching materials

ROLLER-COMPACTED CONCRETE

Roller-compacted concrete is compacted by a roller when the concrete hasn't yet hardened. This type of concrete is best suited for large placement areas where there is little to no reinforcement or embedded metals. New construction of dams and pavement is a suitable application for roller-compacted concrete. It is also used to repair dams. Added strength and stability is an expected result with roller-compacted concrete.

ROUTING AND SEALING

Routing and sealing is a common method of repairing dormant cracks. This procedure should not be used on active cracks. A minimum surface width for a crack to be routed and sealed is one-quarter of an inch. When you are dealing with pattern cracks or narrow cracks, the routing will enlarge the cracks to make them suitable for sealants. Sealants are used to prevent water infiltration.

SHOTCRETE

Shotcrete is mortar that is pneumatically projected at high velocity onto a surface. It can contain coarse aggregate, fibers, and admixtures. When shotcrete is used, the result can be an excellent bond and a viable repair. It can be used with the following types of concrete structures:

- Bridges
- Buildings
- Lock walls
- Dams
- Hydraulic structures

SHRINKAGE-COMPENSATING CONCRETE

Shrinkage-compensating concrete is an expansive cement concrete used to repair cracks. The concrete expands in volume after setting and during hardening. With the proper reinforcement and containment, good strength can be obtained.

CHAPTER

14

Maintenance Matters

OUTLINE

The maintenance of concrete is the answer to reducing expenses associated with concrete repairs. Prevention is the best medicine. As obvious as this should be, it is too often ignored. Routine maintenance can prevent, or at least postpone, the need for costly and time-consuming rehabilitation and repairs. Typical maintenance can be expensive, but it is far less expensive than major repairs. Money invested in advance may not be a desirable option, but it is often better than the alternative.

STAINS

Stains on concrete can be a sign of trouble. They can penetrate the concrete if it is porous and absorbent. For this reason, not to mention eye appeal, stains should be removed. However, there are many different types of stains:

- Iron rust
- Oil
- Grease
- Dirt
- Mildew
- Asphalt
- Efflorescence
- Soot
- Graffiti

Not all stains can be treated equally. Removal methods vary. Choosing a cleaning method depends on what type of stain is being cleaned. Identifying the stain is preferred, but when identification is not possible, test different types of cleaning options. The cleaning components can be tested on small parts of the stained concrete. If you don't know what type of stain you are dealing with, test with the following substances in the order in which they are listed:

- Organic solvents
- Oxidizing bleaches
- Reducing bleaches
- Acids

Protective coatings can be applied to concrete to prevent chemical attacks. Not all coatings perform all jobs. See Table 14.1 for advice on what coatings to use for various purposes.

TABLE 14.1 Guidance on Selection of Concrete Coatings to Prevent Chemical Attack and Reduce Moisture Penetration

Coating	Water Repellency	Cleanability	Aesthetic	Concrete Dusting	Mild Chemical	Severe Chemical	Moderate Physical	Severe Physical
Silicones/silanes/siloxane	R	NR	NR	NR	NR	NR	NR	NR
Cementitious	R	NR	R	NR	NR	NR	NR	NR
Thin-film polyurethane	R	R	R	R	R	NR	R	NR
Epoxy polyester	R	R	R	R	R	NR	NR	NR
Latex[1]	R	R	R	R	NR[2]	NR	NR[2]	NR
Chlorinated rubber	R	R	R	R	R	NR	R	NR
Epoxy	R	R	R	R	R	R	R	NR
Epoxy phenolic	R	R	R	R	R	R	R	R
Aggregate-filled epoxy	R	R	R	R	R	R	R	R
Urethane elastomers	R	R	R	R	R	R	R	R
Epoxy or urethane coal tar	R	R	NR	R	R	R	R	R
Vinylester/polyester	R	R	NR	R	R	R	R	R

[1] *Excluding vinyl latices.*
[2] *Certain latices may be suitable for service.*
R = recommended; NR = not recommended.
Note: The recommendations provided are general. Candidate coating systems must be thoroughly evaluated to ensure that they are appropriate for the intended service conditions and meet other desired characteristics. The above list is not necessarily all-inclusive.
Reprinted with permission from NACE International. The complete edition of NACE Standard RP0591-91 is available from NACE International, P. O., Houston, Texas 77218-8340, phone: 713/492-0535, fax: 713/492-8254. Courtesy of United States Army Corps of Engineers.

STAIN REMOVAL

Stain removal can be accomplished with various methods. Brushing and washing is a common option for simple stains. Steam cleaning is frequently used to remove stubborn stains. Additional cleaning options include water blasting, abrasive blasting, flame cleaning, mechanical cleaning, and chemical cleaning. Each method has its pros and cons.

Cleaning with water is usually not very invasive, because it is done with a fine mist. Too much pressure can drive stains deeper into concrete so the water is applied from the top down. When water alone is not getting the job done, brushing might be the answer. Soap, ammonia, and vinegar are sometimes used with water as a cleaning solution.

Code Consideration

All external tendons and tendon anchorage regions are required to be protected from corrosion.

Water blasting can remove some of the concrete surface. If too much pressure is used, it can be destructive. When dealing with dirt or chewing gum, steam cleaning is effective, but expensive. Abrasive blasting is likely to remove some of the surface area of concrete. When this form of cleaning is used, the nozzle used to deliver the pressure should not be held too close to the concrete surface.

Organic stains may not be able to be removed with solvents. Flame cleaning is capable of cleaning these types of stains; however, the procedure may cause concrete to scale and the fumes can be objectionable.

Another way to remove stains is by using mechanical equipment ranging from a chisel to a grinder. Damage to concrete is not unusual when mechanical cleaning is performed. Many believe that chemical cleaning is the best bet for most types of stains. When used properly, the chemicals do not harm concrete. The downside to chemical cleaning is the health risk associated with the process. When chemical cleaning is used, close attention must be paid to proper safety procedures.

Code Consideration

Unbonded construction that can be affected by repetitive loads must be protected from the possibility of fatigue in anchorages and couplers.

CLEANING DETAILS

The cleaning details for concrete depend on the type of stains being cleaned. Once you know the type of stain, the cleaning details for the job can be determined. Take iron rust as an example. Let's say that you have an iron rust stain that is light or shallow. What will be your course of action? In this type of situation, start by mopping the stained area with a solution of oxalic acid and water. Wait 2–3 hours then scrub the surface with a stiff brush. Rinse as needed with clean water. This should do the trick, but if it doesn't, move to the next level of cleaning.

Deep stains can be treated with a poultice by mixing sodium citrate, glycerol, and diatomaceous earth or talc with water. Trowel the poultice over the stain. Leave the poultice in place for up to 3 days. If the stain is still evident, repeat the process. Just keep working at it until the stain disappears.

Oil Stains

Freshly spilled oil stains should not be rubbed. They should be soaked up with absorbent paper. Do not wipe the spill. Cover the spill area with an absorbent material that could be something as simple as kitty litter or a more professional absorbent. Wait a full day and then sweep up the absorbent material to remove the oil. Next, scrub the stained area with a scouring powder or strong soapy solution.

Code Consideration

Experimental and numerical analysis procedures are allowed when the procedures offer a reliable basis for design.

Old oil stains require a different method of attack. A mixture of equal parts of acetone and amyl acetate will be needed. This is mixed with some material, like flannel, to be laid over the stain. Once the cloth is over the stain, cover the cloth with a glass panel for about 15 minutes. Repeat the process as needed. When done, rinse the work area with clean water.

Grease

Grease can be scraped from a concrete surface. Scouring powder can be scrubbed on a grease stain, or you can scrub the area with a strong soap solution or sodium orthophosphate. If these methods fail, make a poultice of chlorinated solvents. Repeat the process as needed. Once the stain is under control, rinse the affected area with clean water.

Code Consideration

The specified compressive strength of concrete is 3,000 lb/PSI.

Dirt

Dirt is normally pretty simple to remove from a concrete surface. Clean water is often all that is required. Soap and water may be needed in some cases. If you encounter difficult dirt stains, apply a solution of 19 parts water and 1 part hydrochloric acid to remove the stains.

Steam cleaning can be used, but it can also be expensive. Dirt with a clay density can be removed with hot water that contains sodium orthophosphate and a scrub brush.

Mildew

Remove mildew stains with a mix of powdered detergent and sodium orthophosphate with commercial sodium hypochlorite solution and water. This is done by applying the mixture and waiting a few days. After the waiting period, use a brush to scrub the affected area. Rinse with clean water. Be careful when working around metal as sodium hypochlorite may corrode it.

Code Consideration

The specified yield strength of non-prestressed reinforcement must not be in excess of 60,000 lb/PSI.

Asphalt

How do you remove asphalt from a concrete surface? If the temperature is hot, start by chilling molten asphalt with ice. Scrape or chip off the asphalt while it is brittle and scrub the area with an abrasive powder and rinse completely with clean water. Then scrub with scouring powder and rinse the area. You can use a poultice of diatomaceous earth or talc and a solvent to remove cutback asphalt. Once the poultice is dry, you should be able to brush it off. Applying solvents to emulsified asphalt can carry the emulsions deeper into concrete.

Efflorescence

Water and a scrub brush will remove most recent efflorescence stains. Older stains may require water blasting or sandblasting. Chemical removal with hydrochloric or phosphoric acid will generally work for difficult stains, but the solution can discolor the affected area. General practice calls for applying the chemical solution to all visible concrete in the cleaning zone to maintain a consistent appearance.

Code Consideration

When the direction of reinforcement varies more than 10 degrees from the direction of principal tensile membrane force, an assessment must be done to prevent cracking.

Soot

Sometimes soot can be removed with water, scouring powder, and a scrub brush. A powdered pumice or grit can also be used for normal stains. Tough stains can be treated by soaking sections of cotton material in trichloroethylene and applying it to the stain.

Horizontal stains can be covered with the saturated material and the material can be held in place with a heavy object. Vertical stains require the material to be braced in place against the stain.

Monitor the saturated material and remove it periodically. Wring the material out. Soak the material in fresh trichloroethylene and reapply the material to the stain. Continue the process until the stain is removed.

Toxic gases may be created when trichloroethylene is put into contact with fresh concrete or other strong alkalis. Use all proper safety precautions when working with any tools or chemicals.

COATINGS AND SEALING COMPOUNDS

Coatings and sealing compounds are applied to concrete surfaces to protect against chemical attacks. They are also used to control water penetration into concrete. Some thick coatings can be used to protect concrete from physical damage, but before any coating is used, it must be determined if the concrete needs protection. If a protective coating or sealant should be used, the concrete surface must be prepared properly; it must be sound, clean, and dry. The types of coatings and sealants available include:

- Silicones
- Siloxanes
- Silanes
- Cementitious coatings
- Urethanes
- Epoxy polyesters
- Latexes
- Chlorinated rubbers
- Epoxies
- Epoxy phenolics
- Aggregate-filled epoxies
- Thick-film elastomers
- Thin-film polyurethane
- Latex
- Urethane elastomers
- Epoxy coal tar
- Urethane coal tar
- Vinylesters
- Polyesters

Once you have chosen the best coating or sealant for your needs, follow the manufacturer's recommendations for application. Surface temperatures and other variables come into play with coatings and sealants, which is why you need to read and heed the working instructions from each product manufacturer.

CHAPTER

15

Specialized Repairs

OUTLINE

There are many types of specialized concrete repairs. For example, a residential building contractor will generally think of foundation walls or concrete slabs, and a civil engineer may think of a bridge deck or a lock wall. Architects might turn their thoughts to ornamental retaining walls. If you put your mind to it, you can come up with a wide variety of specialized concrete repairs.

Rehabilitation of existing concrete requires different tactics for different working conditions. A concrete pier submerged in water will not be repaired with the same procedures used for a parking deck. In this chapter we will look at specialized repairs with lock walls.

REHABBING LOCK WALLS

The rehabbing of lock walls can involve basic scaling to deeper damage. It is common to remove anywhere from 1–3 feet of concrete from the face of a lock wall. New Portland cement concrete is then used as a conventional concrete to resurface the wall. Other methods to be considered are shotcrete, preplaced-aggregate concrete, and precise concrete that is in stay-in-place forms. Thin overlays are sometimes all that is required for a simple repair.

Code Consideration

When the dimensions of structural elements are evaluated, the assessment should be on critical sections.

CAST-IN-PLACE

Cast-in-place concrete can be economically feasible. Much of the cost factor is related to the depth of a repair. Shotcrete is a good option when repair sections range in thickness from 6 to 12 in. Either of these two options is usually cost-effective. Thicker sections of concrete require traditional concrete forming to maintain the cost-effectiveness. The minimum width that normally dictates the need for a form is 12 in. Conventional cast-in-place concrete offers a number of advantages over other rehab materials. Some of these advantages include:

- It can be proportioned to stimulate existing concrete substrate.
- It can minimize strains resulting from material incompatibility.
- Admixtures can be used in freezing and thawing temperatures.
- Conventional concrete using proven methods that equipment and skilled workers are available for.

BLASTING LOCK WALLS

Blasting of lock walls is a common method of surface preparation before a repair is made. (See Figure 15.1 for compression and tension factors that may be considered during blasting.) Workers often drill small-diameter holes along the top of a lock wall in a direction parallel to the removal face. Then a light explosive, such as detonating cord, is loaded into the holes and cushioned by stemming the holes. Detonation is generally accomplished with electric blasting caps. Proper planning must go into this type of work. Engineers and other experts may use test results or historical reviews of previous jobs to determine a course of action for the blasting.

Code Consideration

The code requires a set of final response measurements to be made within 24 hours of the time that a test load is selected for testing.

When blasting is complete, the remaining concrete has to be checked for loose or defective concrete. This is done with sounding. When bad concrete is discovered, it can be removed with chipping, grinding, or water blasting. Before a repair is made, the concrete surface must be clean and free of debris that might compromise the bonding process.

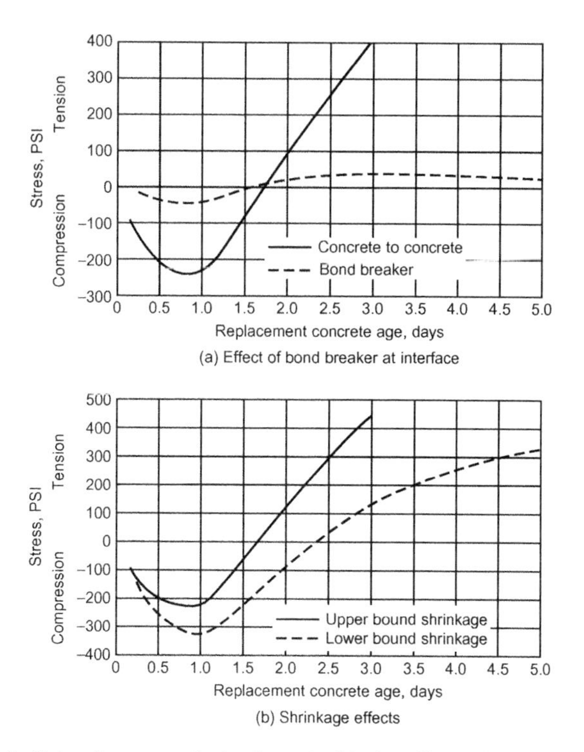

FIGURE 15.1 Results of a finite element analysis of a typical lock wall resurfacing. *Courtesy of United States Army Corps of Engineers.*

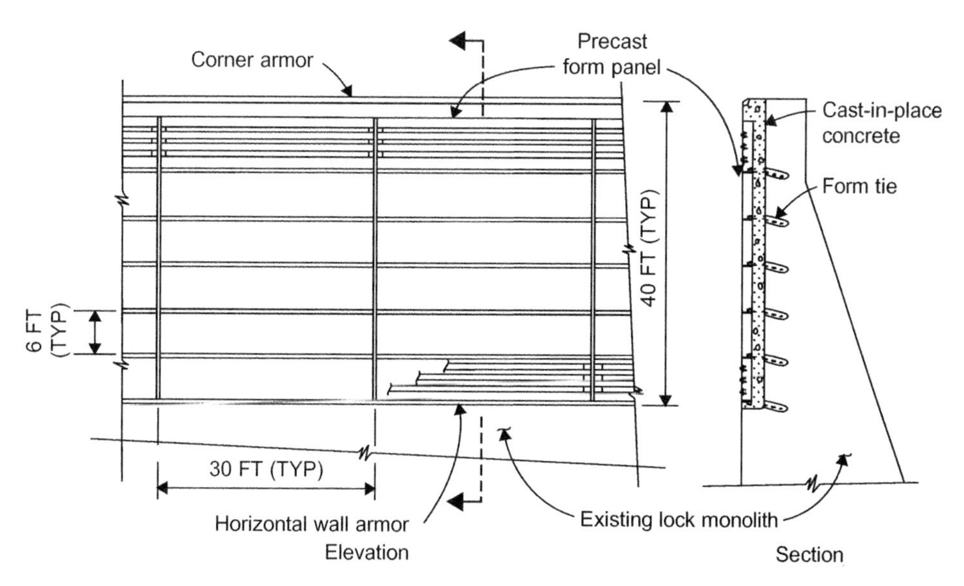

FIGURE 15.2 Pre-cast concrete stay-in-place forming system for lock wall rehabilitation. *Courtesy of United States Army Corps of Engineers.*

ANCHORS

Anchors are needed for repairing the face of a lock wall. The type of anchors used can vary, but dowels are the typical choice. They provide anchor points for reinforcing steel and for the bonding of new concrete to old concrete.

The spacing of dowels is determined by engineers or existing test data. A typical spacing for dowels is 4 feet measured center to center. When openings exist in a wall, the spacing of dowels should be limited to 2 feet on center.

Core samples are taken as a way of determining what type of repair materials will be best suited for the job. In the case of dowels, the core samples help to determine how deeply a dowel needs to be embedded in existing concrete. For example, a wall that has an average compressive strength of 3,000 PSI or more requires dowels to be embedded to a depth that is at least 15 times the nominal diameter of the dowel. Field tests may prove that shorter holes can be used. It is also common practice to test a percentage of dowels installed to confirm a proper repair. One such requirement calls for 3 out of every 1,000 dowels to be tested.

Code Consideration

The code contains special provisions for seismic designs.

Once the dowels are put in place, reinforcing steel can be installed. This normally consists of either number 5 or number 6 bars placed on 12 in. centers in every direction. The reinforcing is hung vertically over the anchors. Occasionally, reinforcing mats, wall armor, or other wall appurtenances are installed on anchors before concrete is placed.

CONCRETE PLACEMENT

Concrete placement for the repair of lock walls is pumped or poured into forms. This may be done with flexible piping. The heights of walls vary. Some full-face pours may be as tall as 50 feet. General procedure is to pour concrete on alternating monoliths, and concrete forms are normally removed anywhere from one to three days after the concrete is placed. Membrane-curing compounds are often applied to formed concrete surfaces.

Cracking is a persistent problem with repair concrete on lock walls. The reasons thin layers of repair concrete crack are

- Shrinkage
- Thermal gradients
- Autogenous volume changes

SHOTCRETE

Shotcrete is a cost-effective way of repairing walls when the repair thickness does not exceed 6 in. Overall, shotcrete is a durable material that does an adequate job of structural repair. There are, however, some potential problems associated with this material. The following list includes examples of the possible shortcomings of shotcrete.

- Moisture may be trapped between the existing lock wall and the shotcrete repair. This can be a serious problem when freezing and thawing conditions exist.
- Spalling can happen over time.
- Delamination may occur.

PREPLACED-AGGREGATE CONCRETE

Preplaced-aggregate concrete is more resistive to shrinkage and creep than conventional concrete because of the aggregate. The net result is more protection against cracking. This type of concrete can be used on numerous types of structures. Using preplaced-aggregate concrete is more expensive, but it may be worth it in the long run.

Code Consideration

When longitudinal reinforcement is required by design the welding of stirrups, ties, inserts, and similar elements is not allowed.

PRE-CAST CONCRETE

Pre-cast concrete has a lot to offer when it comes to concrete repairs. When compared to cast-in-place concrete, the advantages of using pre-cast concrete are numerous. Listed next are some of the key advantages:

- Minimal cracking
- Durability
- Rapid construction
- Lower maintenance costs
- Improved appearance
- Lower impact from on-site weather conditions
- Ability to inspect the concrete before it is used
- May eliminate the need for dewatering a lock chamber during repairs

See Figure 15.2 for an illustration of using pre-cast concrete with a lock wall.

CUTOFF WALLS

Concrete cutoff walls are cast-in-place structures. They are used to provide a positive cutoff of the flow of water under or around a hydraulic structure, such as a dam. Geotechnical monitoring and review prior to making a decision to build a cutoff wall is prudent.

An average cutoff wall is somewhere between 2 and 4 feet thick. The procedure for preparing for and creating a cutoff wall can be tricky. A concrete-lined guide trench is made along the axis of the wall to be repaired. This type of trench is usually only a few feet deep, but it gives a working surface on both sides of the wall and helps to maintain the alignment of the wall.

Code Consideration

Plain concrete is not allowed to be used in footings on piles.

Concrete is placed in the trench to create the cutoff wall. It must be screened for discontinuities as they can cause serious performance problems. The concrete mixture is very important, because it will be used for tremie placement, and this requires strict adherence to required specifications. You are not looking for compressive strength; instead the key elements are flowability and cohesion.

Test panels should be created in noncritical locations near the wall location. These panels can then be tested to determine if the concrete mixture and the placement process are providing the desired result.

PRE-CAST CONCRETE APPLICATIONS

Pre-cast concrete has been used extensively. We have already discussed the advantages to this type of repair process. The following list details some of the different concrete structures that use pre-cast concrete:

- Navigation locks
- Dams
- Channels
- Flood walls
- Levees
- Coastal structures
- Marine structures
- Bridges
- Culverts
- Tunnels
- Retaining walls
- Noise barriers
- Highway pavement

UNDERWATER REPAIRS

Underwater repairs are sometimes necessary. As you might imagine, an underwater repair requires special tactics and techniques, which make dewatering extremely expensive. Naturally, the surface to be repaired must be clean, and specialized equipment is essential for these types of repairs to be done underwater.

How do you excavate underwater? Common methods include air lifting, dredging, and jetting. Air lifts are often used when the repair is in water up to 75 feet deep. When greater depths are encountered, dredging and jetting are the prime options.

Code Consideration

A strut this is wider at mid-length than at its ends is known as a bottle-shaped strut.

Debris removal is normally needed. This can involve the removal of cobbles, sediment, or reinforcing steel. Cobbles and sediment can be dislodged and removed with basic excavation techniques. Steel is another matter. It is usually removed with one of two methods: mechanical and thermal. Mechanical removal is often accomplished with either hydraulically powered shears.

There are three potential thermal techniques for underwater cutting: oxygen-arc cutting, shielded-metal-arc cutting, and gas cutting. Most people prefer oxygen cutting. A newer type of cutting now available and evolving is abrasive-jet cutting.

Cleaning is always of prime importance when working with concrete repairs. This is no different when working underwater. The hard way uses hand tools, such as chisels, brushes, abrasive disks, and so forth. Powered cleaning tools make the job easier, and on large jobs, a self-propelled cleaning vehicle can be used.

Code Consideration

An abrupt change in geometry or loading is discontinuity.

When mixing concrete for submerged use, the mixture must be workable, cohesive, and protected from water until it is in place. A tremie pipe has long been used for underwater concrete placement. Vertical placement is generally done with the tremie method. Tremie pipe is placed at up to a 45 degree angle to slow the flow of concrete, because a straight drop of free-falling concrete is not always desirable.

Other placement methods include the hydrovalve method and the Kajima double tube tremie method. These are both variations of the traditional tremie method. A flexible hose that collapses under hydrostatic pressure and carries a controlled amount of concrete down the hose in slugs is used, which helps to prevent segregation. These methods are reliable, inexpensive, and can be used by any contractor who has workers skilled at working underwater.

Pumping is frequently preferred over the tremie method for placing concrete below water. When thin layers are needed, pumping offers multiple advantages:

- Fewer transfer points.
- No gravity-feeding problems.
- Ability to use a boom for placement allows for better control.

If you encounter a large void that needs to be filled underwater, consider using preplaced-aggregate concrete. With this method concrete is placed in a form and grout is injected from the bottom of the preplaced aggregate.

You must prevent loss of material from the top of the concrete forms. This is normally done by placing a permeable fabric next to the concrete, backed with a wire mesh. This is supported by a stronger backing of perforated steel and plywood. Pressure is put out as grout is injected, which can raise the concrete forms. Dowels can be used to protect against excessive form movement.

Pre-cast concrete panels and modular sections have been used in underwater repairs for dams, stilling basins, and lock walls. (See Figure 15.3 for data on stilling basins.) Successful underwater repairs and concrete forms have been made with prefabricated steel panels, which are also used to control erosion. Both of these options have pros and cons. Consider the underwater application at hand and consider both options when choosing an appropriate repair procedure. Figure 15.4 shows underwater repair of concrete spalling.

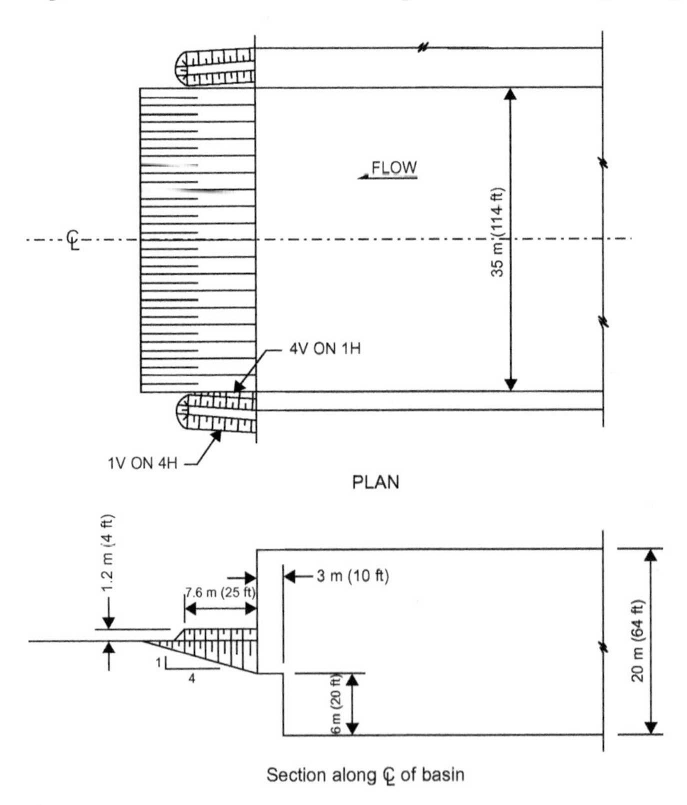

FIGURE 15.3 Stilling basin wall extension fills, Dworshak Dam. *Courtesy of United States Army Corps of Engineers.*

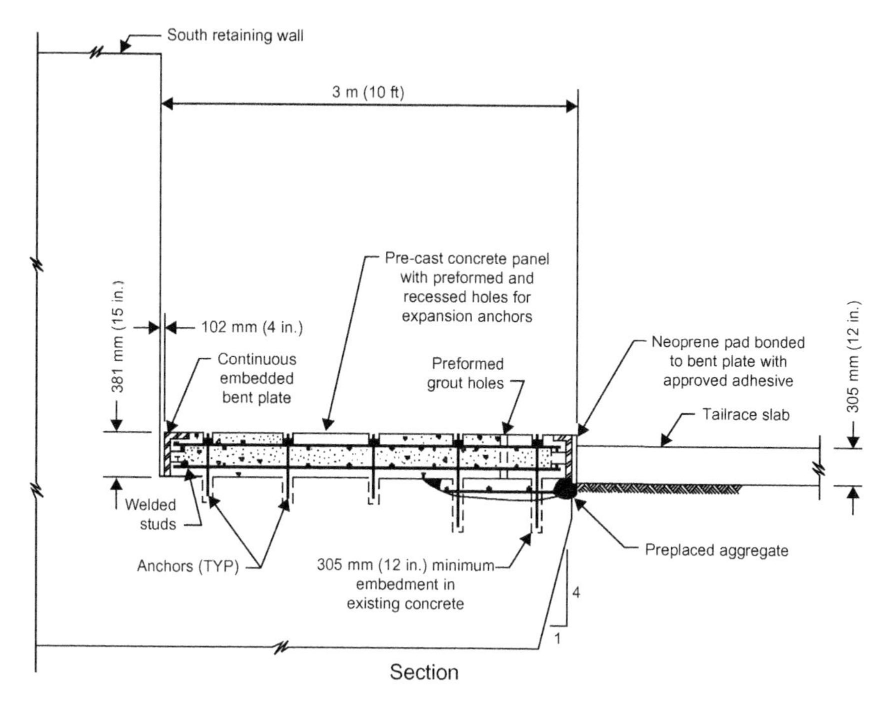

FIGURE 15.4 Underwater repair of concrete spalling, Gavins Point Dam. *Courtesy of United States Army Corps of Engineers.*

GEOMEMBRANE WORK

Geomembrane work is very useful when dealing with dams. Seepage control is just one use. They can also be used when working with canals, reservoirs, storage basins, dams, and tunnels. In Europe, geomembranes have been used to repair old concrete and masonry dams.

What is a geomembrane? It is a polymeric membrane that constitutes a flexible, watertight material with a thickness of 0.5 mm or more. The manufacture of geomembranes is done with a wide range of polymers such as plastics, elastomers, and blends of polymers.

Geomembranes used to be installed on the upstream side of structures with either nails or adhesives. More recently, stainless steel profiles are anchored. This type of system uses two vertical U-shaped anchors with one anchor larger than the other. The smaller anchor is fastened to the face of the structure first and then the larger anchor is placed over the smaller one and connected to the face. This creates two voids to be filled with geomembranes.

ROLLER-COMPACTED CONCRETE

Roller-compacted concrete (RCC) is a low-cost, fast method for concrete construction and repair. When working with dams, RCC can be used to repair damaged overflow structures, to protect embankment dams during overtopping, and to raise the crest of a dam.

Another use is to build a buttress on the downstream side of dams to increase dam strength.

There are a number of specialized repairs that use concrete in one form or another. The information in this chapter can be applied to many of them. There are more samples and examples that could be presented. The goal here is to bring various options to light. Each job will require its own evaluation when determining a suitable repair method.

CHAPTER

16

Problem Solving

OUTLINE

Troubleshooting defects in concrete is required to make a proper repair. In this chapter we are going to concentrate on the troubleshooting process. Some defects can be difficult to troubleshoot, because one defect might mimic the cause of another defect. This can lead to an improper repair. Knowing what to look for and where to look for it can save you time, money, and frustration.

The use of too much water in the concrete mix is probably the most common cause of concrete failure. When excessive water is added to the mix, many conditions can occur. Here are some examples of what would indicate a defect relevant to a bad mixture:

- Reduced strength
- Reduced abrasion resistance
- Increased curing time
- Increased drying shrinkage
- Increased porosity
- Increased creep

When concrete with excessive water in the mixture is installed, it may have to be removed and replaced. A repair may not be feasible. If the damage is shallow, say less than 1.5 in., a concrete sealer is one repair option. If a sealant is used, expect to reapply the sealer on a routine basis. Repairs to concrete damaged by excessive water in the mix are sometimes able to extend the useful life of the concrete, but they are rarely a permanent solution.

Damage that is 1.5 to 6 in. deep might be able to be repaired with an epoxy-bonded replacement concrete. Deeper flaws can be filled with replacement concrete. The best bet is to monitor concrete mixes as they are being used to avoid these problems.

BAD DESIGN DATA

Bad design data can result in a host of concrete problems. This problem can affect any element of an installation or repair. We would have to cover every potential defect to list all of the risks of bad designs. However, there are some design and/or installation factors to consider as common causes of damaged concrete.

How many times have you seen cracks, humps, flaking, or similar surface defects located near encased conduits and pipes? If electrical conduits or outlets are close to the concrete surface, you can expect problems. The same is true when the bases of handrails and similar surface-mounted elements are placed too close to the exterior corners of concrete surfaces.

Some pipe used for handrails can experience thermal expansion and contraction. What does this have to do with concrete? Since the handrails are secured to the concrete, stress from the movement of the handrail can damage the concrete. Slip joints need to be installed in the concrete to absorb this movement without cracking. Once cracking occurs, moisture can enter the surface. When freezing and thawing is a factor, the cracks and damage will accelerate.

A common cause of concrete damage on a bridge or hydroelectric structure is insufficient cover over reinforcing steel. It is normally acceptable to add a 3 in. layer of new concrete over an affected area. However, if corrosion is probable, the thickness of the repair layer should be no less than 4 in.

Concrete slabs need to be equipped with expansion and contraction joints. If they are not installed, cracking is likely. This can be a very big problem for bridge decks, dam roadways, floors of spillways, and so on. Delamination is common under these conditions.

Bad designs are bound to lead to problems. Repairs are costly. Try to ensure good design to avoid a multitude of potential problems.

CHEMICAL ATTACK

Concrete can be damaged from chemical attacks. Sodium, magnesium, calcium sulfates, and salts that can be found in alkali soils and groundwater could cause expensive concrete damage. A chemical reaction can occur between these natural elements and the hydrated line and hydrated aluminate in cement paste. Calcium sulfate and calcium sulfoaluminate may form. When this occurs, concrete may experience expansion damage. Type V Portland cement has low calcium aluminate content and is very resistant to sulfate reaction.

Concrete being damaged by chemical reactions can benefit from a thin polymer concrete overlay. Wetting and drying the concrete can slow the rate of deterioration. When damage gets out of control, the concrete should be replaced with Type V Portland cement.

ALKALI–AGGREGATE REACTION

Alkali–aggregate reaction can damage concrete. The cause of this type of reaction can be related to sand and aggregates, such as opal, chert, flint, and volcanic ash that have a high concentration of high silica. Calcium, sodium, and potassium hydroxide alkalies in Portland cement could have a reaction with these components, which can result in destructive expansion. A low alkali Portland cement and fly ash pozzolan used in new construction can reduce this risk.

Once a reaction has damaged the concrete, repairs are futile until all of the damage is done. In some cases, cutting relief slots in the concrete can extend the useful life of a structure, and in many cases the destruction will cease after enough time has passed. When the damage is no longer active, normal repairs can be made.

FREEZING

Freezing and thawing damages concrete. Concrete sealers can be used to minimize the risk of this type of destruction. The key to preventing damage from freezing and thawing is protecting concrete from excessive moisture content. When repairs are needed, they are normally done with replacement concrete.

MOVING WATER

Moving water channeled by concrete structures can contain destructive elements. Abrasion from waterborne sand and rocks can extensively damage a concrete surface. Generally, the water must be moving quickly for this to be a significant problem. This type of defect will be evident by the polished appearance of the concrete surface. When repairs are required, they are usually made with either polymer concrete or silica fume concrete.

CAVITATION

Cavitation can damage concrete when fast-moving water encounters discontinuities on the flow surface. When flowing water is lifted off the concrete surface, bubbling can occur. This creates a negative pressure zone. When the bubbles come into contact with concrete and burst, there is a very high pressure impact. It seems strange that bubbles could damage concrete, but as they pop, they remove particles from concrete. Over time, this damage can become serious. Common locations for this type of damage are water control gates and gate frames. It is so intense that not even cast iron or stainless steel can stand up against it. Repair of cavitation damage normally calls for full replacement of the damaged concrete. There are, however, some situations where a repair can be made with an epoxy-bonded replacement.

THE ROUNDUP

The roundup of all possible concrete defects is not feasible in the space available here. We have covered troubleshooting damage, causes, and repairs that can occur. If concrete is suffering from damage of being overloaded, the damage is usually obvious and noted over time.

Cracking can come from a variety of causes. Corrosion in reinforcing steels is not considered a cause for concrete damage, but rather an effect of damage from some relevant cause.

When you are troubleshooting a cause of damage for repair options, be thorough. Consider all viable options before making a repair decision. Attempting to correct a problem with the wrong repair procedure is simply going to cause more trouble and cost more money. Time spent in evaluation and research is time well spent.

CHAPTER

17

Code Requirements

OUTLINE

It is impossible to cover the entire 430-page concrete codebook in this chapter. Here you will find tips, tidbits, and commentary on many elements of the concrete code. This chapter cannot replace your codebook, but it can give you a jump start on some of the more commonly used code requirements.

CONCRETE SELECTION

First, proportions of concrete must be selected. Before this can be done properly, you must confirm that your selection is in compliance with the concrete code. How will the workability and consistency of the concrete be? Will you be able to work the mixture into forms and around reinforcement under conditions of placement without segregation or excessive bleeding? The mixture must be workable and have a quality consistency.

Before a concrete material can be used, it must have a code-approved resistance to special exposures. The mixture must also conform to strength test requirements. When different materials are planned for use in different portions of proposed work, the combination of the materials must be considered and determined effective.

Code Consideration

The codebook provides tables to be used as reference points in establishing various conditions and requirements.

COMPRESSIVE STRENGTH

Test records used to demonstrate proposed concrete proportions are required to represent materials and conditions similar to those expected. When there are changes in materials, conditions, and proportions within the test records, they must be as restrictive, or more restrictive, than the conditions used in the test samples.

Tests for compressive strength require no less than 30 test records. Ten consecutive tests are acceptable when the test records cover a period of time for no less than 45 days. When an acceptable record of field test results is not available, there are alternative methods to compensate for this.

Trial mixtures with proportions and consistencies required for the proposed work require the use of a minimum of three different water-cementitious materials contents that provide a range of strengths encompassing compressive strength.

Code Consideration

The slump of a trial mixture must be designed not to exceed ±0.75 in. of the maximum permitted and for air-entrained concrete, within ±0.5% of the maximum allowable air content.

When cylinder testing is done, the test must be done for 28 days or at the test age designated for the proper calculation of compressive strength. Results from these tests can be used to plot a curve to show the relationship between the water–cementitious materials ratio or cementitious materials content and compressive strength at a set test age.

FIELD-CURED SPECIMENS

Test results from field-cured specimens tested in cylinders may be required for the inspection of code officers. All field-cured test cylinders are required to be molded at the same time and from the same samples as test cylinders that have been tested under laboratory conditions.

The requirements for protecting and curing concrete must be improved when strength of field-cured cylinders at test age designated for determination of compressive strength is less than 85% of that found in companion laboratory-cured cylinders. This rule is not required when the field-cured strength exceeds compressive strength requirements by more than 500 PSI.

If test results indicate low-strength concrete and show that the load-carrying capacity is heavily reduced, test cores drilled for the area in question are allowed.

Core samples must be wiped dry prior to transport. They are to be placed in watertight bags or containers immediately after drilling. Core samples must not be tested sooner than 48 hours after being taken nor more than 7 days after coring. The only exception to this is if a registered design professional delivers an acceptable alternative.

What makes concrete in an area represented by coring acceptable? There must be three cores that are equal to at least 85% of the compressive strength required. It is also necessary that no more than one core sample is less than 75% of compressive strength. It is acceptable to take core samples from locations represented by erratic core strength.

PUTTING CONCRETE IN PLACE

There are rules and regulations that apply to the installation of concrete. For example, all equipment used to mix or transport concrete must be clean. Any ice present on the work section must be removed. Concrete forms are required to be coated prior to the pouring process. If you use masonry filler units that will be in contact with concrete, the fillers must be drenched with water. All reinforcement material must be free of ice and deleterious coatings. Water in the pour site must be removed as well as any loose material in the pour area prior to concrete installation.

MIXING CONCRETE

Ready mixed concrete must be mixed until there is a uniform distribution of materials. The full contents of the mixer must be emptied before the mixer is recharged.

Conditions change when concrete is mixed on a jobsite. A batch mixer of an approved type must be used for the mixing process. The operator of the mixer must know and operate the mixer at a speed recommended by the manufacturer of the mixer. Once all materials are in the drum of the mixer the mixing process must continue for a minimum of 90 seconds.

Detailed records are required to be kept of the mixing and placement procedures for job-mixed concrete. These records must include:

- Number of batches of concrete used
- Proportions of materials used
- Approximate location of final deposit in structure
- Time of mixing and placing
- Date of mixing and placing

Code Consideration

Concrete must be conveyed in such a manner as to prevent separation or loss of materials.

PLACING CONCRETE

Concrete should be placed as nearly as practical to its final position. This helps to avoid segregation due to re-handling or flowing. All concrete being placed must be plastic and must flow readily into spaces between reinforcements. Any bad concrete mix, such as a mixture that has begun to harden, must be disposed of properly and not used in the construction process.

Once concreting begins it must flow continuously until a suitable stopping point is reached. Top surfaces of vertically formed lifts must be level. Concrete must be thoroughly consolidated by suitable means during placement and must be completely worked around reinforcement and embedded fixtures and into the corners of forms.

CURING

The curing of most concrete should be done at an air temperature of at least 50°F in moist conditions. This curing process should continue for a minimum of seven days. High-early-strength concrete can be cured in just three days under the same curing conditions.

High-pressure steam can be used to accelerate the curing process. Steam at atmospheric pressure, heat, and moisture can be used for curing concrete quickly. These circumstances can also be used to accelerate strength gain. When these techniques are used, the concrete must be equal to required design strength at the load stage.

Concrete curing in cold weather requires equipment to be available that keeps materials from freezing. Concrete materials and all reinforcement, forms, fillers, and ground that will come into contact with concrete must be free of frost and ice. No frozen materials may be used in concreting.

When the weather is particularly hot, newly poured concrete must be protected against water evaporation. This can be done by dampening the concrete as needed. Tenting may be required to prevent excessive temperatures that alter the curing of concrete.

FORM DESIGN

Form design results in a form that is made to create an encasement for concrete that holds it in place to produce the shapes and sizes desired during the curing process. All forms must be built to prevent leakage. Appropriate bracing and restraints are required to keep forms in their proper placement. When you are rehabbing existing concrete, the forms for new concrete must be constructed in such a way that the existing concrete will not be harmed.

A few factors come into play with form design. For instance, all forms must be built with consideration of the rate and method in which concrete will be placed in the form. Vertical, horizontal, and impact loads must be considered.

Forms used for prestressed concrete members must be designed and constructed to permit movement of the member without damage during application of prestressing force.

FORM REMOVAL

Safety and serviceability must not be compromised during the removal of forms. Concrete contained in a form must be cured to a strength suitable for exposure prior to the removal of any form work from the concrete.

A procedure must be developed that will show the procedure and schedule for the removal of shores and installation of reshores and for calculating the loads transferred to a structure during the process. Code officers may require documentation of the plans and procedures for form removal and shoring.

Form supports for prestressed concrete members are not to be removed before the members can support the dead load. Structural analysis is required to determine when concrete is strong enough to carry its required loads.

EMBEDDED ITEMS

Conduits, pipes, sleeves, and similar items may be embedded in concrete. Aluminum conduits must not be embedded in concrete unless the conduit is effectively coated or covered to prevent aluminum–concrete reaction or electrolytic action between aluminum and steel. When a conduit or pipe penetrates a wall or beam, impairing the strength of construction must be avoided during installation.

Columns may be used to conceal embedded items. However, the items must not displace more than 4% of the area of cross section on which strength is calculated or which is required for fire protection.

Code Consideration

Commonly embedded items must not have an outside dimension that is more than one-third the overall thickness of the slab, wall, or beam in which the embedding occurs.

Conduits and similar embedded items are not to be installed closer than three diameters or widths on center. Embedded items must not be exposed to rusting or other deterioration. Pipes used under concrete are uncoated or galvanized iron or steel no thinner than standard Schedule 40 steel pipe. The inside diameters of pipes must not exceed 2 in. All embedded items must be designed to resist effects of the material, pressure, and temperature to which they will be subjected.

Until concrete reaches its design strength, embedded conduits are not allowed to convey liquid, gas, or vapor. One exception to this is water that does not exceed 90°F. The pressure of the water being conveyed must not exceed 50 PSI.

Piping in solid slabs must be placed between the top and bottom reinforcement, unless the piping is to be used for radiant heat for snow melting. Concrete cover over embedded items must not be less than 1.5 in. when concrete is exposed to weather or the earth. When concrete is not exposed to exterior conditions, the minimum cover required is three-quarters of an inch.

Code Consideration

Reinforcement with an area of no less than 0.002 times the area of the concrete section should be provided, as standard procedure for piping.

CONSTRUCTION JOINTS

The surface of all construction joints must be cleaned and all laitance must be removed. Before new concrete can be placed all construction joints have to be wetted and all standing water removed. The strength of a structure must not be impaired by construction joints. All construction joints located within the middle third of spans of slabs, beams, and girders must be clean and dry. Vertical support members that are still plastic must not be used to support beams, girders, or slabs. Except when shown otherwise in design drawings or specifications, beams, girders, haunches, drop panels, and capitals are to be placed monolithically as part of a slab system.

Code Consideration

Reinforcement partially embedded in concrete is not allowed to be bent in the field, with one exception: if the bending is shown on design drawings or is permitted by an engineer.

REINFORCEMENT

Reinforcement material must not be covered with mud, oil, ice, or any other element that will decrease the reinforcement's ability to bond. With the exception of prestressing steel, steel reinforcement with rust or mill scale or both can be determined to be satisfactory for use. If the reinforcement has the proper minimum dimensions and can be cleaned with a hand brush, the material should be acceptable. A light coating of rust is acceptable on prestressing steel.

Code Consideration

Reinforcement, including tendons and post-tensioning ducts, must be accurately placed and adequately supported before concrete is installed.

Minimum clear spacing between parallel bars in a layer is not allowed to be less than 1 in. Reinforcement bars placed in layers must be placed directly above any bars below with a minimum clear distance of 1 in. Spirally reinforced or tied reinforced compression members require a minimum clear distance of 1.5 in. between reinforcements.

Code Consideration

Welding of crossing bars is not permitted for assembly of reinforcement unless authorized by an engineer.

When bundled bars are used as a reinforcement they make a unit, and this unit must not contain more than four bars. Stirrups and ties are used to house bundled bars. No bar larger than a number 11 is to be used in a bundle to be used in a beam.

Code Consideration

Bundling or post-tensioning ducts are permitted if concrete can be satisfactorily placed and if a provision is made to prevent the prestressing steel, when tensioned, from breaking through the duct.

COLUMN REINFORCEMENT

Column reinforcement often includes the use of offset bars. The offset bent longitudinal bars must meet minimum requirements for code compliance. The slope of an inclined portion of an offset bar with axis of column is not allowed to exceed 1 in 6. The portions of bars above and below an offset are required to be parallel to the axis of the column. Offset bars that require bending must be bent prior to placing them in a concrete form.

Offset bends must be provided horizontal support by lateral ties, spirals, or parts of the floor construction. The support must be designed to resist 1.5 times the horizontal component of the computed force in the inclined portion of an offset bar. When lateral ties or spirals are used for support they must be placed no more than 6 in. from the points of bend.

Code Consideration

Longitudinal bars must not be offset bent where a column face is offset by three or more inches.

Ends of structural steel cores should be accurately finished. They must bear at end-bearing splices with positive provision for alignment of one core above the other in concentric contact. End-bearing splices are considered to be effective when they do not transfer more than 50% of the total compressive stress in the steel core. Transfer of stress between column base and footing must be designed in compliance with the concrete code.

CONNECTIONS

Enclosure at connections of principal framing elements, such as beams, must be provided for splices of continuing reinforcement and for anchorage of reinforcement terminating in such connections. External concrete, internal closed ties, spirals, or stirrups can be used for the enclosure at connections.

SPIRALS

Spirals are to consist of evenly spaced continuous bar or wire of such size and so assembled to permit handling and placing without distortion from designed dimensions. Spirals for cast-in-place construction require a minimum diameter of 3/8 in. The clear spacing between spirals is not to exceed 3 in. The minimum clear spacing between spirals is not to be less than 1 in. The anchorage of spiral reinforcement must be provided by 1.5 extra turns of spiral bar or wire at each end of a spiral unit. Spirals are required to extend from the top of a footing or slab in any story to the level of the lowest horizontal reinforcement in members that are supported above.

Code Consideration

Spirals must be held in place and to be true to line.

TIES

Beams and brackets are not always framed into all sides of a column. When this is the case, ties are required. The ties are to extend from above the termination of spiral to the bottom of the slab or drop panel. Columns that have capitals require spirals that extend to a level at which the diameter or width of capital is twice that of the column.

Non-prestressed bars are to be enclosed by lateral ties, at least number 3 in size for longitudinal bars number 10 or smaller, and at least number 4 in size for numbers 11, 14, 18, and bundled longitudinal bars. Deformed wire or welded wire reinforcement, or an equivalent area, will be permitted.

Code Consideration

The spacing between vertical ties is not to exceed 16 longitudinal bar diameters, 48 tie bar or wire diameters, or least dimension of the compression member.

Ties require specific arrangement. All corner and alternate longitudinal bars require lateral support provided by the corner of a tie with an included angle of no more than 135 degrees, and no bar is allowed to be more distant than 6 in. clear on each side along the tie from such a laterally supported bar. Longitudinal bars located around the perimeter of a circle may use a complete circular tie.

Ties are to be located vertically no more than one-half a tie spacing above the top of a footing or slab. Spacing must be no more than one-half a tie spacing below the lowest horizontal reinforcement in a slab or drop panel.

For beams and brackets that frame from four different directions into a column, termination of ties no more than 3 in. below the lowest reinforcement in the shallowest of such beams or brackets is permitted.

Code Consideration

Lateral reinforcement for flexural framing members subject to stress reversals or to torsion at supports must consist of closed ties, closed stirrups, or spirals extending around the flexural reinforcement.

SHRINKAGE

Reinforcements for shrinkage are to be spaced no more than five times the slab thickness nor farther apart than 18 in. Tendons have to be proportioned so that a minimum average compressive stress of 100 PSI on gross concrete area using effective prestress after losses is calculated. The maximum allowable distance for spacing tendons is 6 feet.

STRUCTURAL INTEGRITY REQUIREMENTS

Cast-in-place construction requires certain minimum requirements; for example, beams along the perimeter of a structure require continuous reinforcement. At least two bars are required. Support will consist of at least one-sixth of the tension reinforcement required for negative moment at the support. You must provide at least one-quarter of the tension reinforcement required for positive moment at midspan. When splices are needed to provide continuity, the top reinforcement is to be spliced at or near midspan. Bottom reinforcement must be spliced at or near the support. Class “A” tension splices are normally used.

Code Consideration

Pre-cast construction calls for the use of tension ties. They are to be provided in the transverse, longitudinal, and vertical directions and around the perimeter of the structure to tie everything together.

Code Consideration

With the exception of prestressed concrete, approximate methods of frame analysis are permitted for typical construction.

LIVE LOADS

Some assumptions are allowed when considering live loads. Live loads are to be applied only to the floor or roof under consideration. The far ends of columns built integrally with the structure are considered to be fixed. Common practice is to assume that the arrangement of live loads is limited to combinations of factored dead loads on all spans with full factored live load on two adjacent spans and factored dead load on all spans with full factored live load on alternate spans.

T-BEAMS

When working with T-beam construction the flange and web are to be built integrally or otherwise effectively bonded together. The width of slab effective as a T-beam flange must not exceed one-quarter of the span length of the beam and the effective overhanging flange width on each side of the web. It is not to exceed eight times the slab thickness and one-half the clear distance to the next web.

There are times when beams are placed with a slab on only one side. When this is the case the effective overhanging flange width will not exceed one-twelfth the span length of the beam or six times the slab thickness or one-half the clear distance to the next web.

Isolated beams where T-shapes are used to provide a flange for additional compression area require a flange thickness of no less than one-half the width of the web and an effective flange width no more than four times the width of the web.

JOIST CONSTRUCTION

What is joist construction? It is a monolithic combination of regularly spaced ribs and a top slab arranged to span in one direction or two orthogonal directions. Ribs are required to have a minimum width of 4 in. The depth of the ribs cannot be more than 3.5 times the minimum width of the rib. Thirty inches of clear space is required between ribs.

If fillers are used, there must be a minimum of 1/12 the clear distance between ribs and not less than 1.5 in. Slab thickness requires a minimum clear distance of 1/12 the distance between the ribs and not less than 2 in.

Code Consideration

Concrete floor finishes can be considered part of the required cover or total thickness for nonstructural considerations.

FIRST STEP

What you have just read is a good first step toward understanding the general basis of the concrete code. There is a lot more in the code than what has been covered here, but this chapter shows a good cross section of what to expect when you dig into code requirements.

Math plays a substantial role in the concrete code. There are many equations used. Your codebook provides a strong background for you to work with. Many of the symbols and abbreviations are defined in the front section, and definitions of terms are clear. It will take time and some work to master the code, but you can pick up the codebook and begin to understand and use it immediately. I suggest that you do so.

During my 30 years in the construction business I have seen numerous people fall far short of their potential in the trades. There have been countless reasons for this. Alcohol and drugs certainly contributed to many failed careers, and old-fashioned laziness has cut many construction careers short. The code and licensing requirements are also responsible for holding many people back.

Being fluent in the code makes you much more valuable on the job. Consider the plumbing and electrical trades. How much does an apprentice make per year? What is the expected income of a journeyman? Masters of the trade can expect the highest income and have the ability to open their own businesses. What separates each level? Experience is one factor. Being willing to learn their respective codes and pass the required licensing tests is what separates the highest earners from the lowest paid employees.

Where do you want to be in the food chain? Are you happy raking gravel in slab beds? Does rolling reinforcement wire out and placing rebar excite you? All of this is necessary in a learning process, but most people would prefer to wear the white hat and call the shots. If you want to be a supervisor or a business owner, concentrate on the code. Knowledge is what makes you more valuable.

CHAPTER

18

How Much Do You Know?

OUTLINE

How much do you know about the code pertaining to concrete work? Are you someone who feels that knowing the code should be someone else's job? If you are going to make a career in concrete you will have to understand the code. This chapter will give you a broad look at the types of requirements specified in the code. Take the time to test your existing knowledge and you will gain insight into how much effort you should put forth to learn various elements of the code.

What follows are quizzes on concrete code facts. Some are in true–false format; others are in a multiple-choice form. Answers are found at the back of this chapter. I am going to work though the codebook and pose questions for you to answer. Some will be simple. Many of the questions will pertain to day-to-day concrete work. Some of the questions will be more removed from daily life.

No one expects you to memorize the concrete code. It is sufficient to be able to look up answers to your questions in the codebook when you are on the job. However, this can make customers a bit nervous about your knowledge and ability. At the very least, you need to know what questions to ask yourself when doing a job. Let's start with some true–false questions and see how you do.

TRUE OR FALSE QUIZ

1. It is necessary for all equipment for mixing and transporting concrete to be clean.
   True False
2. Areas to be filled with concrete must be free of ice and debris.
   True False

3. Concrete forms must be protected from being coated at all times.
   True False
4. Masonry filler units are not allowed to come into contact with concrete.
   True False
5. Reinforcement must be thoroughly clean of ice.
   True False
6. Reinforcement must be thoroughly clean of deleterious coatings.
   True False
7. Concrete must be deposited within 50 feet of its final destination to avoid segregation during handling.
   True False
8. Concreting is to be done at such a rate that concrete is at all times plastic and flows readily into spaces between reinforcement.
   True False
9. Preferred curing conditions and time for concrete requires the curing to be done at temperatures above 62°F and in moist conditions for at least nine days.
   True False
10. Curing concrete with high-pressure steam is prohibited.
    True False
11. When concrete is cured with an accelerated process the compressive strength of the concrete at the load stage must be considered at least equal to the required design strength at that load stage.
    True False
12. Many installers are aware of the need for special provisions when concreting in cold temperatures. Hot temperature concreting offers its own challenges and requires special consideration for water evaluation and related risks.
    True False
13. When repairing existing concrete, forms for concreting must be installed and supported to avoid any damage to existing concrete.
    True False
14. The rate and method of placing concrete must be factored into the design of formwork for concreting.
    True False
15. Conduits and pipes, with their fittings, embedded within a column will not displace more than 7.5% of the area of cross section on which strength is calculated or which is required for fire protection.
    True False
16. Pipes embedded in concrete will not hold or convey any liquid, gas, or vapor, except water not exceeding 90°F or 50 PSI pressure until design strength is obtained.
    True False

17. The surface of concrete construction joints must be cleaned without removing any existing laitance.
True False

18. Construction joints must be wetted and standing water removed prior to placing new concrete.
True False

19. Construction joints in girders will be offset a minimum distance of two times the width of intersecting beams.
True False

20. Construction joints in floors will be located within the middle third of spans of slabs, beams, and girders.
True False

21. All bending of reinforcement material is to be done while the material is hot.
True False

22. When dealing with bundled bars the groups of parallel reinforcing bars bundled in contact to act as a unit will be limited to six in one bundle.
True False

23. Bars larger than number 11 will not be bundled in beams.
True False

24. Bundled bars will be enclosed within stirrups or ties.
True False

25. In walls and slabs other than concrete joist construction, primary flexural reinforcement will not be spaced farther apart than three times the wall or slab thickness, nor farther apart than 18 in.
True False

How do you think you have done so far? Do the correct answers jump out at you, or do you have to scratch your head from time to time to come up with an answer? Are you curious about your score? Take a moment and check your answers against the correct answers at this back of this chapter. See how you did. Take a break, and then begin the multiple-choice section of the test.

So, how did you do? How much do you really know about the concrete code? Are you good to go or should you brush up on the code? Are you ready to try the multiple-choice questions? Let's do it.

MULTIPLE-CHOICE QUESTIONS

1. Generally speaking, bundled bars must be covered by a minimum amount of concrete equal to the diameter of the bundle. The maximum cover, under general conditions is
A) 2 in. B) 3 in. C) 4 in. D) 6 in.

2. When bundled bars exist in concrete that is going to be cast against and permanently exposed to earth, the minimum amount of coverage allowed is
   A) 2 in. B) 3 in. C) 4 in. D) 6 in.

3. When working with offset bars the slope of inclined portion of an offset bar with axis of column will not exceed:
   A) 1 in 6 B) 3 in 6 C) 4 in 6 D) 5 in 6

4. Horizontal support at offset bends will be provided by lateral ties, spirals, or parts of the floor construction. Horizontal support provided will be designed to resist ________ times the horizontal component of the computed force in the inclined portion of an offset bar.
   A) 1 B) 1.5 C) 2.25 D) 3

5. Lateral ties or spirals, if used, are to be placed no more than ________ in. from points of bend.
   A) 2 B) 4 C) 6 D) 8

6. Offset bars are bent ________ placement in forms.
   A) Before B) After C) During D) In

7. Enclosure at connections will consist of external concrete or internal closed ties, ________, or stirrups.
   A) Racks B) Leverage C) Spirals D) Concaves

8. For cast-in-place construction, size of spirals will not be less than ________ of an inch in diameter.
   A) 1/8 B) 1/4 C) 1/2 D) 3/8

9. Clear spacing between spirals will not exceed ________ in.
   A) 2 B) 3 C) 6 D) 9

10. Vertical spacing of ties will not exceed ________ longitudinal bar diameters.
    A) 4 B) 10 C) 12 D) 16

11. Vertical spacing of ties will not exceed ________ tie bar or wire diameters, or least dimension of the compression member.
    A) 12 B) 24 C) 48 D) 62

12. Shrinkage and temperature reinforcement are to be spaced not farther apart than five times the slab thickness, nor farther apart than ________ in.
    A) 4 B) 6 C) 12 D) 18

13. The spacing of tendons must not exceed ________ feet.
    A) 3 B) 6 C) 9 D) 12

14. When dealing with joist construction, clear spacing between ribs will not exceed ________ in.
    A) 16 B) 22 C) 24 D) 30

15. Slab thickness in joist construction must not be less than one-twelfth the clear distance between ribs, nor less than _____ in.
A) 2 B) 6 C) 24 D) 30

16. Design yield strength of structural steel core will be the specified minimum yield strength for the grade of structural steel used but not to exceed __________ PSI.
A) 100 B) 10,000 C) 50,000 D) 75,000

17. A shearhead will not be deeper than _____ times the web thickness of the steel shape.
A) 10 B) 20 C) 50 D) 70

18. Mechanical and welded splices ________ permitted.
A) Will be B) Will not be C) May be D) Are never

19. Walls more than _______ in. thick, except basement walls, will have reinforcement for each direction placed in two layers parallel with faces of the wall.
A) 4 B) 6 C) 8 D) 10

20. Thickness of nonbearing walls will not be less than ______ in., or less than 1/30 the least distance between members that provide lateral support.
A) 4 B) 6 C) 8 D) 10

21. The depth of footing above bottom reinforcement will not be less than _______ in. for footings on soil.
A) 4 B) 6 C) 8 D) 10

22. The depth of footing above bottom reinforcement will not be less than _______ in. for footings on piles.
A) 6 B) 8 C) 10 D) 12

23. Structural integrity ties around the perimeter of each floor and roof, within 4 feet of the edge, will provide a nominal strength in tension not less than ________ pounds.
A) 10,000 B) 12,000 C) 16,000 D) 24,000

24. The use of an entire composite member or portions thereof for resisting shear and moment will ________.
A) Be permitted B) Not be permitted C) Exist D) Be discouraged

25. Three-dimensional effects will be considered in design and analyzed using three-dimensional procedures or approximated by considering the summation of effects of _____ orthogonal planes.
A) 2 B) 3 C) 4 D) 6

This amounts to 50 questions and answers that pertain to the concrete code. Check your answers to see what your natural knowledge of the code is. Remember, the code is a big part of your career in concrete. If you don't know or understand the code requirements, spend some time studying the code. It will be a good investment of your time.

CORRECT ANSWERS FOR TRUE–FALSE QUESTIONS

1. T
2. T
3. F
4. F
5. T
6. T
7. F
8. T
9. F
10. F
11. T
12. T
13. T
14. T
15. T
16. T
17. F
18. T
19. T
20. T
21. T
22. F
23. T
24. T
25. T

CORRECT ANSWERS FOR MULTIPLE-CHOICE QUESTIONS

1. A
2. B
3. A
4. B
5. C
6. A
7. C
8. D
9. C
10. D
11. C
12. D
13. B
14. D
15. A
16. C
17. D
18. A
19. D
20. A
21. B
22. D
23. C
24. A
25. A

CHAPTER

19

Working with Code Requirements

OUTLINE

Using the code in the real world is a little different from reading and understanding a codebook. In theory, the codebook should hold all the answers and you should be able to apply them without incurring any unexpected problems. But reality is not always so simple. Putting the concrete code to practical use doesn't have to be difficult, but it can be.

There are two extremes to applying code requirements on a job. On one side, you have situations where the code enforcement is lax. Then there are times when the code enforcement is extremely strict. Most jobs are somewhere between the two extremes. Whether the job you are working on is lax or strict, you could have problems dealing with code requirements.

LAX JOBS

There are some workers who welcome lax jobs. These people enjoy not having their work scrutinized too tightly. I guess everyone might enjoy an easy job, but there are risks to lax jobs.

In my opinion professional concrete workers should perform professional services, even if they can get away with less-than-credible work. This should be an ethical commitment. But there are workers who cut corners in jurisdictions where they know that they can get away with it. This is not fair to the customer, and it can put the installer at risk.

Inspectors provide a form of protection for installers and contractors. Cutting corners can put you at great risk for a lawsuit. Leaving the financial side of such a suit out of the picture, try to imagine how you would deal with the guilt of people injured as a result of your deviation from the code.

How many times have you installed new work without a permit? Did you know that a permit and inspection were required? Many workers have skipped the permit and inspection element of a job at one time or another, but they probably never thought of the risk. What would happen if the concrete installation failed? Would people be hurt? Could anyone be killed? What is the most extreme risk that you could be subject to? Do you really want to lose sleep at night over cutting corners on permits? It is not worth it. When a permit is required, get one.

Code Consideration

Your liability insurance may not be willing to cover any claim made for work that you did without following code requirements. If you had done the job by the book, you would have some foundation for defending yourself. Additionally, you would have an approved inspection certificate that would validate the fact that you did the work properly and within code requirements. This could be very helpful in a lawsuit.

The small amount of money you save by cheating on permits and inspection could come back as a massive lawsuit and wipe out your career or your business. When you work in an area prone to lax inspections, you can be at risk. It is in your best interest to do all work to code requirements and to have that work inspected and approved.

STRICT CODE ENFORCEMENT

Strict code enforcement can be a blessing. It helps get you off the hook if something goes wrong. On the other hand, it can be a real pain when you are being ripped apart on a simple job. Again, you have to use common sense in your response to such situations.

Should an inspector fail your job because you are missing one anchor? Technically, yes. In reality, especially if the code officer knows that you are reputable and professional, what would it hurt to pass the job with a notice for you to install the missing anchor?

SAFETY

The concrete code exists for a reason. Safety is what the code is focused on, and this is a valid reason for code regulations. Most installers have felt that code regulations are too strict or not needed for certain elements of the business. This may be true on occasion, but the basic foundation of the code is sound, and it should be observed. There are elements of code regulations that I feel are too detailed for practical purposes, but I always attempt to comply with code requirements.

FEES

The fees charged for permits are often complained about by both contractors and homeowners. These fees pay for the administration of the code and the protection of consumers. Indirectly, contractors receive protection from inspections. When you are feeling taken advantage of due to costly fees charged by a code enforcement office, consider the benefits that our society is receiving. I am not trying to be a proponent of code fees, but I do believe that people in the trades should look at the fees in a reasonable fashion.

KNOW YOUR INSPECTORS

To make your life easier, get to know your inspectors. Knowing the inspectors in your area and what they expect will make jobs run more smoothly. Most inspectors are accessible and helpful as long as you don't approach them with a bad attitude.

Code Consideration

When doing work in a new location for the first time it is often helpful to meet with the local code officers to introduce yourself and to see what they will expect.

LOCAL JURISDICTIONS

Local jurisdictions adopt a code and have the right to amend it for local requirements. In simple terms, this can mean that a local code may vary from the primary code. Even if you know the primary code by heart, you might have a conflict with a local jurisdiction. A visit with your local code officers will clear up these differences before they become a problem on a job.

COMMON SENSE

Common sense goes a long way in the installation of concrete systems. It is not uncommon for a mixture of common sense and code requirements to be used to achieve a successful installation. Code regulations must be followed, but they are not always as rigid as they may appear to be. Don't be afraid to consult with inspectors to arrive at a reasonable solution to difficult problems.

In closing, using the code in the real world is largely a matter of proper communication between contractors and code officers. This is an important fact to keep in mind. Learn the code and use it properly. If you feel a need to deviate from it, talk with a code officer and arrive at a solution that is acceptable to all parties affected. If you keep a good attitude and don't scrap the code because you think you know better, you should do well in your trade.

CHAPTER

20

Avoiding On-the-Job Injuries

OUTLINE

Avoiding on-the-job injuries should be a priority. Working with concrete can be dangerous. From overhead accidents to falling on rebar, the risks are always there. Workers are injured on the job every year. Most of the injuries could be prevented, but they aren't. Why is this? People are in a hurry to make a few extra bucks, so they cut corners. This happens with employees, contractors, and piece workers. It even affects hourly installers who want to shave 15 minutes off their workday so that they can head back to the shop early.

Based on my field experience, most accidents occur as a result of negligence. Workers try to cut corners and they wind up getting hurt. This has proved true with my personal injuries. I have only suffered two serious on-the-job injuries, and both of them were a direct result of my carelessness. I knew better than to do what I was doing when I was hurt, but I did it anyway. Sometimes you don't get a second chance, and the life you affect may not always be your own. So, let's look at some sensible safety procedures that you can implement in your daily activity.

VERY DANGEROUS

Construction can be a very dangerous trade. The tools of the trade can be potential killers. Requirements of the job can place you in positions where a lack of concentration could result in serious injury or death. The fact that working with concrete can be dangerous is no reason to rule out the trade as your profession. Driving can be extremely dangerous, but few people never set foot in an automobile out of fear.

Fear is generally a result of ignorance. When you have a depth of knowledge and skill, fear begins to subside. As you become more accomplished at what you do, fear is forgotten. There is a huge difference between fear and respect. It is advisable to learn to work without fear, but you should never work without respect.

If, as an installer, you are afraid to climb up high enough to set a pour form, you are not going to last long in the plumbing trade. However, if you scurry up recklessly, you could be injured severely, perhaps even killed. You must respect the position you are putting yourself in. If you are using a ladder you must respect the outcome of a mistake.

Being afraid of heights could limit or eliminate your career. Respect is the key. If you respect the consequences of your actions, you are aware of what you are doing and the odds for a safe trip improve.

Code Consideration

Safe working conditions are a good place to start when building a career in the concrete business. If you are injured your ability to do your job is likely to be limited.

Many young installers are fearless in the beginning. They think nothing of darting around on a roof or jumping down in a trench. As their careers progress, they usually hear about or see on-the-job accidents. Someone gets buried in the cave-in of a trench, somebody falls off a roof, or a metal ladder being set up hits a power line. The list of possible job-related injuries is a long one.

Millions of people are hurt every year in job-related accidents. Most of these people were not following solid safety procedures. Sure, some of them were victims of unavoidable accidents, but most were hurt by their own hand. You do not have to be one of these statistics.

In over 30 years in construction work, I have only been hurt seriously on the job twice, and both times were my fault. I got careless. In one of the instances, I let loose clothing and a powerful drill work together to chew up my arm. In the other incident, I tried to save myself the trouble of repositioning my stepladder while drilling holes in floor joists. My desire to save a minute cost me torn stomach muscles and months of pain brought on from a twisting drill.

My accidents were not mistakes; they were stupidity. Mistakes are made through ignorance. I was not ignorant of what could happen to me. I knew the risk I was taking, and I knew the proper way to perform my job. Even with my knowledge, I got hurt. Luckily, both of my injuries healed, and I did not pay a lifelong price for my stupidity.

During my long career I have seen a lot of people get hurt. Most of these people have been helpers and apprentices. Of all the on-the-job accidents I have witnessed, every one of them could have been avoided. Many of the incidents were not extremely serious, but a few were.

As a concrete worker, you will do dangerous work. Hopefully, your employer will provide you with quality tools and equipment. If you have the right tool for the job, you are off to a good start in staying safe.

Code Consideration

Safety training is something you should seek from your employer. Some contractors fail to tell their employees how to do their jobs safely. It is easy for someone who knows a job well to forget to inform an inexperienced person of potential danger.

For example, a supervisor might tell you to break up the concrete around a pipe to allow the installation of new plumbing and never consider telling you to wear safety glasses. The supervisor will assume you know the concrete is going to fly up in your face as it is chiseled up. However, as a rookie, you might not know about the reaction concrete has when hit with a cold chisel. One swing of the hammer could cause extreme damage to your eyesight.

Simple jobs, like the one in the example, are all it takes to ruin a career. You might be really on your toes when asked to scoot across an I-beam, but how much thought are you going to give to carrying a few bags of concrete mix to a mixer? The risk of falling off the I-beam is obvious. Hurting your back by carrying heavy loads the wrong way may not be so obvious. Either way, you can have a work-stopping injury.

Safety is a serious issue. Some jobsites are very strict in maintaining safety requirements, but a lot of jobs have no written rules of safety. If you are working on a commercial job, supervisors are likely to make sure you abide by the rules of the Occupational Safety and Health Administration (OSHA). Failure to comply with OSHA regulations can result in stiff financial penalties. However, if you are working residential jobs, you may never set foot on a job where OSHA regulations are observed.

In all cases, you are responsible for your own safety. Your employer and OSHA can help you to remain safe, but in the end, it is up to you. You are the one who has to know what to do and how to do it. Not only do you have to take responsibility for your own actions, you also have to watch out for the actions of others. It is likely that you could be injured by someone else's carelessness. Now that you have had the primer course, let's get down to the specifics of job-related safety.

As we move into specifics, you will find the suggestions in this chapter broken down into various categories. Each category will deal with specific safety issues related to the category. For example, in the section on tool safety, you will learn procedures for working safely with tools. As you move from section to section, you may notice some overlapping of safety tips. For example, in the section on general safety, you will see that it is wise to work without wearing jewelry. Then you will find jewelry mentioned again in the tool section. This duplication is done to point out definite safety risks and procedures. We will start with general safety.

GENERAL SAFETY

General safety covers a lot of territory. It starts from the time you get into the company vehicle and carries you right through to the end of the day. Much of the general safety recommendations involve the use of common sense.

Code Consideration

If you are unloading heavy items, don't put your body in awkward positions. Learn the proper way to lift, and never lift objects inappropriately.

Vehicles

Many construction workers are given company trucks to go back and forth from jobs. You will probably spend a lot of time loading and unloading company trucks as well as riding in or driving them. All of these areas can threaten your safety.

If you will be driving the truck, take the time to get used to how it handles. Loaded work trucks do not drive like the family car. Remember to check the vehicle's fluids, tires, lights, and related equipment. Many company trucks are old and have seen better days. Failure to check the vehicle's equipment could result in unwanted headaches. Also, remember to use the seat belts; they do save lives.

Apprentices are normally charged with the duty of unloading the truck at the jobsite. There are many ways to get hurt doing this job. Many trucks use roof racks to haul supplies and ladders. If you are unloading these items, make sure they do not come into contact with low-hanging electrical wires. Aluminum ladders make very good electrical conductors, and they will carry the power surge through you on the way to the ground. If you are unloading heavy items, do not put your body in awkward positions. Learn the proper ways to lift, and never lift objects inappropriately. If the weather is wet, be careful climbing on the truck. Step bumpers get slippery, and a fall can impale you on an object or bang up your knee.

When it is time to load the truck, observe the same safety precautions you did in unloading. In addition to these considerations, always make sure your load is packed evenly and is well secured. Be especially careful of any load you attach to the roof rack, and always double check the cargo doors on trucks with utility bodies. If you are carrying a load of forms in the bed of your truck, make very sure that they are strapped in securely.

Code Consideration

Eye and ear protection is often overlooked. An inexpensive pair of safety glasses can prevent you from spending the rest of your life blind. Ear protection reduces the effect of loud noises from equipment such as jackhammers and drills. You may not notice much benefit now, but in later years you will be glad you wore safety equipment.

CLOTHING

Clothing is responsible for a lot of on-the-job injuries. Sometimes it is the lack of clothing that causes the accidents, and there are many times when too much clothing creates the problem. Generally, it is wise not to wear loose fitting clothes. Shirttails should be tucked in, and short-sleeve shirts are safer than long-sleeved shirts when operating some types of equipment.

Caps can save you from minor inconveniences and hard hats provide some protection from potentially damaging accidents, such as having a steel fitting dropped on your head. If you have long hair, keep it up and under a hat.

Good footwear is essential in the trade. Normally a strong pair of hunting-style boots are best. The thick soles provide some protection from nails and other sharp objects you may step on. Boots with steel toes can make a big difference in your physical well-being. If you are going to be climbing, wear foot gear with a flexible sole that grips well. Gloves can keep your hands warm and clean, but they can also contribute to serious accidents. Wear gloves sparingly, depending upon the job you are doing.

JEWELRY

On the whole, jewelry should not be worn in the workplace. Rings can inflict deep cuts in your fingers. They can also contribute to finger amputation. Chains and bracelets are equally dangerous, and probably more so.

EYE AND EAR PROTECTION

Eye and ear protection is often overlooked. An inexpensive pair of safety glasses can prevent you from spending the rest of your life blind. Ear protection reduces the effect of loud noises from equipment such as jackhammers and drills. You may not notice much benefit now, but in later years you will be glad you wore it.

PADS

Kneepads make your job more comfortable and help to protect the knees. Some workers spend a lot of time on their knees, and pads should be worn to ensure that they can continue to work for many years.

The embarrassment factor plays a significant role in job-related injuries. People, especially young people, feel the need to fit in and make a name for themselves. Concrete work is considered a macho trade. There is no secret that construction workers often fancy themselves as strong human specimens. The work can be hard and in doing it, becoming strong is a side benefit. But you cannot allow safety to be pushed aside for the purpose of making a macho statement.

Too many people believe that working without safety glasses, ear protection, and so forth makes them tough. That is just not true; it may make them appear dumb, and it may get them hurt, but it does not make them look tough. If anything, it makes them look stupid and inexperienced.

Don't fall into the trap so many young tradesmen do. Never let people goad you into bad safety practices. Some people are going to laugh at your kneepads. Let them laugh, you will still have good knees when they are hobbling around on canes. I am very serious about this issue. There is nothing sissy about good safety precautions. Wear your gear in confidence, and don't let the few jokesters get to you.

TOOL SAFETY

Tool safety is a big issue. Anyone in the trades will work with numerous tools. All of these tools are potentially dangerous, but some of them are especially hazardous. This section is broken down by the various tools used on the job. You cannot afford to start working without the basics in tool safety. The more you can absorb on tool safety, the better off you will be.

The best starting point is reading all of the literature from the manufacturers of your tools. The people that make the tools provide some good safety suggestions. Read and follow the manufacturers' recommendations.

The next step in working safely with your tools is to ask questions. If you do not understand how a tool operates, ask someone to explain it to you. Do not experiment on your own, the price you pay could be much too high.

Common sense is irreplaceable in the safe operation of tools. If you see an electrical cord with cut insulation, you should have enough common sense to avoid using it. In addition to this type of simple observation, you will learn some interesting facts about tool safety. Now, let me tell you what I have learned about tool safety over the years.

Code Consideration

Know your tools well and keep them in good repair. Be especially sure to check all electrical cords for damage.

There are some basic principles you should apply to all of your work with tools. We will start with the basics, and then move on to specific tools.

- Keep body parts away from moving parts.
- Do not work with poor lighting conditions.
- Be careful of wet areas when working with electrical tools.
- If special clothing is recommended for working with your tools, wear it.
- Use tools only for their intended purposes.
- Get to know your tools well.
- Keep your tools in good condition.

Now, let's take a close look at the tools you are likely to use.

Drills and Bits

Drills have been my worst enemy. The two serious injuries I have received were both related to my work with a drill. The drills used by many construction workers are not the little pistol-grip, hand-held types of drills most people are familiar with. The day-to-day drilling done in concrete work involves the use of large drills that have enormous power when they get in a bind. Hitting an obstruction while drilling can do a lot of damage. You can break fingers, lose teeth, suffer head injuries, and so on. As with all electrical tools, you should always check the electrical cord before using the drill. If the cord is not in good shape, don't use the drill.

Always know what you are drilling into. If you are working new construction, it is fairly easy to look before you drill. However, drilling in a remodeling job can be much more difficult. You cannot always see what you are getting into. If you are unfortunate enough to drill into a hot wire, you can get a considerable electrical shock.

The bits you use in a drill are part of the safe operation of the tool. If your drill bits are dull, sharpen them. Dull bits are much more dangerous than sharp ones. If you will be drilling metal, be aware that the metal shavings will be sharp and hot.

Power Saws

The most common types of power saws used by concrete workers are concrete saws, reciprocating saws, and circular saws. These saws are used to cut concrete, form material, pipe, plywood, floor joists, and a whole lot more. All of the saws have the potential for serious injury.

Reciprocating saws are reasonably safe. Most models are insulated to help avoid electrical shocks if a hot wire is cut. The blade is typically a safe distance from the user, and the saws are pretty easy to hold and control. However, the brittle blades do break, and this could result in an eye injury.

Circular saws are occasionally used by concrete workers. The blades on these saws can bind and cause the saws to kick back. If you keep your body parts out of the way and wear eye protection you can use these saws safely. Concrete saws are heavy, noisy, and they generate a lot of dust. Protect your eyes and your respiratory system from the flying chips of concrete and dust.

POWER MIXERS

Power mixers are often used on small jobs. These tools make the mixing of concrete simple and require far less work than manually mixing concrete. Whenever there are moving parts, as there are in mixers, there are safety risks. Workers must be careful not to get body parts or clothing caught in the moving parts of tools. Also, keep your back in mind when lifting bags of material to dump into a mixer.

Air-Powered Tools

Air-powered tools are not used very often by concrete workers. Jackhammers are probably the most used air-powered tools for individuals rehabbing concrete. When using tools with air hoses, make sure to check all connections carefully. If you experience a blow-out, the hose can spiral wildly out of control. The air hose can also create a tripping hazard that must be avoided. Any type of power washer, sand blaster, or related equipment can cause injuries.

Powder-Actuated Tools

Powder-actuated tools are used to secure objects to hard surfaces, like concrete. If the user is properly trained, these tools are not too dangerous. However, good training and eye and ear protection are all necessary. Misfires and chipping hard surfaces are the most common problem with these tools.

Ladders

Both stepladders and extension ladders are used frequently by construction workers, and ladder accidents are a possibility. You must always be aware of what is around you when handling a ladder. If you brush against a live electrical wire with a ladder you are carrying, your life could be over. Ladders often fall over when the people using them are not careful. Reaching too far from a ladder can be all it takes to fall.

Code Consideration

Rebar placed prior to a concrete pour should be installed with protector cups to reduce the risk of injury if someone falls onto the reinforcement bars.

When you set up a ladder or rolling scaffolds, make sure they are set up properly. The ladder should be on firm footing, and all safety braces and clamps should be in place. When using an extension ladder, many plumbers use a rope to tie rungs together where the sections overlap. The rope provides an extra guard against the ladder's safety clamps failing and the ladder collapsing. When using an extension ladder, be sure to secure both the base and the top. I had an unusual accident on a ladder that I would like to share with you.

I was on a tall extension ladder, working near the top of a commercial building. The top of my ladder was resting on the edge of the flat roof. There was metal flashing surrounding the edge of the roof, and the top of the ladder was leaning against the flashing. There was a picket fence behind me and electrical wires entering the building to my right. The entrance wires were far enough away, so I was in no immediate danger. As I worked on the ladder, a huge gust of wind blew around the building. I don't know where it came from; it had not been very windy when I went up the ladder.

The wind hit me and pushed me and the ladder sideways. The top of the ladder slid easily along the metal flashing, and I could not grab anything to stop me. I knew the ladder was going to go down, and I did not have much time to make a decision. If I pushed off of the ladder, I would probably be impaled on the fence. If I rode the ladder down, it might hit the electrical wires and fry me. I waited until the last minute and jumped off of the ladder.

I landed on the wet ground with a thud, but I missed the fence. The ladder hit the wires and sparks flew. Fortunately, I was not hurt and electricians were available to take care of the electrical problem. This was a case where I was not really negligent, but I could have been killed. If I had secured the top of the ladder, my accident would not have happened.

Screwdrivers and Chisels

Eye injuries and puncture wounds are common when working with screwdrivers and chisels. When the tools are used properly and safety glasses are worn, few accidents occur.

The key to avoiding injury with most hand tools is simply to use the right tool for the job. If you use a wrench as a hammer or a screwdriver as a chisel, you are asking for trouble.

There are other types of tools and safety hazards found in the concrete trade. However, this list covers the ones that result in the most injuries. In all cases, observe proper safety procedures and utilize safety gear, such as eye and ear protection.

CO-WORKER SAFETY

Co-worker safety is the last segment of this chapter. I am including it because workers are frequently injured by the actions of their co-workers. This section is meant to protect you from others and to make you aware of how your actions might affect other around you.

Most installers find themselves working around other people. This is especially true on construction jobs. When working around other people, you must be aware of their actions, as well as your own. If you are walking out of a house to get something off the truck and a roll of roofing paper gets away from a roofer, you could get an instant headache.

If you don't pay attention to what is going on around you, it is possible to wind up in all sorts of trouble. Sometimes cranes lose their loads, and such a load landing on you is likely to be fatal. Equipment operators don't always see the concrete worker kneeling down to drive in a section of rebar. It is not hard to have a close encounter with heavy equipment. While we are on the subject of equipment, let me share another one of my on-the-job experiences.

One day I was in a ditch. The section of ditch that I was working in was only about 4 feet deep. There was a large pile of dirt near the edge of the trench that had been created when the ditch was dug. The dirt was not laid back like it should have been; it was piled up.

As I worked in the ditch, a backhoe came by. The operator had no idea I was in the ditch. When he swung the backhoe around to make a turn, the small scorpion-type bucket on the back of the equipment hit the dirt pile.

I had stood up when I heard the hoe approaching, and it was a good thing I had. When the equipment hit the pile of dirt, part of the mound caved in on me. I tried to run, but it caught both of my legs and the weight drove me to the ground. I was buried from just below my waist. My head was okay, and my arms were free. I was still holding my shovel.

I yelled, but nobody heard me. I must admit, I was a little panicked. I tried to get up and couldn't. After a while, I was able to move enough dirt with the shovel to crawl out from under the dirt. I was lucky. If I had been on my knees working I might have been smothered. As it was, I came out of the ditch no worse for wear. But, boy, was I mad at the careless backhoe operator.

This is a prime example of how other workers can hurt you and never know they did it. You have to watch out for yourself at all times. As you gain field experience, you will develop a second nature for impending co-worker problems. You will learn to sense when something is wrong or is about to go wrong. But you have to stay alive and healthy long enough to gain that experience.

Always be aware of what is going on over your head. Avoid working under other people and hazardous overhead conditions. Let people know where you are, so you won't get stranded on a roof or in an attic when your ladder is moved or falls over.

You must also remember that your actions could harm co-workers. If you are on a roof to flash a pipe and your hammer gets away from you, somebody could get hurt. Open communication between workers is one of the best ways to avoid injuries. If everyone knows where everyone else is working, injuries are less likely. Primarily, think and think some more. There is no substitute for common sense. Try to avoid working alone, and remain alert at all times.

CHAPTER

21

Worksite Survival

OUTLINE

You probably don't think about worksite survival when you grab your lunch and your hardhat, but you should be ready to help yourself and others in the event of an injury. You never know when having first-aid skills may save your life. Construction work can be a dangerous life. On-the-job injuries are common. Most injuries are fairly minor, but they often require treatment. Do you know the right way to get a sliver of copper out of your hand? If your helper suffers from an electrical shock when a drill cord goes bad, do you know what to do? Well, many people don't know how to help in these situations.

Before we get too far into this chapter, there are a few points I want to make. First of all, I am not a medical doctor or any type of trained medical-care person. I have taken first-aid classes, but I am certainly not an authority on medical issues. The suggestions that I will give you in this chapter are for informational purposes only. This book is not a substitute for first-aid training offered by qualified professionals.

My intent here is to make you aware of some basic first-aid procedures that can make life on the job much easier. But I am not telling you to use my advice to administer first-aid. Hopefully, this chapter will show you the advantage of taking first-aid classes. Before you attempt first-aid on anyone, including yourself, you should attend a structured, approved first-aid class. I am going to give you information that is as accurate as I can make it, but don't assume that my words are enough. You may never use what you learn, but the one time it is needed, you will be glad you made the effort to learn what to do.

OPEN WOUNDS

Open wounds are a common problem on construction sites. Many tools and materials used by workers can create open wounds. What should you do if you or one of your workers is cut?

- Stop the bleeding as soon as possible.
- Disinfect and protect the wound from contamination.
- You may have to take steps to avoid shock symptoms.
- Once the patient is stable, seek medical attention for severe cuts.

When a bad cut is encountered, the victim may slip into shock. A loss of consciousness could result from a loss of blood. Death from extreme bleeding is also a risk. As a first-aid provider, you must act quickly to reduce the risk of serious complications.

Bleeding

To stop bleeding, direct pressure is normally a good tactic. This may be as crude as clamping your hand over the wound, but a cleaner compression is desirable. Ideally, a sterile material should be placed over the wound and secured, normally with tape (even if it is duct tape). Whenever possible, wear rubber gloves to protect yourself from possible disease transfer if you are working on someone else. Thick gauze used as a pressure material can absorb blood and begin to allow the clotting process to start.

Bad wounds may bleed right through the compress material. If this happens, don't remove the blood-soaked material. Add a new layer of material over it. Keep pressure on the wound. If you are not prepared with a first-aid kit, you could substitute gauze and tape with strips cut from clothing that can be tied in place over the wound.

When you are dealing with a bleeding wound, it is usually best to elevate it. If you suspect a fractured or broken bone in the area of the wound, elevation may not be practical. When we talk about elevating a wound, it simply means to raise the wound above the level of the victim's heart. This helps the blood flow to slow, due to gravity.

Code Consideration

Tourniquets can do a great deal of harm if not used properly. They should only be used in life-threatening situations.

Super Serious Bleeding

Super serious bleeding might not stop even after a compression bandage is applied and the wound is elevated. When this is the case, you must put pressure on the main artery that is producing the blood. Constricting an artery is not an alternative for the steps that we previously discussed.

Putting pressure on an artery is serious business. First, you must be able to locate the artery. Don't keep the artery constricted any longer than necessary. You may have to apply pressure for awhile, release it, and then apply it again. It is important that you do not restrict the flow of blood in arteries for long periods of time. I hesitate to go into too much detail on this process, as I feel it is a method that should be taught in a controlled, classroom situation. However, I will hit the high spots. But remember, these words are not a substitute for professional training from qualified instructors.

Open arm wounds are controlled with the brachial artery. The location of this artery is in the area between the biceps and triceps, on the inside of the arm, about halfway between the armpit and the elbow. Pressure is created with the flat parts of your fingertips. Basically, you are holding the victim's wrist with one hand and closing off the artery with your other hand. Pressure exerted by your fingers pushes the artery against the arm bone and restricts blood flow. Again, don't attempt this type of first-aid until you have been trained properly in the execution of the procedure.

Severe leg wounds may require the constriction of the femoral artery. This artery is located in the pelvic region. Normally, bleeding victims are placed on their backs for this procedure. The heel of a hand is placed on the artery to restrict blood flow. In some cases, fingertips are used to apply pressure. Don't rely solely on what I am telling you. It is enough that you understand that knowing when and where to apply pressure to arteries can save lives and that you should seek professional training in these techniques.

Code Consideration

When someone is impaled on an object, resist the urge to remove the object. Call professional help and leave the object in place, as removing it could result in more excessive bleeding.

Tourniquets

A tourniquet should only be used in a life-threatening situation. When a tourniquet is applied, there is a risk of losing the limb to which the restriction is applied. This is obviously a serious decision and one that must be made only when all other means of stopping blood loss have been exhausted.

Unfortunately, concrete workers might run into a situation where a tourniquet is the only answer. For example, if a worker allowed a power saw to get out of control, a hand might be severed or some other type of life-threatening injury could occur. This would be cause for the use of a tourniquet. Let me give you a basic overview of what is involved when a tourniquet is used.

Tourniquets should be at least 2 in. wide. A tourniquet should be placed at a point that is above a wound, between the bleeding and the victim's heart. However, the binding should not encroach directly on the wound area. Tourniquets can be fashioned out of many materials. If you are using strips of cloth, wrap the cloth around the limb that is wounded and tie a knot in the material. Use a stick, screwdriver, or whatever else you can lay your hands on to tighten the binding.

Once you have made a commitment to apply a tourniquet, the wrapping should be removed only by a physician. It is a good idea to note the time that a tourniquet is applied, as this will help doctors later in assessing their options. As an extension of the tourniquet treatment, you will most likely have to treat the patient for shock.

Infection

Infection is always a concern with open wounds. When a wound is serious enough to require a compression bandage, don't attempt to clean the cut. Keep pressure on the wound to stop bleeding. In cases of severe wounds, be on the lookout for shock symptoms, and be prepared to treat them. Your primary concern with a serious open wound is to stop the bleeding and gain professional medical help as soon as possible.

Lesser cuts, which are more common than deep ones, should be cleaned. Regular soap and water can be used to clean a wound before applying a bandage. Remember, we are talking about minor cuts and scrapes at this point. Flush the wound generously with clean water. A piece of sterile gauze can be used to pat the wound dry, and then a clean, dry bandage can be applied to protect the wound while in transport to a medical facility.

Fast Facts

- Use direct pressure to stop bleeding.
- Wear rubber gloves to prevent direct contact with a victim's blood.
- When feasible, elevate the part of the body that is bleeding.
- Extremely serious bleeding can require putting pressure on the artery supplying the blood to the wound area.
- Tourniquets can do more harm than good.
- Tourniquets should be at least 2 in. wide.
- Tourniquets should be placed above the bleeding wound, between the bleeding and the victim's heart.
- Tourniquets should not be applied directly on the wound area.
- Tourniquets should only be removed by trained medical professionals.
- If you apply a tourniquet, note the time you did it.
- When a bleeding wound requires a compression bandage, don't attempt to clean the wound. Simply apply compression quickly.
- Watch victims with serious bleeding for symptoms of shock.
- Lesser bleeding wounds should be cleaned before being bandaged.

SPLINTERS AND FOREIGN OBJECTS

Splinters and other foreign objects are a problem for construction workers. Getting these items out cleanly is best done by a doctor, but there are some on-the-job methods that you might want to try. A magnifying glass and a pair of tweezers work well together when removing embedded objects, such as splinters and slivers of copper tubing. Ideally, tweezers should be sterilized either over an open flame, such as the flame of your torch, or in boiling water before use.

Splinters and slivers beneath the skin can often be lifted out with the tip of a sterilized needle. The use of a needle in conjunction with a pair of tweezers is very effective in the removal of most simple splinters. If you are dealing with something that has gone extremely deep into tissue, it is best to leave the object alone until a doctor can remove it.

EYE INJURIES

Eye injuries are very common on construction and remodeling jobs. Most of these injuries could be avoided if proper eye protection was worn, but far too many workers don't wear safety glasses and goggles. This sets the stage for eye irritations and injuries.

Before you attempt to help someone who is suffering from an eye injury, you should wash your hands thoroughly. I know this is not always possible on construction sites, but cleaning your hands is advantageous. In the meantime, keep the victim from rubbing the injured eye as it can make matters much worse.

Never attempt to remove a foreign object from someone's eye with the use of a rigid device, such as a toothpick. Wet cotton swabs can serve as a magnet to remove some types of invasion objects. If the person you are helping has something embedded in an eye, get the person to a doctor as soon as possible. Don't attempt to remove the object yourself.

When you are investigating the cause of an eye injury, you should pull down the lower lid of the eye to determine if you can see the object causing trouble. A floating object, such as a piece of sawdust trapped between an eye and an eyelid, can be removed with a tissue, a damp cotton swab, or even a clean handkerchief. Don't allow dry cotton material to come into contact with an eye.

If looking under the lower lid doesn't produce a source of discomfort, check under the upper lid. Clean water can be used to flush out many eye contaminants without much risk of damage to the eye. Objects that cannot be removed easily should be left alone until a physician can take over.

- Wash your hands, if possible, before treating eye injuries.
- Don't rub an eye wound.
- Don't attempt to remove embedded items from an eye.
- Clean water can be used to flush out some eye irritants.

SCALP INJURIES

Scalp injuries can be misleading. What looks like a serious wound can be a fairly minor cut. On the other hand, what appears to be only a cut can involve a fractured skull. If you or someone around you sustains a scalp injury, such as having a hammer fall on your head from an overhead worker, take it seriously. Don't attempt to clean the wound. Expect profuse bleeding.

If you don't suspect a skull fracture, raise the victim's head and shoulders to reduce bleeding. Try not to bend the neck. Put a sterile bandage over the wound, but don't apply excessive pressure. If there is a bone fracture, pressure could worsen the situation. Secure the bandage with gauze or some other material that you can wrap around it. Seek medical attention immediately.

FACIAL INJURIES

Facial injuries can occur on jobs. I have seen helpers let their right-angle drills get away from them resulting in hard knocks to the face. On one occasion, I remember a tooth being lost, and split lips and tongues that have been bitten are common when a drill goes on a rampage.

Extremely bad facial injuries can cause a blockage of the victim's air passages. This is a very serious condition. If the person's mouth contains broken teeth or dentures, remove them. Be careful not to jar the individual's spine if you have reason to believe there may be injury to the back or neck. It is critical that air passages are open at all times.

Conscious victims should be positioned, when possible, so that secretions from the mouth and nose will drain out. Shock is a potential concern in severe facial injuries. For most on-the-job injuries, workers should be treated for comfort and sent for medical attention.

NOSE BLEEDS

Nose bleeds are not usually difficult to treat. Typically, pressure applied to the side of the nose where bleeding is occurring will stop the flow of blood. Applying cold compresses can also help. If external pressure does not stop the bleeding, use a small, clean pad of gauze to create a dam on the inside of the nose. Then apply pressure on the outside of the nose. This usually works. If it doesn't, get to a doctor.

Code Consideration

Falls result in all sorts of hidden injuries. These could include concussions or a broken back. Don't attempt to move fall victims. Call in the professionals and keep the victim stable.

BACK INJURIES

There is really only one thing that you need to know about back injuries. Don't move the injured party. Call for professional help and see that the victim remains still until help arrives. Moving someone who has suffered a back injury can be very risky. Don't do it unless there is a life-threatening cause for your action, such as a person is trapped in a fire or some other type of deadly situation.

LEGS AND FEET

Legs and feet sometimes become injured on jobsites. The worst case that I can remember was when a plumber knocked a pot of molten lead over on his foot. It sends shivers up my spine just to recall that incident. When someone suffers a minor foot or leg injury, you should clean and cover the wound. Bandages should be supportive without being constrictive. The appendage should be elevated above the victim's heart level when possible. Prohibit the person from walking. Remove boots and socks so that you can keep an eye on the person's toes. If the toes begin to swell or turn blue, loosen the supportive bandages.

Code Consideration

Signs of shock: skin turns pale or blue and is cold to the touch, moist and clammy skin, and general weakness.

Blisters

Blisters may not seem like much of an emergency, but they can sure take the steam out of a hard worker. In most cases, blisters can be covered with a heavy gauze pad to reduce pain. It is generally recommended to leave blisters unbroken. When a blister breaks, the area should be cleaned and treated as an open wound. Some blisters tend to be more serious than others. For example, blisters in the palm of a hand or on the sole of a foot should be looked at by a doctor.

HAND INJURIES

Hand injuries are common in the trades. Little cuts are the most frequent complaint. Serious hand injuries should be elevated as this tends to reduce swelling. You should not try to clean really bad hand injuries. Use a pressure bandage to control bleeding. If the cut is on the palm of a hand, a roll of gauze can be squeezed by the victim to slow the flow of blood. Pressure should stop the bleeding, but if it doesn't, seek medical assistance. As with all injuries, use common sense on whether or not professional attention is needed after first-aid is applied.

SHOCK

Shock is a condition that can be life-threatening even when the injury is not otherwise fatal. In this case we are talking about traumatic shock, not electrical shock. Many factors can lead to a person going into shock, but the most common is a serious injury. There are certain signs that indicate someone is going into shock.

If a person's skin turns pale or blue and is cold to the touch, it is a likely sign of shock. Skin that becomes moist and clammy and a general weakness are also signs of shock. When a person is going into shock, the individual's pulse is likely to exceed 100 beats per minute, and breathing is usually increased, but it may be shallow, deep, or irregular. Chest injuries usually result in shallow breathing. Victims who have lost blood may be thrashing about as they enter into shock. Vomiting and nausea can also signal shock.

As a person slips into deeper shock, the individual may become unresponsive. Look at the eyes; they may be widely dilated. Blood pressure can drop, and in time, the victim will lose consciousness. Body temperature will fall, and death can occur if treatment is not rendered.

There are three main goals when treating someone for shock: get the person's blood circulating well, make sure an adequate supply of oxygen is available to the individual, and maintain the person's body temperature.

When you have to treat a person for shock, you should keep the victim lying down so the blood will circulate better. Cover the individual to minimize loss of body heat. Get medical help as soon as possible. Remember, if you suspect back or neck injuries, don't move the person.

People who are unconscious should be placed on one side so that fluids will run out of the mouth and nose. It is also important to make sure that air passages are open. A person with a head injury may be laid out flat or propped up, but the head should not be lower than the rest of the body. It is sometimes advantageous to elevate a person's feet when they are in shock. However, if there is any difficulty in breathing or if pain increases when the feet are raised, lower them.

Body temperature is a big concern with shock patients. You want to overcome or avoid chilling. However, don't attempt to add additional heat to the surface of the person's body with artificial means. This can be damaging. Use only blankets, clothes, and other similar items to regain and maintain body temperature.

Avoid the temptation to offer the victim fluids, unless medical care is not going to be available for a long time. Avoid fluids completely if the person is unconscious or is subject to vomiting. Under most jobsite conditions, fluids should not be administered.

CHECKLIST OF SHOCK SYMPTOMS

- ☐ Skin that is pale, blue, or cold to the touch
- ☐ Moist and clammy skin
- ☐ General weakness
- ☐ Pulse rate in excess of 100 beats per minute
- ☐ Increased breathing
- ☐ Shallow breathing
- ☐ Thrashing
- ☐ Vomiting and nausea
- ☐ Unresponsive action
- ☐ Widely dilated eyes
- ☐ A drop in blood pressure

BURNS

Burns are not common among concrete installers, but they can occur in the workplace. There are three types of burns: first degree, second degree, and third degree. First-degree burns are the least serious. These burns typically come from overexposure to the sun, which construction workers often suffer from; quick contact with a hot object, like the tip of a torch; and scalding water, which could happen when working with a boiler or water heater.

Second-degree burns are more serious. They can occur with deep sunburn or from contact with hot liquids and flames. A person who is affected by a second-degree burn may have a red or mottled appearance, blisters, and a wet appearance of the skin within the burn area. This wet look is due to a loss of plasma through the damaged layers of skin.

Third-degree burns are the most serious. These can be caused by contact with open flames, hot objects, or immersion in very hot water. Electrical injuries can also result in third degree burns. This type of burn can look similar to a second degree burn, but the difference is the loss of all layers of skin.

Treatment

Treatment for most job-related burns can be administered on the jobsite and will not require hospitalization. First-degree burns should be washed with or submerged in cold water. A dry dressing can be applied if necessary. These burns are not too serious. Eliminating pain is the primary goal with first-degree burns.

Second-degree burns should be immersed in cold (but not ice) water. The soaking should continue for at least one hour and last up to two hours. After soaking, the wound should be layered with clean cloths that have been dipped in ice water and wrung out. Then the wound should be dried by blotting, not rubbing. Dry, sterile gauze should then be applied. Don't break open any blisters. It is also not advisable to use ointments and sprays on severe burns. Burned arms and legs should be elevated and medical professionals called.

Bad burns, the third-degree type, need quick medical attention. First, don't remove a burn victim's clothing as skin might come off with it. A thick, sterile dressing can be applied to the burn area. Personally, I would avoid this if possible. A dressing might stick to the mutilated skin and cause additional skin loss when the dressing is removed. When hands are burned, keep them elevated above the victim's heart. The same goes for feet and legs. You should not soak a third-degree burn in cold water as it could induce more shock symptoms. Don't use ointments, sprays, or other types of treatments. Get the burn victim to competent medical care as soon as possible.

Code Consideration

You owe it to yourself, your family, and the people you work with to learn first-aid techniques. This can be done best by attending formal classes in your area. Most towns and cities offer first-aid classes on a regular basis.

HEAT-RELATED PROBLEMS

Heat-related problems can include heat stroke and heat exhaustion. Cramps are also possible when working in hot weather. There are people who don't consider heat stroke to be serious. They are wrong. Heat stroke can be life-threatening. People affected by heat stroke can develop body temperatures in excess of 106°F. Their skin is likely to be hot, red, and dry without sweat. Pulse is rapid and strong, and victims can sink into an unconscious state.

If you are dealing with heat stroke, you need to lower the person's body temperature quickly. There is a risk, however, of cooling the body too quickly once the victim's temperature is below 102°F. You can lower body temperature with rubbing alcohol, cold packs, cold water on clothes, or in a bathtub of cold water. Avoid the use of ice in the cooling process. Fans and air-conditioned space can be used to achieve your cooling goals. Get the body temperature down to at least 102°F and then seek medical help.

Cramps

Cramps are fairly common among workers during hot spells. A simple massage can be all it takes to cure this problem. Salt-water solutions are another way to control cramps. Mix one teaspoonful of salt per glass of water and have the victim drink half a glass about every 15 minutes.

Exhaustion

Heat exhaustion is more common than heat stroke. A person affected by heat exhaustion is likely to maintain a fairly normal body temperature, but the person's skin may be pale and clammy. Sweating may be very noticeable, and the individual will probably complain of being tired and weak. Headaches, cramps, and nausea may accompany the symptoms. In some cases, fainting might occur.

The salt-water treatment described for cramps will normally work with heat exhaustion. Victims should lie down and elevate their feet about a foot off the floor or bed. Clothing should be loosened, and cool, wet cloths can be used to add comfort. If vomiting occurs, get the person to a hospital for intravenous fluids.

We could continue talking about first-aid; however, the help I can give you here for medical procedures is limited. You owe it to yourself, your family, and the people you work with to learn first-aid techniques. This can be done best by attending formal classes in your area. Most towns and cities offer first-aid classes on a regular basis. I strongly suggest that you enroll in one. Until you have some hands-on experience in a classroom and gain the depth of knowledge needed, you are not prepared for emergencies. Don't get caught unaware. Prepare now for the emergency that might never happen.

APPENDIX

I

Background Facts and Issues Concerning Cement and Cement Data

OUTLINE

Background Facts and Issues Concerning Cement and Cement Data

By Hendrik G. van Oss

Open-File Report 2005-1152

U.S. Department of the Interior

U.S. Geological Survey

U.S. Department of the Interior

Gale A. Norton, Secretary

U.S. Geological Survey

P. Patrick Leahy, Acting Director

For product and ordering information:
World Wide Web: http://www.usgs.gov/pubprod
Telephone: 1-888-ASK-USGS

For more information on the USGS—the Federal source for science about the Earth, its natural and living resources, natural hazards, and the environment:
World Wide Web: http://www.usgs.gov
Telephone: 1-888-ASK-USGS

Preface

This report is divided into two main parts. Part 1 first serves as a general overview or primer on hydraulic (chiefly portland) cement and, to some degree, concrete. Part 2 describes the monthly and annual U.S. Geological Survey (USGS) cement industry canvasses in general terms of their coverage and some of the issues regarding the collection and interpretation of the data therein. The report provides background detail that has not been possible to include in the USGS annual and monthly reports on cement. These periodic publications, however, should be referred to for detailed current data on U.S. production and sales of cement. It is anticipated that the contents of this report may be updated and/or supplemented from time to time.

Because some readers will choose to access only specific sections of this report, the individual sections have been written on a more stand-alone basis than might have been otherwise. A variety of technical terms related to cement (in this report and elsewhere) have definitions that may differ from the same terms as used in other fields. Most terms in this report have been defined at first usage in the text, but a brief glossary of technical terms has been provided to limit sectional repetition of definitions. A short table of units of measure has also been included.

Units of measurement

For the most part, this report makes use of metric units despite the fact that the U.S. cement (and concrete) industry, at the time of writing, continues to use nonmetric units for commerce and a mix of units for internal accounting. Nonmetric units are also commonly encountered in the literature on cement. Conversion of units is provided below for the convenience of the reader. It is important to note that some of the conversion factors are given here to more significant figures than can generally be justified in reporting cement data.

The cement industry tends not to distinguish between units of weight and mass but almost invariably conducts its business in the sense of weight. Thus the metric ton, a unit of mass, is treated as if it were a unit of weight; strictly, the conversions to weight are applicable only at mean sea level.

Barrel (bbl) as a unit of cement weight (not volume) is no longer used by the cement industry but is frequently encountered in the historical literature. Its weight conversion has varied over the years by type of cement, reporting source, and by reporting region. For original (not re-converted) U.S. data reported by the U.S. Government, equivalences are as follows:

Cement Type	Years	Weight (lbs) of 1 barrel
Portland cement:	through 1919	380
	1920 onwards	376
Natural cement:	through 1920	300 (imposed*)
	1921 onwards	276
Pozzolanic and slag cements:	through 1915	380
	1916–1920	300
	1921 onwards	376
Masonry cement:	through 1960	376
	1961 onwards	280
Generic cement data:	All years	376

*USGS-imposed (in original reports) conversion so as to standardize industry-submitted data that ranged from 265 to 300 pounds per barrel

Thus the general conversions:

1 metric ton = 5.80163 bbl (of 380 lbs);
5.86335 bbl (of 376 lbs);
7.87365 bbl (of 280 lbs).

1 bag (in USA) of portland cement = 94 lbs (this could change if the industry switches to kilogram-denominated bags).

1 bag (in USA) of masonry cement = 70 lbs (this could change if the industry switches to kilogram-denominated bags).

Units of measurement and their equivalents

Units of Weight and Mass		
1 metric ton	1,000 kilograms	1.10231 short tons
1 short ton	2,000 pounds avoirdupois (lbs)	0.907185 metric tons
1 kilogram	2.20462 lbs	
Units of Length		
1 meter	3.28084 feet	39.3701 inches
Units of Volume		
1 cubic meter	1,000 liters	1.3079 cubic yards
1 cubic meter	264.172 U.S. gallons	35.3147 cubic feet
1 barrel (volume) of fuel	42 U.S. gallons	158.9878 liters
Note: When initially reported in metric tons, the conversion of fuels to volume units (e.g., to use volume-related heat or energy equivalences) varies by type of fuel. For petroleum-based liquid fuels, a rough (first approximation) equivalence is: 1 metric ton of liquid fuel = 7 barrels (range is about 6–9 barrels).		
Units of Energy		
1 kilowatt hour (electricity)	3,412.14 British thermal units (Btu)	3.6000 megajoules (MJ)
1 British thermal unit (Btu)	1.055056 kilojoules (kJ)	
1 million Btu (Mbtu)	1.055056 gigajoules (GJ)	
Temperature		
°F	32 + 9/5 °C	
°C	5/9 (°F – 32)	
Prefixes		
kilo (k)	$= 10^3$	
mega (M)	$= 10^6$	
giga (G)	$= 10^9$	
tera (T)	$= 10^{12}$	
peta (P)	$= 10^{15}$	
exa (E)	$= 10^{18}$	

Contents

Figures

Tables

Part 1: Overview of Hydraulic Cements

Introduction

Hydraulic cements are the binding agents in concretes and most mortars and are thus common and critically important construction materials. The term *hydraulic* refers to a cement's ability to set and harden under, or with excess, water through the hydration of the cement's constituent chemical compounds or minerals. Hydraulic cements are of two broad types: those that are inherently hydraulic (i.e., require only the addition of water to activate), and those that are pozzolanic. Loosely defined, the term *pozzolan* (or pozzolanic) refers to any siliceous material that develops hydraulic cementitious properties when interacted with hydrated lime [$Ca(OH)_2$], and in this overview includes true pozzolans and *latent cements*. The difference between true pozzolans and latent cements is subtle: true pozzolans have no cementitious properties in the absence of lime, whereas latent cements already have some cementitious properties but these are enhanced in the presence of lime. Pozzolanic additives or extenders may be collectively termed *supplementary cementitious materials* (SCM).

Concrete is an artificial rock-like material (in effect, an artificial conglomerate) made from a proportioned mix of hydraulic cement, water, fine and coarse aggregates, air, and sometimes additives. The cement itself can either be a pure hydraulic cement or a mix of hydraulic cement and SCM. Concrete mix recipes vary, but most have compositions (in volumetric terms) in the range of about: 7%–15% cement powder, 15%–20% water, 0.5%–8% air, 25%–30% fine aggregates (e.g., sand), and 30%–50% coarse aggregates (e.g., gravel or crushed stone) (Kosmatka and others, 2002). The cement powder and water together form *cement paste*. The aim in a good concrete mix is to have a completely unsorted (by size) mix of the aggregate particles, all bound together with just enough cement paste to completely coat all of the aggregate grain surfaces and fill all unintentional voids. For a typical concrete mix, 1 metric ton (t) of cement (powder) will yield about 3.4–3.8 cubic meters (m^3) of concrete weighing about 7–9 t (that is, the density is typically in the range of about 2.2–2.4 t/m^3). Although aggregates make up the bulk of the mix, it is the hardened cement paste that binds the aggregates together and contributes virtually all of the strength of the concrete, with the aggregates serving largely as low cost fillers. Hydraulic *mortars* are similar to concrete, except that only fine aggregates are incorporated and the cement used is formulated so as to be somewhat more plastic in character. A few mortars use nonhydraulic binders such as lime. Mortars are used to bind together bricks, blocks, or stones in *masonry* construction. In the older literature, "mortar" can sometimes also be a casual term for cement itself.

Current (2004) world total annual production of hydraulic cement is about 2 billion t (Gt), and production is spread very unevenly among more than 150 countries. This quantity of cement is sufficient for about 14–18 Gt/yr of concrete (including mortars), and makes concrete the most abundant of all manufactured solid materials. The current yearly output of hydraulic cement is sufficient to make about 2.5 metric tons per year (t/yr) of concrete for every person on the planet.

Brief history of hydraulic cement

The use of various cements in construction is a very ancient practice but some of the details as to who first made, or used, what form of cement are still uncertain. Nevertheless, a general sequence of developments can be described; much of the following synopsis is derived from Lea (1970, ch. 2), Bogue (1955, ch.1), Klemm (2004), Lesley (1924, ch. 1–2), and Wilcox (1995).

Ancient use of cements

The earliest binder used in masonry construction was plain mud (with or without straw) and mud binders are still used, mainly for adobe construction, in many parts of the world today. Bitumen (natural asphalt) was used as a binder in some parts of ancient Mesopotamia. Some of the earliest use of true mortars was in ancient Egypt, Greece, and Crete. In Greece and Crete, lime mortars were made, as today, from the burning of limestone. In contrast, the Egyptians mainly used crude gypsum (plaster) mortars, although they had access to abundant limestone and did make some lime mortars. The Egyptian practice of using gypsum (a much rarer material than limestone) appears primarily to have been because of a shortage of fuels; the conversion of limestone to lime requires significantly higher temperatures (hence more fuel) than that of gypsum to plaster. Some foundations for ancient Egyptian buildings even made use of gypsum concrete (Kemp, 1994), in which limestone quarry and/or construction debris comprised the coarse aggregates. Some of the Greek mortars used hydraulic lime made from impure limestones, and still other hydraulic mortars were made by mixing lime with certain volcanic ashes, most notably that from the island of Thíra (or Thera; now called Santorin or Santorini).

The ancient Romans learned about various types of mortars from the Greeks, but because the Romans improved the quality and methods of application of hydraulic mortars and made far more extensive use of them, it is the Romans who are commonly given the lion's share of the credit for the development of hydraulic cement. Most significant was the Roman use of pozzolan-lime cements incorporating volcanic ash. A volcanic ash particularly favored by the Romans for this purpose was that quarried in large quantities from the distal slopes of Mt. Vesuvius near the village of Pozzuoli. This material became known as *pozzolana* (also spelled puzzolana, pouzzolana, pozzuolana) based on this village's name; likewise, the more general terms pozzolan and pozzolanic. Pozzolana has come to be applied to all volcanic ashes (such as santorin earth and trass) having pozzolanic character. Where pozzolana was unavailable, the Romans made use of crushed tiles or potshards as an artificial pozzolan; the Greeks may have earlier made similar use of crushed ceramics.

The Roman pozzolan-lime cements were so strong that, in practice, the proportion of aggregates in the mortars could be significantly increased over that used in unmodified lime mortars. Even coarse aggregates (commonly demolition debris) could be incorporated to form a bulk building material in its own right, i.e., concrete. Roman engineers used pozzolan-lime mortars and concrete throughout the Roman Empire, not only in a multitude of buildings (perhaps the best known surviving example of which is the Pantheon in Rome), but also in applications such as the waterproof lining of aqueducts and the construction of sea walls for artificial harbors.

Post-Roman use of cements

Hydraulic mortars and concretes made in the first few centuries following the fall of the Roman Empire were of lower quality, owing either to poor understanding of the techniques of cement and lime manufacture or a lack of pozzolans in some places, or both. Regular lime mortars appear to have fared better, however. Sporadic interest remained in reproducing the quality of the old Roman cements, and in the 18th century active research into improving the quality of cements was becoming fairly widespread in Western Europe.

A major breakthrough in the understanding of hydraulic cement resulted from the research of John Smeaton following his being awarded, in 1756, the contract to build a replacement lighthouse on the Eddystone Rocks, offshore from Plymouth, England. For this project, Smeaton conducted experiments to make a mortar that could withstand especially severe marine conditions. Smeaton discovered that strong hydraulic lime mortar could be made from calcining limestones that contained appreciable amounts of clay—his key insight being the link between the clay component and the development of hydraulic character. He found that an even better mortar could be made by combining this hydraulic lime with pozzolans, and it was a hydraulic lime-pozzolan mortar (specifically, hydraulic lime made from Aberthaw, Wales limestone and Italian pozzolana) that he used to build the new Eddystone lighthouse, which then stood for 126 years before needing replacement. Smeaton published the results of his research on hydraulic limes in 1791.

In 1796, a patent was granted to James Parker (Joseph Parker in some writeups) in England for hydraulic cement made from argillaceous limestone nodules (septaria). Within a few years, cements derived from a variety of argillaceous limestones were being marketed under the misleading but persistent term Roman cement. The name was based on the claim that the cement was as good as, and of similar reddish color to, the ancient Roman product, but was despite the fact that the new material contained no pozzolana and was compositionally quite unlike its namesake. Because the new cement (better termed *natural cement*) possessed good strength and hydraulic properties and set and hardened fairly quickly, its popularity grew rapidly, and natural cement remained the dominant cement type produced in England and most of the rest of Europe until the mid-19th century.

Production of cement of any type came later in the United States than in Europe, with the initial impetus being a need for waterproof mortars for the lining and lockworks for the Erie Canal, New York. Construction of the canal began in 1817 and, in the following year, deposits of argillaceous limestone suitable for the manufacture of natural cement were discovered near the canal. Manufacture of natural cement began shortly thereafter. The argillaceous limestone was locally called *cement rock.* The natural cement industry in the United States grew steadily as cement rock deposits were subsequently discovered in eastern Pennsylvania and elsewhere. Until the early 1870s, the only cements made in the United States were natural cements and slag-lime cements (chiefly based on quenched iron furnace slag).

Natural cements in the United States and in Europe exhibited significant regional variations in quality, owing to differences in processing methods and in the composition of the cement rock raw material. Smeaton's discovery of the importance of clay to the development of hydraulic character in lime mortars inspired research into ways to improve the quality, and/or reduce the variability, of natural cements. An ultimately more important avenue of research was that into the making of so-called artificial cement. Probably the most influential researcher in this area of investigation was the French engineer Louis J. Vicat. His research was first published in 1818 and showed how, in the absence of argillaceous limestones or cement rock, high quality hydraulic limes could be made with ordinary limestone, provided that controlled amounts of clay or shale were added. A more comprehensive review was published in 1828, and was translated into English in 1837 (Vicat, 1837).

The first portland cement

The findings of Smeaton, Vicat, and other researchers appear to have been fairly well disseminated, and a number of patents were issued in England and France for various types of hydraulic limes and cements. In 1810, Edgar Dobbs received a patent in England for a cement that utilized clay, road dust (from limestone), and lime. The Dobbs patent expired in 1824, and this may have inspired Joseph Aspdin, a brick mason and experimenter from Leeds, England, to file later that year a patent for what superficially appears to be a similar product. Aspdin's December 15, 1824 patent "An improvement in the modes of producing an artificial stone" (British patent 5022) was for a product that he called *Portland Cement.* This name alluded to Portland stone, a well-regarded, very tough, dimension stone quarried on the Isle of Portland along the South Dorset coast. The Portland stone comparison also had been used, much earlier, by Smeaton in extolling the qualities of his own cement. Aspdin's patent is interesting by modern standards in that it reveals remarkably little specific information about the product or its manufacture; indeed it would seem to be a rather tenuous foundation for a future major industry! Omitting the introductory remarks and salutations, the patent's technical description reads:

> My method of making a cement or artificial stone for stuccoing buildings, waterworks, cisterns, or any other purpose to which it may be applicable (and which I call Portland Cement) is as follows: I take a specific quantity of limestone, such as that generally used for making or repairing roads, and I take it from the roads after it is reduced to a puddle, or powder; but if I cannot procure a sufficient quantity of the above from the roads, I obtain the limestone itself, and I cause the puddle or powder, or the limestone, as the case may be, to be calcined. I then take a specific quantity of argillaceous earth or clay, and mix them with water to a state approaching impalpability, either by manual labor or machinery. After this proceeding I put the above mixture into a slip pan for evaporation, either by the heat of the sun or by submitting it to the action of fire or steam conveyed in flues or pipes under or near the pan until the water is entirely evaporated. Then I break the said mixture into suitable lumps, and calcine them in a furnace similar to a lime kiln till the carbonic acid is entirely expelled. The mixture so calcined is to be ground, beat, or rolled to a fine powder, and is then in a fit state for making cement or artificial stone. This powder is to be mixed with a sufficient quantity of water to bring it to the consistency of mortar, and this applied to the purposes wanted.

The first portland cement (the name, commonly, is no longer capitalized) plant was set up at Wakefield, England by Aspdin shortly after receiving his patent. His son, William Aspdin, later set up plants on the Thames and at Gateshead-on-Tyne. In the early years of the British portland cement industry the expensive new cement found difficulty in capturing market share from the well-regarded natural cements. Marketing of portland cement received a major boost in 1838 when it was chosen for the prestigious project to construct the Thames River tunnel.

Early efforts by others to duplicate the quality of Aspdin's portland cement were largely unsuccessful and led to speculation that Aspdin's patent lacked (perhaps deliberately) critical details regarding the cement's manufacture. The duplication efforts, however, yielded by the mid-1840s important ideas as to what those missing details were, most notably the need to heat the raw materials in the kiln to much higher temperatures than that merely needed to calcine the limestone. Empirical evidence for this was the discovery that over-burned, partially vitrified, material (hitherto discarded) from the kilns yielded better quality cement; a rival cement maker, Isaac C. Johnson, is generally credited with making public this need for high-temperature burning. It was eventually determined that the improvement in cement quality was because of the presence of hydraulic dicalcium silicate in the vitrified material (clinker) and, even more important, the presence of tricalcium silicate (discussed in more detail later). There remains debate as to whether Aspdin's original portland cement contained either of these minerals or whether it was merely a well-made hydraulic lime, relying on heat-activated clay pozzolans. It is now known that both dicalcium silicate and (minor) tricalcium silicate were present in at least some of the later portland cements made by Joseph Aspdin (Stanley, 1999; Campbell, 1999, p. 1–3), and were certainly present in cements made by his son. Through trial and error, numerous improvements to the manufacturing process and mineralogic composition of portland cement were made in England and elsewhere over the course of the following decades. Improvements still continue and there remains only superficial similarity between modern portland cements and those manufactured in the mid- to late-19th century. The portland cement name has been retained by the industry largely because the modern material is still an artificial cement made from limestone and argillaceous raw materials and because the name has an unrivaled cachet.

Early use of portland cement in the United States

The first portland cement plants outside of England were constructed in Belgium and Germany around 1855. Portland cement imports into the United States began in the late 1860s and reached a peak level of about 0.5 million metric tons (Mt) in 1895 before declining owing to increased local production. The first portland cement plant in the United States was established in 1871 by David Saylor at Coplay, Pennsylvania; natural cement had been made there since 1850. The initial portland cement made at Coplay was not very satisfactory. However, by 1875, Saylor had more or less overcome problems of low kilning temperatures and inadequate mixing of raw materials and was making a high-quality product. The superiority of portland cement led to a proliferation of portland cement plants elsewhere in the United States; a detailed review of the history of the early (through about 1920) portland cement industry in the United States is provided by Lesley (1924). Output of portland cement in the United States grew rapidly from about 1890 onwards. By 1900, U.S. production had reached 1.46 Mt, exceeding for the first time the combined U.S. output (1.22 Mt) of natural and slag-lime cements.

Growth of the market for portland cement

Overall growth in demand for concrete was given a great boost by the invention of reinforced concrete in the late 1860s. Because concrete is strong in compression but relatively weak in tension, this innovation of incorporating reinforcing steel bar to provide tensile strength opened up a number of new uses for the concrete, such as for high-rise buildings and suspended slab structures (such as bridge decks).

All of the early portland and natural cements were made on a batch basis in vertical shaft or chimney-type kilns. A major advance in output capacity and in thoroughness of mixing and heating of the raw materials came with the invention in 1873 of the rotary kiln. The English engineer F. Ransome patented an improved version of the rotary kiln in 1885 that allowed continuous throughput of materials. Thomas Edison developed the first high capacity rotary kilns in 1902; Edison's kilns were about 46 meters (m) long, as opposed to just 18–24 m for the rotary kilns up to that time.

In the first half of the 20th century, world use of concrete, and hence production of cement, grew erratically, experiencing major ebbs during the two World Wars and the Great Depression. World hydraulic cement production does not appear to have reached 100 Mt/yr until 1948 (see figure 1). However, since World War II, cement production has experienced steady, strong growth, with output in 2004 at about 2 Gt, as noted earlier. Today, there are very few countries that do not have at least one cement plant. As can be seen in figure 1, the greatest growth in cement output since 1950 (and especially since 1980) has been in Asia, which currently accounts for more than half of total world production. Much of Asia's overall growth has been by China. In 1950, China produced only about 2 Mt of cement, whereas output in 2004 was about 950 Mt. Currently, the top five world producers are, in descending order, China, India, the United States, Japan, and the Republic of Korea. The vast majority of cement produced in the world today is portland cement or closely related cements having portland cement as a base.

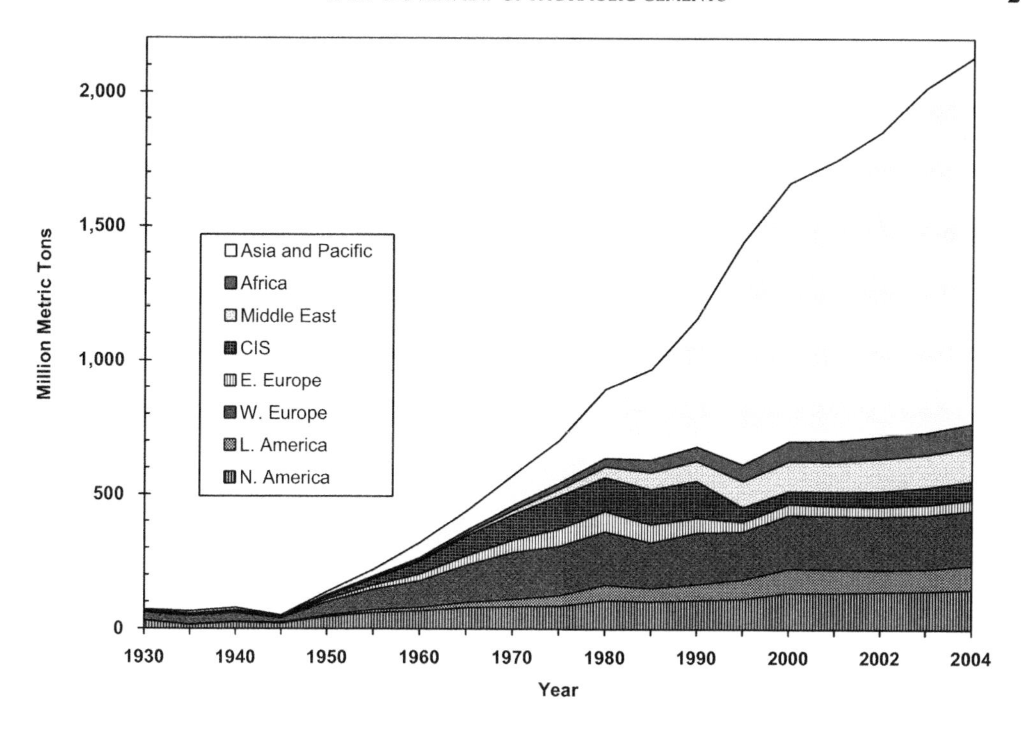

Figure 1. World production of hydraulic cement, by region.

Because of its important use in construction, cement production and consumption trends in developed countries tend to mirror those for the general economy. The historical graph for the United States (see figure 2) is illustrative; most of the major national economic perturbations, such as the World Wars, Great Depression, and the energy crises, can be seen as short-term disruptions to a generally increasing demand for cement. The general increase is largely owing to the combination of a shift in construction preferences to concrete and the rapidly rising population. The very rapid growth beginning in the early 1950s is due to the commencement of construction of the interstate highway network. Until 1970, U.S. production and consumption of portland cement and production of clinker were in balance. Since that time, however, U.S. production of clinker has fallen behind production of portland cement, and cement output has fallen greatly behind consumption. The production shortfalls have been met by imports of clinker and cement. Details of U.S. cement production, consumption, and trade can be found in the USGS Minerals Yearbooks and monthly Mineral Industry Surveys.

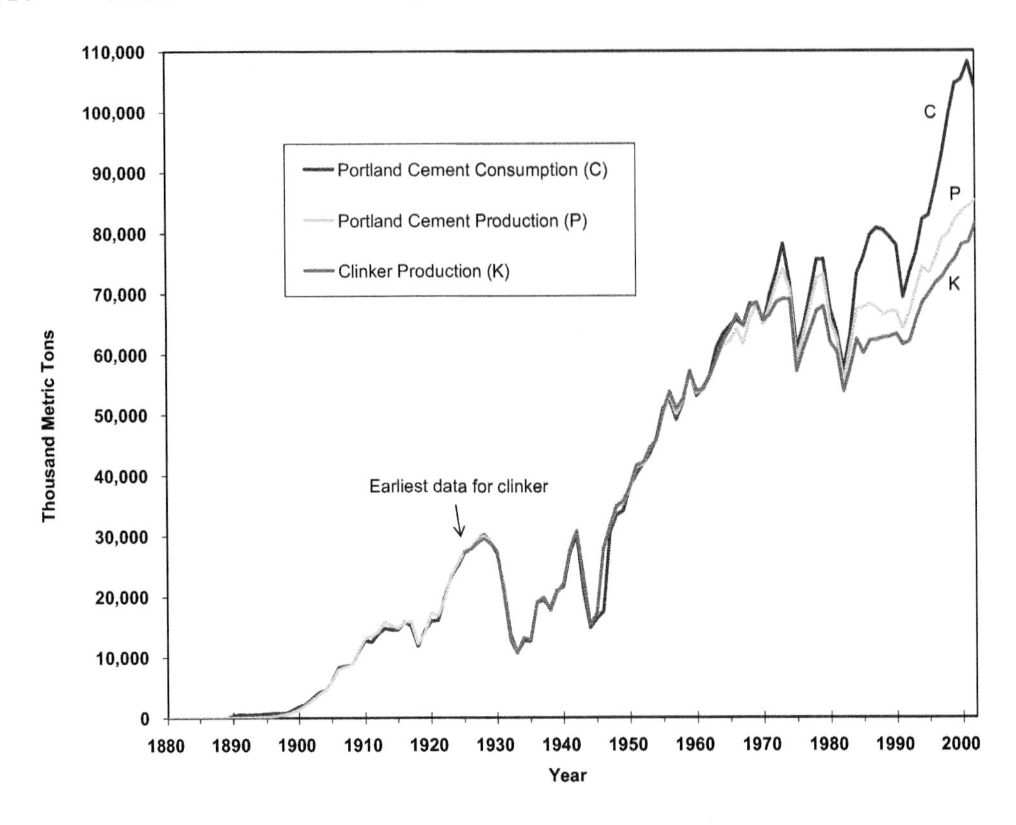

Figure 2. U.S. production and consumption of portland cement.

Types of hydraulic cement

As partly enumerated in the historical review above, there are a number of different types of hydraulic cement, most of which are still in at least occasional use today. The main ones are:

Hydraulic lime

This lime contains certain variable amounts of hydraulic silicates (mostly as calcined clay pozzolans but some varieties could contain small amounts of dicalcium silicate) as well as free lime, and is made from calcination of clay-rich limestones. Hydraulic lime was the active ingredient in natural cements. Today, hydraulic lime is used in some specialized mortars, but is not widely produced.

Natural cements

Commonly and misleadingly called Roman cement in some older U.S. and English literature, natural cements are made from argillaceous limestone or cement rock or intimately interbedded limestone and clay or shale, with few, if any, additional raw materials. Some natural cements were made at relatively low temperatures, and the cementing properties were substantially due to the formation of clay pozzolans. Better quality natural cements were made at much higher temperatures, and some may thereby have included hydraulic dicalcium silicate. Because significant blending of raw materials is not done, the quality of natural cements is dependent (assuming similar manufacturing parameters) on the composition of the local cement rock and so can vary regionally. Although of great historical importance, natural cements were generally found to be inferior to portland cement, and most natural cement plants worldwide eventually switched to the production of portland cement.

Portland cement

Portland cement is an artificial cement in that its manufacture involves the mixing of raw materials (details on its manufacture and mineralogy are given in a later section), which allows for a uniform composition of the raw material feed to the kiln and which, in turn, allows for uniform properties of the finished cement, regardless of where it was made. The blending of raw materials and the fact that portland cements are made at higher temperatures are key differences between portland and natural cements. Today, straight (i.e., fitting within the specifications of ASTM C-150—see below) portland cement is defined as a finely interground mixture of portland cement clinker (an intermediate product in cement manufacture described in more detail later) and a small quantity (typically 3%–7% by weight) of calcium sulfate, usually in the form of gypsum. The ASTM C-150 standard for portland cement was revised recently to allow the incorporation of up to 5% of ground limestone as a filler.

By modifying the raw material mix and, to some degree, the temperature of manufacture, slight compositional variations in the clinker can be achieved to produce portland cements with slightly different properties. During the 20th century, the mineralogical ratios and particle size distribution in portland cement have been changed to favor faster and greater development of strength; this has been achieved principally by increasing the proportion of tricalcium silicate relative to dicalcium silicate, and grinding the clinker more finely.

Similar varieties of portland cement are made in many parts of the world, but go by different names. In the United States, the different varieties of straight portland cement are denoted per the American Society for Testing and Materials (ASTM) standard C-150, as:

Type I: general use portland cement. In some countries, this type is known as *ordinary* portland cement.

Type II: general use portland cement exhibiting moderate sulfate resistance and moderate heat of hydration.

Type III: high early strength portland cement.

Type IV: portland cement having a low heat of hydration.

Type V: portland cement having high sulfate resistance.

For Types I, II, and III, the addition of the suffix A (e.g., Type IA) indicates the inclusion of an *air-entraining* agent. Air entraining agents impart myriad tiny bubbles to the concrete containing the hydrated cement, which can offer certain advantages to the concrete, especially improved resistance to freeze-thaw cracking. In practice, many companies market hybrid portland cements; Type I/II is a common hybrid and meets the specifications of both Types I and II. Another common hybrid is Type

II/V. Because the uses and mineralogic composition (see table 3 below) of modern Type I and II portland cements are so similar, many statistical compilations (including those of the USGS) do not distinguish between them.

Portland cements are usually gray, but a more expensive *white* portland cement (generally within the Type I or II designations) can be obtained by burning only raw materials of very low contents of iron and transition elements. Both gray and white portland cement can be the basis of colored portland cements through the incorporation of pigments.

Notwithstanding the ASTM C-150 specifications, it is important to note the common industry practice, and that of the USGS, to include within the portland cement designation a number of other cements not within ASTM C-150 that are composed largely of portland cement and which are used for similar applications (e.g., concrete). These include blended cement (see below), block cement, expansive cement, oil well cement, regulated fast setting cement, and waterproof cement. However, plastic cements and portland-lime cements are not part of the portland cement umbrella, being instead grouped by the USGS and by most cement companies within the *masonry cement* designation (see below). In Europe and some other regions, the term portland cement, used loosely, may also include cement more properly termed *portland-limestone* cement (see below); this cement is not currently in use in the United States.

Some statistical compendia, including USGS Minerals Yearbook chapters and pre-1998 monthly Mineral Industry Surveys reports, also include *blended* cement (see below) within the portland cement designation. Blended cements have the same uses as Types I, II, IV, and V portlands, but generally not instead of Type III portland. Some jurisdictions allow certain (usually Type I) portland cements to contain small amounts (1%–3%) of inert and/or cementitious extenders while retaining the portland cement name. A relatively common example of this is where (unground) granulated blast furnace slag has been added in the finish mill to act as a clinker-grinding aid in the production of Type I cement. In this instance, the slag is converted (through grinding) into an SCM component to the finished portland cement. Because the use of SCM is increasing, interest is growing in having the cement requirements in construction project specifications change from a compositional basis (e.g., based on ASTM C-150) to a performance basis. The current performance-basis standard is ASTM C-1157, which identifies the following basic performance types of cement (essentially irrespective of composition) for concrete and similar applications:

Type GU: general use cement (performance equivalent to Type I in ASTM C-150)

Type HE: high early strength cement (performance equivalent to Type III)

Type MS: a cement providing moderate sulfate resistance (i.e., equivalent to Type II)

Type HS: a cement providing high sulfate resistance (i.e., equivalent to Type V)

Type MH: a cement providing moderate heat of hydration (i.e., equivalent to Type II)

Type LH: a cement providing low heat of hydration (i.e., equivalent to Type IV)

Any of the ASTM C-1157 cements can be further specified as requiring *Option R*, which indicates a requirement that the cement exhibit low reactivity with alkali-reactive concretes. This is to avoid alkali-silica or alkali-aggregate reactions between the cement paste and the aggregates in the concrete.

For low heat of hydration applications, Type IV cement has now been virtually entirely replaced with blended cements (generally incorporating fly ash) meeting the requirements of ASTM C-1157 Type LH .

Portland-limestone cements

These are cements wherein relatively large amounts (6% to 35%) of ground limestone have been added as a filler to a portland cement base. Although not in use, as yet, in the United States, portland-limestone cements are reported to be in common use in Europe for certain relatively low strength general construction applications (Moir, 2003).

Blended cements

Blended cements (called *composite* cements in some countries) are intimate mixes of a portland cement base (generally Type I) with one or more SCM extenders. The SCM commonly makes up about 5%–30% by weight of the total blend (but can be higher). In many statistical compendia (including some by the USGS), blended cements are included within the term portland cement; however, beginning with the January 1998 edition, the USGS monthly Mineral Industry Surveys on cement list blended cements as a separate category.

In blended cements, the SCM (or pozzolans) are activated by the high pH resulting from the lime released during the hydration of portland cement (see chemistry of hydration discussion below). The most commonly used SCM are pozzolana, certain types of fly ash (from coal-fired powerplants), ground granulated blast furnace slag (GGBFS—now increasingly being referred to as *slag cement*), burned clays, silica fume, and cement kiln dust (CKD). In general, incorporation of SCM with portland cement improves the resistance of concrete to chemical attack, reduces the concrete's porosity, reduces the heat of hydration of the cement (not always an advantage), may improve the flowability of concrete, and produces a concrete having about the same long-term strength as straight portland cement-based concretes. However, SCM generally reduce the early strength of the concrete, which may be detrimental to certain applications.

Blended cement, strictly, refers to a finished blended cement product made at a cement plant or its terminals, but essentially the same material can be made by doing the blending within a concrete mix. Indeed, most of the SCM consumption by U.S. concrete producers is material purchased directly for blending into the concrete mix; concrete producers buy relatively little finished blended cement. Increasingly, cement and concrete companies would prefer the flexibility of offering their products on a performance basis rather than on a recipe (specific type of portland cement) basis. Performance specifications for cements are covered under ASTM Standard C-1157-02 as noted above. A recent detailed review of blended cements is provided by Schmidt and others (2004).

The designations for blended cements vary worldwide, but those currently in use in the United States meet one or the other of the ASTM Standards C-595 and C-1157. The definitions of blended cements given in ASTM Standard C-595 are summarized below.

Portland blast furnace slag cement (denoted IS [pronounced "one-S"]) contains 25%–70% GGBFS. Type IS cements are for general purpose uses, and can be designated for special properties: Type IS(MS) has moderate sulfate resistance; Type IS(A) is air-entrained; and Type IS(MH) has moderate heat of hydration.

Portland-pozzolan cement contains a base of portland and/or IS cement and 15%–40% pozzolans. The pozzolan type is not specified. There are two main types of portland-pozzolan cement: Types IP ("one-P") and P.

Type IP cements are for general use; and Types IP(MS), IP(A), and IP(MH) share the same modifiers as the IS cements above.

Type P cements are for general uses not requiring high early strength. Again, there are the same P(MS), P(A), and P(MH) varieties, as well as Type P(LH); the LH designates low heat of hydration.

Pozzolan-modified portland cement: Designated I(PM); the base is portland and/or Type IS cement, with a pozzolan addition of less than 15%. Type I(PM) is for general use; and Types I(PM)(MS); I(PM)(A); and I(PM)(MH) share the same modifiers as the IS cements above.

Slag-modified portland cement: Designated I(SM), these cements contain less than 25% GGBFS. Type I(SM) is for general use; and Types I(SM)(MS), I(SM)(A), and I(SM)(MH) share the same modifiers as the IS cements above.

Slag cement: Under C-595, slag cement is designated Type S, defined as having a GGBFS content of 70% or more. The slag can be blended with portland cement to make concrete or with lime for mortars; the latter combination would make the final cement a pozzolan-lime cement. Type S(A) is air-entrained.

It is important to note that true Type S cements are no longer commonly made in the United States. Instead, the name *slag cement* (but with no abbreviation) is now increasingly given to the unblended 100% GGBFS product; this use of the term has led to some confusion because the material is not used directly (i.e., on its own) as a cement. The GGBFS is sold as an SCM either to cement companies to make blended or masonry cements, or to concrete manufacturers as a partial substitute for portland cement in the concrete mix (in effect making a IS or I(SM) blended cement paste as the binder).

Pozzolan-lime cements

These include the original Roman cements (c.f., the natural cements above) and are an artificial mix of one or more pozzolans with lime (whether or not hydraulic). Little, if any, of this material is currently manufactured in the United States (but slag-lime cements were once popular) and these cements are now relatively uncommon elsewhere as well.

Masonry cements

These are hydraulic cements based on portland cement to which other materials have been added primarily to impart plasticity. The most common additives are ground limestone (unburned) and/or lime, but others, including pozzolans, can be used. In a true masonry cement, the typical mix will incorporate about 50%–67% portland cement (or its clinker plus gypsum equivalent) and the remainder will be the additives. Masonry cements are used to make hydraulic mortars for binding blocks together, and controlling their spacing, in masonry-type construction.

In addition to true masonry cements, USGS and many other statistical compendiums include within this category two other cements of similar application. These are *plastic cements*, which are portland cements to which only small amounts (generally <12%) of plasticizers have been added, and *portland-lime* cement. Portland-lime cements can be, and commonly are, made at the construction site by combining purchased portland (generally Type I) cement with hydrated lime.

Aluminous cements

Also known as *calcium-aluminate* cement, high-aluminous cement or, for some versions, *ciment fondu*, aluminous cements are made from a mix of limestone and bauxite as the main raw materials. Aluminous cements are used for refractory applications (such as for cementing furnace bricks) and in certain rapid-hardening concrete applications. These cements are much more expensive than portland cements and are made in relatively tiny quantities by just a few companies worldwide. Currently, there is only one production facility in the United States. Data on production and sales of aluminous cement are almost always proprietary and hence unavailable.

Mechanisms of cementation, and the chemistry and mineralogy of portland cement

As noted earlier, portland cement and some other cements based on it are overwhelmingly the dominant hydraulic cements produced today. The chemistry and mineralogy of portland cement can and will be described simply and qualitatively in the following overview. It should be stressed, however, that the simple descriptions and explanations provided below are mere representations of more complex reactions into which research is still ongoing. The same comment applies to the mechanisms of binding or cementation.

Mechanisms of cementation

A number of binding mechanisms can occur in a mix of a cementing agent and aggregate particles. Qualitatively, one can envision the ultimate strength of the cemented material being due primarily to any or a combination of the mechanical strength of the hardened cement paste itself, the degree to which the hardened paste physically interlocks with the aggregate particles or chemically binds or otherwise reacts with the particles, the mechanical strength of the aggregate particles themselves, and the degree to which the aggregate particles and paste have been properly mixed.

Geologically speaking, a binding agent may be thought of as a material that has cooled from a melt, settled from suspension, precipitated from aqueous solution, or is a preexisting material that has undergone a change (perhaps through diagenesis) in morphology, mineralogy, or chemical composition. It is this last aspect that is of chief importance in artificial cements, and the changes of interest are *carbonation* and *hydration*. Hydraulic cements rely on hydration as their cementing mechanism.

Carbonation

Ordinary (nonhydraulic) lime mortars harden through carbonation (rarely, but incorrectly, called carbonisation) of free lime. Lime (CaO) is formed by heating a source of calcium carbonate ($CaCO_3$) such as calcite (the main mineral in limestone) to high temperatures (for practical purposes, about 950°C or more). The resulting *calcination* reaction to drive off carbon dioxide (CO_2) is simple: $CaCO_3 + heat \rightarrow CaO + CO_2\uparrow$. The carbonation reaction for lime is simply the slow reversal of this reaction by the lime's gradual absorption of atmospheric carbon dioxide. In lime mortars, the actual lime species present is hydrated or slaked lime (or *portlandite*) and is formed simply by the hydration reaction $CaO + H_2O \rightarrow Ca(OH)_2$. The carbonation reaction for hydrated lime is: $Ca(OH)_2 + CO_2 \rightarrow CaCO_3 + H_2O$. The ultimate strength of the lime mortar will depend on the completeness of the carbonation; the hydrated lime particles develop a shell of $CaCO_3$ which, to one degree or another, slows CO_2 diffusion to the residual lime particle cores and thus slows further carbonation. But when the carbonation process is complete (months to decades or more), lime mortars can be quite strong.

It is important to note that binders that rely on carbonation to harden and gain strength are not hydraulic cements. Ordinary lime, therefore, is not a hydraulic cement, but hydraulic lime is, although it may also develop some strength through carbonation.

Hydration

A simple example of cementation by hydration is that of gypsum or plaster mortars. The common mineral gypsum ($CaSO_4 \cdot 2H_2O$), when heated to about 150°C, partially dehydrates to the hemihydrate phase ($CaSO_4 \cdot ½H_2O$), called *plaster of Paris* or simply plaster. If either gypsum or hemihydrate is heated to about 190°–200°C, all the structural water is lost and "soluble" *anhydrite* ($CaSO_4$) is formed. In the presence of water, soluble anhydrite fairly readily rehydrates to hemihydrate. Hemihydrate very quickly rehydrates to gypsum. However, if anhydrite is formed at temperatures above about 200°C, its ability to rehydrate slowly diminishes and, at about 600°C, is lost entirely. The rehydration of hemihydrate to gypsum is the binding reaction in plaster mortars. In addition, it is the defining reaction in the manufacture of gypsum wallboard.

Most hydraulic cements rely on far more complex hydration reactions to set and develop strength than those of plaster. The hydration reactions of portland and related cements will be discussed below, following an introduction to the chemistry and mineralogy of portland cement.

Chemical composition of portland cement and clinker

Modern straight portland cement is a very finely ground mix of portland cement clinker and a small amount (typically 3%–7%) of gypsum and/or anhydrite. The chemical composition and mineralogy of the portland cement itself can be described fairly simply, but first the shorthand notation used in cement chemistry needs to be introduced, as it simplifies the presentation of hydration and other reactions and is commonly encountered in the cement literature.

Cement chemistry is generally denoted in simple stoichiometric shorthand terms for at least the major constituent oxides. This notation can be initially confusing to readers familiar only with standard forms of chemical notation (geologists, however, use an almost identical shorthand for AFM diagrams and similar compositional representations). Consider, for example, a non-cement example such as the familiar feldspar mineral *anorthite*. The formula for anorthite is generally shown in the geological literature using standard chemical notation: $CaAl_2Si_2O_8$. This can be recast in terms of oxide groupings: $(CaO)(Al_2O_3)(SiO_2)_2$ for mass balance computational purposes, or in such a way as to show structural or crystal chemistry relationships. Cement shorthand merely abbreviates the oxides; in this shorthand, anorthite's formula becomes CAS_2.

The shorthand notation for the major oxides in the cement literature is given in table 1 below, which also shows the chemical composition of a typical modern portland cement and its clinker. For clinker, the oxide compositions would generally not vary from the rough averages shown in table 1 by more than 2%–4% absolute (i.e., relative to 100% total oxides). Likewise, the oxide composition of portland cement would vary slightly depending on its actual gypsum fraction (5% is shown but the range is generally 3%–7%), or if anhydrite substitutes for any of the gypsum, or if any other additives are present.

Table 1. Typical chemical composition of clinker and portland cement

Oxide formula	Shorthand notation	Percentage by mass in clinker [1]	Percentage by mass in cement [1,2]
CaO	C	65	63.4
SiO_2	S	22	20.9
Al_2O_3	A	6	5.7
Fe_2O_3	F	3	2.9
MgO	M	2	1.9
$K_2O + Na_2O$	K + N	0.6	0.6
Other (incl. SO_3^-)	...(... $\bar{S}$)	1.4	3.6
H_2O	H	"nil"	1
	Total:	100	100

[1]Values shown are representative to only 2 significant figures.

[2]Based on clinker shown plus 5% addition of gypsum ($CaSO_4 \cdot 2H_2O$).

Because the shorthand just conveys the oxide chemistry, a given shorthand formula will apply to all minerals or compounds having the same formula. Accordingly, many cement chemists confine the use of shorthand notation to the common or major cement minerals or chemical phases, and may use conventional chemical notation for less common cement minerals or compounds or for others (such as the mineral anorthite) not generally associated with cement. So a mix of notations may be encountered in a single equation. Cement shorthand is particularly convenient for denoting stoichiometric balances.

Mineralogy of portland cement and its clinker

The major oxides in clinker are combined essentially into just four cement or clinker minerals, denoted in shorthand: C_3S; C_2S; C_3A; and C_4AF. It is important to note that these mineral formulas are not completely fixed; they are more or less averages and ignore the fact that, in real clinker, the minerals are commonly somewhat impure. For example, C_4AF is actually the mean value of a solid solution with end members C_6A_2F and C_6AF_2. The ratios among these four minerals (and gypsum) in typical modern portland cements, and major functions of the minerals, are shown in table 2.

Table 2. Typical mineralogical composition of modern portland cement

Chemical formula	Oxide formula	Shorthand notation	Description	Typical percentage	Mineral function (see notes)
Ca_3SiO_5	$(CaO)_3SiO_2$	C_3S	Tricalcium silicate ('alite')	50–70	(1)
Ca_2SiO_4	$(CaO)_2SiO_2$	C_2S	Dicalcium silicate ('belite')	10–30	(2)
$Ca_3Al_2O_6$	$(CaO)_3Al_2O_3$	C_3A	Tricalcium aluminate	3–13	(3)
$Ca_4Al_2Fe_2O_{10}$	$(CaO)_4Al_2O_3Fe_2O_3$	C_4AF	Tetracalcium aluminoferrite	5–15	(4)
$CaSO_4{\cdot}2H_2O$	$(CaO)(SO_3)(H_2O)_2$	$C\bar{S}H_2$	Calcium sulfate dihydrate (gypsum)	3–7	(5)

(1) Hydrates quickly and imparts early strength and set.

(2) Hydrates slowly and imparts long-term strength.

(3) Hydrates almost instantaneously and very exothermically. Contributes to early strength and set.

(4) Hydrates quickly. Acts as a flux in clinker manufacture. Imparts gray color.

(5) Interground with clinker to make portland cement. Can substitute anhydrite ($C\bar{S}$). Controls early set.

Although not shown in table 2, the typical "free" or uncombined lime ($C\bar{S}$) in portland cement or clinker is generally within the range of 0.2%–2.0% (a very low value is sought). As noted in the table 2 footnotes, the minerals in clinker have different functions relating either to the manufacturing process or the final properties of the cement. Thus, for example, the primary function of the "ferrite" mineral (C_4AF) is to lower the temperature required in the kiln to form, in particular, the C_3S mineral, rather than impart some desired property to the cement. In contrast, the proportion of C_3S determines the degree of early strength development of the cement. Accordingly, it is no surprise that the mineralogical ratios differ for the different functional types (I–V) of portland cement defined earlier Table 3 shows typical (not extreme) mineralogical ranges among these cement types.

Table 3. Typical range in mineral proportions in modern portland cements

ASTM C-150	Clinker Mineral (%)*				Properties of cement
Cement Type	C_3S	C_2S	C_3A	C_4AF	
I	50–65	10–30	6–14	7–10	General purpose
II	45–65	7–30	2–8	10–12	Moderate heat of hydration, moderate sulfate resistance
III	55–65	5–25	5–12	5–12	High early strength**
IV	35–45	28–35	3–4	11–18	Low heat of hydration
V	40–65	15–30	1–5	10–17	High sulfate resistance

*Range of minerals is empirical and approximate rather than definitional.

**High early strength is typically achieved by finer grinding of Type I cement.

During the 20th century the mineral ratios in portland cements have been changed in response, mainly, to a growing demand for faster development of strength and greater strength overall. The principal change has been a gradual increase in the ratio of C_3S to C_2S. Typical C_3S to C_2S ratios in the first quarter of the 20th century were more like the reverse of that shown in table 2, and the ratios have continued to evolve in recent years. Table 4 shows the evolution of (empirical) average mineralogical ratios between those made in the 1960–80s (approximately), and those made more recently, as reported by the PCA.

Table 4. Evolution of typical average mineral ratios in modern portland cements

ASTM C-150	Clinker Average Mineral Ratios (%) and Period Ranges[1,2]							
	C_3S		C_2S		C_3A		C_4AF	
Cement Type	Older	Newer	Older	Newer	Older	Newer	Older	Newer
I	55	54	19	18	10	10	7	8
II	51	55	24	19	6	6	11	11
III	56	55	19	17	10	9	7	8
IV[3]	28	42	49	32	4	4	12	15
V	38	54	43	22	4	4	9	13
White[4]	33	63	46	18	14	10	2	1

[1]Except where indicated as white, data are for gray portland cement. Data are empirical averages and do not sum horizontally to 100%.

[2]"Older" is the period of about 1960–80s; data are from Kosmatka and Panarese (1988); "Newer" refers to about 1990 onwards; data are from Kosmatka and others (2002).

[3]Essentially no Type IV is currently made; it has been replaced by IP cement incorporating fly ash.

[4]All forms of white portland cement (i.e., Types I–V).

Although very small changes over time in the averages shown in table 4 (e.g., C_3S in Types I and III) are probably of no statistical significance for these empirical data, some shifts are noteworthy. The average C_3S content of Type IV (rarely made today) and Type V portland cement has been increased significantly, and C_2S decreased; in the case of Type V possibly reflecting the need for faster hydrating sulfate resistant concrete in areas of rapid population growth in the American Southwest (e.g., Arizona, southern California, Nevada). For some portland cement types, not only has the average C_3S content increased (table 4), but the upper end of its proportionality range (table 3) has increased (to about 65%, from about 55% in older Type I cements and from about 50% for Type II). At the upper end of the range, Types I and II cements now can have C_3S contents that used to be more typical of Type III cements, and reflect the fact that many modern Type III cements are merely more finely ground versions of Type I cement. The overall iron content (reflected by C_4AF) has increased modestly, possibly reflecting an increased need for its fluxing role (to increase formation of C_3S) in clinker manufacture, as will be discussed in the process mineralogy section below.

The largest compositional shift is seen for white portland cements, for which the C_3S contents have been more or less doubled, at the expense of the C_2S contents. As will be discussed later, this has interesting process chemistry implications. White cements continue to have extremely low C_4AF contents to avoid the coloring effects of iron.

Hydration of portland cement

Portland cement hydration reactions are complex and not completely understood. Part of the problem is that hydration (hydrated mineral) shells form around the cement mineral particles. The shells shield the remaining cores from easy observation, slow the hydration of the as-yet-unreacted or partly reacted cores, and affect the actual hydration reaction stoichiometries. Nonetheless, it is possible to note a few general "net" equations that are representative of the larger family of reactions that are likely taking place. Much of the following discussion is based on the summary article by Young (1985).

The important strength-developing hydration reactions are those of C_3S and C_2S. Typical hydration reactions (in shorthand notation—see table 1) would be:

for C_3S: $2C_3S + 6H$ (water) $\rightarrow$ $C_3S_2H_3$ ("tobermorite" gel) + 3CH (hydrated lime)

for C_2S: $2C_2S + 4H \rightarrow C_3S_2H_3 + CH$

The formula shown for *tobermorite* is only approximate, and some texts denote it as $C_3S_2H_4$, in which case both hydration equations above would need an additional water (H) to start with. Actually, instead of just tobermorite, a whole family of similar *calcium silicate hydrates* (C-S-H) may be formed, and C-S-H is the preferred general term for these compounds. It is the C-S-H colloid or gel that is the actual binder in hydrated portland cement. The ultimate strength of the hardened cement paste will depend not only on the original total content of C_2S and C_3S but also on the completeness of their hydration.

Although the net hydration reactions for both C_3S and C_2S are similar, the reaction for C_3S is relatively fast, and C-S-H from it is responsible for virtually all of the early (e.g., within 3 days of curing) strength development of the cement. Typically, about 60% (by mass) of the C_3S has hydrated to C-S-H within the first 5 days of curing and about 70% has hydrated within about 10 days. Because of the formation of protective hydration shells, the remaining unreacted C_3S particle cores hydrate much more slowly, reaching about 75% hydration after 20 days of curing, about 80% hydration after 28 days (a standard measurement interval), and 85% after 60 days. Beyond 60 days, the rate of C_3S hydration slows dramatically and the incremental hydration and strength contribution is of little practical importance.

In contrast, the hydration of C_2S is relatively slow, with only about 20% hydration after 5 days of curing, about 30% after 10 days, 35% after 20 days, about 40% after 28 days, and only about 55% at 60 days. Its rate of hydration slows further after 60 days. Accordingly, the C-S-H derived from the hydration of C_2S, while making little contribution to the early strength of the concrete, contributes a significant proportion of the strength gain after the first week or so of curing.

The presence of large amounts of C_3S can be considered a defining characteristic of modern portland cement. In contrast, the very earliest portland cements probably contained little, if any, C_3S, and those of the latter half of the 19th century probably no more than about 15%–20% C_3S.

The 19th century portland cements instead relied primarily on the hydration of C_2S. It is unclear if the original (1824) portland cement even contained much C_2S; the patent description describes only a calcination process. If the patent was indeed a full description of the original process, then the resulting hydraulic species would have been primarily clay pozzolans (as with hydraulic limes and natural cements) although, as described in the process chemistry discussion below, C_2S formation could have been possible given the poor temperature control characteristic of the lime kilns of the day.

As shown above, the C_3S and C_2S hydration reactions release free lime. Based on the typical clinker mineral proportions and their hydration reactions, it can be shown that the net free lime release during clinker hydration, overall, is roughly 25%–33% of the original CaO content of the clinker. Free lime in hardened concrete is not particularly desirable because it increases the chemical reactivity of the surface (including along cracks) and can leach out in an unsightly fashion. On the other hand, by maintaining a high pH in the aqueous phase, free lime can help protect steel reinforcing bars (rebar) in the concrete from corrosion should water and oxygen reach the rebar via cracks. The lime is also available to react with any pozzolans that may have been added to the cement or concrete mix.

Alkalis, particularly sodium (Na_2O, or N in shorthand), can combine with C-S-H to form complex hydrates (e.g., C-S-N-H) that are unstable and prone to swelling compared with regular C-S-H. Alkalis can also react with forms (amorphous, opaline, or very fine grained crystalline) of silica in some aggregates used in concrete, forming highly hygroscopic alkali-silicate hydrates (e.g., N-S-H), and generally weakening the bond between the aggregates and the cement paste and forming higher-volume phases. These and similar reactions, collectively called *alkali-silicate reactions* (ASR) or *alkali-aggregate* reactions, can cause cracking of hardened concrete. The cracks not only weaken the concrete but render its interior susceptible to additional alkali or other chemical attack, and to freeze-thaw damage in cold weather regions. Approaches to controlling ASR reactions include selecting portland cements having lower alkali contents (e.g., ASTM C-150 provides for a low-alkali cement designation if the cement has a total alkali content [defined as $Na_2O + 0.658\ K_2O$] of 0.60% or less), testing of aggregates for reactivity, and the incorporation of pozzolans into the cement paste. Pozzolans contain active silica which "sacrificially" combines with the alkalis in the paste (thus leaving less alkalis available to react with the aggregates), and significantly reduce the hardened concrete's porosity. A review of ASR is provided by Leming (1996).

The other two clinker minerals, C_3A and C_4AF, have complex hydration reaction paths that are similar to each other, but those of C_3A are more important because they are much more rapid and exothermic. Having C_3A in the cement primarily enhances initial set and speeds, via release of heat, the hydration of C_3S (the presence of C_3A also has benefits to the cement manufacturing process because it speeds the overall formation of the clinker). In the absence of significant sulfate, C_3A very rapidly—almost instantaneously—forms C_3A-hydrates, many of which are unstable and may subsequently convert to other forms. One of the many possible sequential hydration reactions is:

$$2C_3A + 21H \rightarrow C_2AH_8 + C_4AH_{13} \rightarrow 2\ C_3AH_6 + 9H.$$

A minor, but lime-consuming, reaction is:

$$C_3A + 12\ H + CH \rightarrow C_4AH_{13}$$

The hydration of C_3A in the absence of sulfate can be so rapid as to cause the undesirable condition known as *flash set*. This is controlled through the addition of sulfate, usually as gypsum and/or anhydrite. Plaster is only rarely used because it hydrates so quickly back to gypsum that its use is rather counterproductive. The use of plaster also increases initial water consumption. A typical hydration reaction of C_3A in the presence of rate-controlling sulfate (here shown as gypsum) would be:

$$C_3A + 3\ C\bar{S}H_2 + 26\ H \rightarrow C_6A\bar{S}_3H_{32}\ (\text{"ettringite"}).$$

Flash set is controlled because ettringite forms a shell around the C_3A particles, which slows water diffusion to, and hence the hydration of, the residual C_3A cores. Ettringite is stable only in the presence of excess sulfate. If this condition is not met (i.e., not enough gypsum present, or in the evolving conditions at the ettringite-residual C_3A core interface), then ettringite reacts with C_3A to form a monosulfate phase:

$$C_6A\bar{S}_3H_{32} + 2C_3A + 4H \rightarrow 3\ C_4A\bar{S}H_{12}\ (\text{"monosulfate"}).$$

Alternatively, C_3A hydration under low sulfate conditions can be expressed by:

$$C_3A + 10H + C\bar{S}H_2 \rightarrow C_4A\bar{S}H_{12}$$

An important property of the monosulfate phase is that, in the presence of sulfate ions, it can re-form ettringite, such as by the reaction:

$$C_4A\bar{S}H_{12} + 2C\bar{S}H_2 + 16H \rightarrow C_6A\bar{S}_3H_{32}$$

Ettringite has a molar volume of about 735 cubic centimeters (cm^3) per mole and monosolfate about 313 cm^3 per mole (Bentz, 1997). Because of this volume difference, re-formation of ettringite from monosulfate can cause expansion of the concrete. This is not much of an issue while the cement paste has yet to harden, but if ettringite re-forms in hardened concrete, the result can be cracking or spalling of the concrete. This process is known as *sulfate attack* and is prevalent in regions (commonly desert areas) having sulfate-rich groundwater, or it can occur if too much gypsum is present in the cement. Thus the proportion of gypsum in the cement is important. Where sulfate attack from groundwater is likely, concretes are better made using a sulfate resistant portland cement, such as Type II, or better yet, Type V; both have low concentrations of C_3A (table 3). A Type IV cement would also show resistance to sulfate attack, but would be less desirable for most applications because its relatively low C_3S content would cause it to develop strength relatively slowly. Alternatively, sulfate resistance is improved by using a blended cement, as addition of pozzolans lowers the overall C_3A content of the cement paste and reduces the porosity (hence sulfate entry potential) of the concrete.

The ferrite mineral C_4AF does not play a critical role in cement hydration. The chief value of ferrite is in its effects on kiln reactions to form C_3S (see process mineralogy discussion below). The hydration of C_4AF is broadly similar to that of C_3A, although the reactions tend to be slower and much less exothermic. The reaction stoichiometries will vary given the fact that, as noted earlier, C_4AF is merely a mean composition for the ferrite solid solution having end members C_6A_2F and C_6AF_2. In the absence of sulfate, the F partially substitutes for some of the A (partial substitution denoted as A,F) in the analogous C_3A hydration products, as shown in the reaction:

$$2C_4AF + 32H \rightarrow 2C_2(A,F)H_8 + C_4(A,F)H_{13} + (A,F)H_3 \quad \text{where total AF = total (A,F)}$$

In the presence of hydrated lime (from C_3S and C_2S hydration), however, the formation of $(A,F)H_3$ is suppressed and a stable AF-hexahydrate ($C_4(A,F)H_6$) is formed that is analogous to C_3AH_6, with a possible net reaction being:

$$C_2(A,F)H_8 + C_4(A,F)H_{13} + (A,F)H_3 + 6CH \rightarrow 3C_4(A,F)H_6 + 12H$$

Even more so than with C_3A hydration, the hydration of C_4AF is slowed in the presence of sulfate by the formation of an ettringite-like phase, with a possible reaction being:

$$3C_4AF + 12C\bar{S}H_2 + 110\,H \rightarrow 4C_6(A,F)\bar{S}_3H_{32} + 2(A,F)H_3$$

And, analogous to C_3A, if the sulfate concentration is insufficient, the "AF" ettringite becomes unstable and forms an "AF" monsosulfate phase:

$$3C_4AF + 2C_6(A,F)\bar{S}_3H_{32} + 14H \rightarrow 6C_4(A,F)\bar{S}H_{12} + 2(A,F)H_3$$

Clinker manufacturing process

Portland cement manufacture involves two main steps: manufacture of clinker followed by fine grinding of the clinker with gypsum and sometimes other materials like pozzolans to make the finished cement product. Integrated plants perform both steps, whereas *grinding plants* make cement by grinding clinker that was made elsewhere. Clinker manufacture itself involves two main steps. First, appropriate raw materials must be quarried, crushed, and then proportioned and blended into a kiln feed called the raw mix or raw meal. Second, the raw mix must be converted into the clinker minerals. This is a thermochemical conversion and because it involves direct flame interaction, the overall procedure is referred to as pyroprocessing. Figure 3 is a generalized flow sheet of cement manufacture.

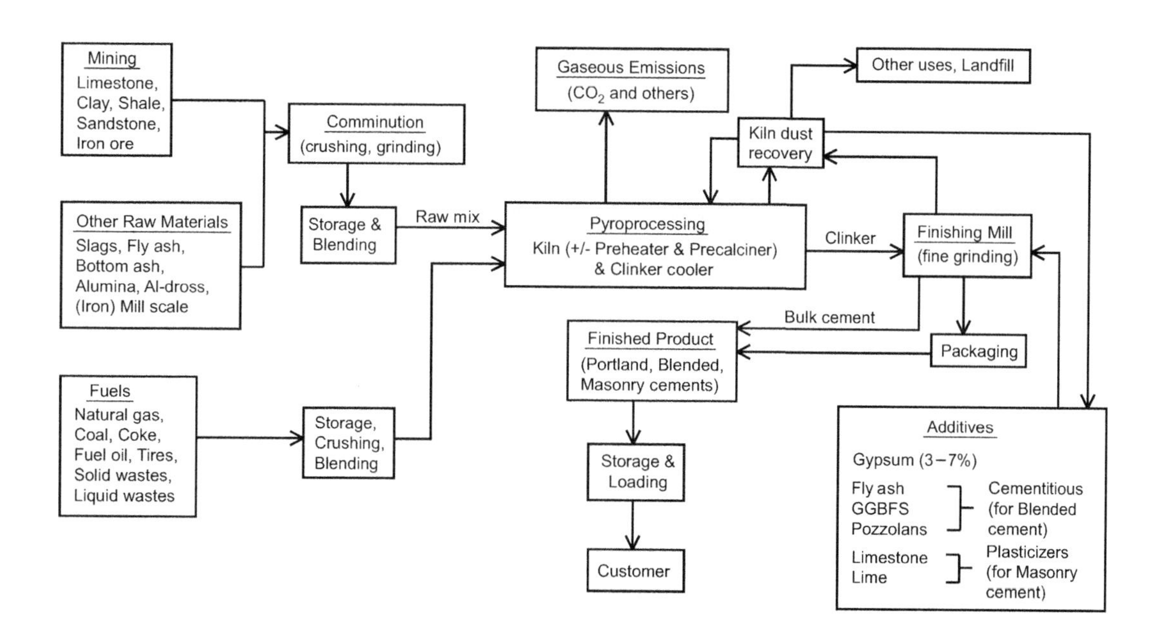

Figure 3. Simplified flow sheet of cement manufacture.

Raw materials for clinker

The nonfuel raw materials for cement must yield the oxides required for clinker in the approximate proportions noted in table 1. Individual raw materials generally provide more than one oxide. Primary raw materials are those that are always used in relatively large quantities by a specific plant. To correct for minor deficiencies in one or more oxides in the primary raw materials, accessory or "sweetener" materials, generally of high purity, may be added. Certain oxides can also be partly supplied by the fuels; for example, the ash in coal supplies a portion of the silica requirements for clinker, and the steel belts in waste tires (a supplementary fuel) supply iron oxide. When a plant evaluates its raw materials, consideration is given not only to each material's potential contribution of major oxides (CaO, SiO_2, Al_2O_3, Fe_2O_3), but also to the content, if any, of undesirable trace elements (e.g., excess MgO, alkalis, toxic species). Consideration also is given to the ease of prepping the material (usually ease of crushing), and the material's "burnability," that is, the heat energy required to break down the material to activate or make accessible its component oxides.

The major oxide requirement for clinker is CaO (table 1) and large amounts of "accessible" CaO-bearing material are thus required for clinker manufacture. In practical terms this means limestone or similar material (e.g., marble). However, because of the need for other oxides, a high purity (>95% $CaCO_3$) limestone is not required, although it is usable. Basically, a cement plant will first analyze its limestone, and then, to the degree that the limestone lacks the other requisite oxides, they will be added from other sources. Argillaceous limestones that supply a more or less complete oxide package are sometimes called *cement rock*; such material is not particularly common but was the key raw material and location determinant for natural cement plants (most of which eventually converted to portland cement production). Preparation of the raw mix for the kiln is a process of constant adjustment based on the frequent chemical testing of the raw materials, the raw mix itself, and the clinker. Generally, most, but not all, of the raw materials are mined adjacent to or within a few miles of the cement plant; the low unit value of most of the raw materials means that few can support the cost of long-distance transport to the plant. Long distance sourcing of major raw materials (i.e., limestone) is generally only economical if waterborne transport of the materials is available.

The USGS Minerals Yearbook chapters on cement show current consumption of raw materials split out by major oxide contribution and (in recent editions) whether the material is used to make clinker or is added to the clinker in the finish (grinding) mill to make finished cement. Table 5 shows an average annual consumption of raw materials for recent years by the U.S. cement industry and also shows some of the variety of materials that can be used. Some accessory raw materials (not necessarily split out in table 5) are waste products of other industries. Examples of these are spent potliners and catalysts from aluminum smelters, lime kiln dust and sludges, mill scale from steel mills, and a variety of ashes and slags from power plants and smelters.

Individual plants will have similar oxide ratios among materials, but may differ significantly in the specific raw materials consumed. Typically, about 1.7 t of nonfuel raw materials (about 1.5 t of which will be limestone or similar calcareous rocks) are required to make 1 t of clinker (table 5). The main reasons for this large ratio are the substantial mass loss (of CO_2) from the calcination of limestone (see environmental discussion below), and the generation of large amounts of cement kiln dust (CKD), which, although captured, are not necessarily returned to the kiln as part of the feed stream.

Table 5. Nonfuel raw materials for clinker and portland cement manufacture in the United States (average for 1995–2000)

(Million metric tons per year)

Major oxide	Materials	Amount
CaO	Limestone, cement rock, marl, marble, CKD, other:	111.0
SiO_2	Sand, sandstone, ferrous slags, fly ash, other ash:	6.0
Al_2O_3	Clay, shale, bauxite, other:	9.1
Fe_2O_3	Iron ore, millscale, other:	1.4
Other:	Gypsum, anhydrite, other:	4.5
	Total raw materials:	132.0
	Raw material content of imported clinker:[1]	6.1
	Total equivalent raw materials:	138.1
	Clinker production:	75.0
	Total raw materials per ton of clinker:[2]	1.7
	Cement production:[3]	84.4
	Total raw materials per ton of cement:	1.6

[1]Calculated as tons of clinker × 1.7.

[2]Excludes gypsum and anhydrite.

[3]Includes cement made from imported clinker.

Pyroprocessing

The heart of the cement manufacturing process is the kiln line, where raw materials undergo pyroprocessing to make clinker (figure 3). Almost all of the raw materials and most of the total energy consumed in cement manufacture are consumed during pyroprocessing.

In the following discussions, the technology of the kiln or pyroprocessing line and the process chemistry of clinker manufacture will be briefly described.

Kiln technology

Early natural and portland cements were made in small vertical chimney-type kilns operating on a batch-process basis. These were slow, labor-intensive, and fuel inefficient, and the quality of the cement was difficult to control. As demand for cement grew, both in terms of quantity and quality, efforts were made to improve the manufacturing technology. The invention of the rotary kiln (1873), its improvement (1885), and significant enlargement (1902) allowed for superior mixing of raw materials, better control of temperature and other processing conditions, and continuous throughput of materials. Further design refinements to rotary kiln lines have been made throughout the 20th century. In most countries today, rotary kilns account for virtually all of the cement made (100% of U.S. production). The most significant exception is China, where a majority of production is still from small vertical shaft kilns (VSK), although the VSK fraction is declining rapidly; Lan (1998) provides a brief review of Chinese VSK technology.

Although VSK are improvements over the old, chimney-type kilns in that some VSK allow for continuous processing, they are considered to be less energy efficient than the rotary kilns, and VSK clinker (and hence cement) is generally considered to be of lower quality. Still, well-operated VSK technology can be appropriate to supply very small (village-scale) markets. The broad thermochemical functions of VSK do not differ significantly from rotary kilns, and the remaining discussion pertaining to rotary kilns will also apply broadly to VSK.

Rotary kilns consist of enormous, gently inclined and slowly rotating steel tubes lined with refractory brick, and are said to be the largest pieces of moving manufacturing equipment in existence. In all rotary kilns, the finely ground raw mix is fed into the upper, "cool" end of the kiln and is gradually heated and transformed into semifused clinker nodules as the material progresses down the kiln. In its simplest form, heating is from a white-hot jet of flame projecting up-kiln from a burner tube at the lower end. The clinker emerges from the lower, discharge end, and falls into a clinker cooler, where the clinker temperature is reduced to a safe handling level (about 100°C) for grinding into cement.

The oldest and largest (dimension) kilns in operation today use *wet* technology; these kilns typically range from about 120–185 m in length (a few are much longer) and about 4.5–7.0 m in internal diameter. Wet kilns derive their name from the fact that they are fed their raw materials in an aqueous slurry; such slurries were an early but effective solution to the problem of achieving a thorough mix of the crushed raw materials. *Dry* (technology) kilns take a dry powder feed and are of three main types. In order of increasing technological advancement, the main dry kiln line types are *long dry* kilns, *preheater* dry kilns, and *preheater-precalciner* dry kilns. Dimensionally, long dry kilns are about 90–120 m or more in length; preheater dry kilns can be the same length but most are somewhat shorter; and preheater-precalciner kilns—the most modern technology—are typically 45–75 m long. Dry kilns typically have internal diameters in the range of 3.5–4.5 m. The main reason for dry kilns having shorter tubes than wet kilns is that dry kiln tubes perform fewer thermochemical functions, as discussed below. Among kilns of the same technology, specific tube length and diameter may reflect throughput capacity (bigger tubes having larger capacity), and whether (any) preheaters and/or precalciners were installed as retrofit upgrades (original kiln tube length being retained or mostly so) or as an integral part of an entirely new kiln line (generally using a short, but large diameter tube). Although the largest (output) capacity kilns today are preheater-precalciner kilns, kiln size and technology, themselves, are not good guides to output capacity.

It must be emphasized that the overall pyroprocessing or thermochemical functions casually ascribed to the kiln generally refer to those of the complete kiln *line*, which consists of the kiln tube itself plus any preheaters and precalciners ahead of the kiln tube, plus the clinker cooler at the discharge end of the kiln tube. All kiln *lines* perform identical pyroprocessing functions, but those performed in the kiln *tube* itself depend on the type of kiln technology.

Pyroprocessing functions of kiln lines

Wet kiln tubes perform the full range of functions of a kiln line, and the basic clinker manufacturing process is most easily described using a wet kiln example. The four major clinker line functions are: drying, preheating, calcining, and sintering (a.k.a. burning or clinkering). Each of these functions is performed sequentially in specific and progressively hotter parts of the wet kiln, described here as *functional zones* (note, the boundaries of adjacent functional zones overlap to some degree) and the temperature ranges shown are for the main functions of the zone (table 6).

Table 6. Sequential functional zones in a wet kiln tube

(Listed in order from upper end to lower end of the kiln)

Functional zone	Approximate range of temperature (°C)	Activity
Drying:[1]	<100 – 200	Drive off water (wet slurry becomes dry powder).
Preheating:[2]	200 – 550	Drive off structurally bound water (from clays etc.).
Calcination:[3]	750 – 1000	Drive off carbon dioxide from carbonate minerals.
Sintering:[4]	1200 – 1450	Form clinker minerals; form clinker nodules.
Cooling:[5]	<1450 – 1300	Cool slightly (no longer in path of flame).

[1]This function is essentially obviated in dry kiln lines.

[2]This function is mostly performed by a separate preheater in a preheater dry kiln line.

[3]This function is mostly performed in a separate precalciner in a preheater-precalciner dry line.

[4]Also known as "burning" or "clinkering."

[5]This cooling happens before the clinker enters the clinker cooler apparatus.

Again, the boundaries of the functional zones are approximate and do not involve complete cutoffs of activity. The drying and preheating zones together occupy roughly the upper one-third of the wet kiln tube, the calcination zone roughly the middle third, and the sintering zone most of the remainder. In the sintering zone, there is not only the thermochemical formation of the individual clinker minerals but also of a liquid phase (i.e., there is partial melting) such that the minerals form an intimate physical mix within semifused nodules or pellets of clinker about 1–10 centimeters in diameter. After their formation, the clinker nodules move into the kiln's short cooling zone—a slightly cooler region beneath the burner tube. When the now slightly cooled clinker reaches the discharge end of the kiln, it drops out of the kiln tube into the clinker cooler apparatus, where the real cooling occurs and the clinker temperature is reduced to about 100°C. The heat from the clinker is recaptured by recycling the hot air from the clinker cooler to the kiln line to be used as combustion air.

The technological progression over the years from wet kilns to preheater-precalciner dry kilns has been accompanied by a successive shortening of the kiln tubes. In succession, the wet kiln's drying zone is obviated in the long dry kiln; the long dry kiln's preheating zone is replaced by a tower-mounted cyclone preheater apparatus in a preheater kiln; and, finally, most of the calcination function is transferred to a (pre)calciner apparatus attached to the preheater in a preheater-precalciner kiln line. Thus, in a modern preheater-precalciner kiln line, the kiln tube itself basically only performs the sintering function, and can be as little as one-quarter the length of a wet kiln tube of similar output capacity.

Residence time for material in kilns ranges from about 2 hours or more in wet kilns to as little as 20 minutes in preheater-precalciner kilns. The residence time in a preheater-precalciner tower itself is only about 20 to 90 seconds. Used alone, preheaters can be heated with hot kiln exhaust gases, but precalciners require their own, separate heating source, and commonly are designed to burn about 60% of the total kiln line fuel supply. Hot air from the clinker cooler apparatus is used as combustion air in the kiln to save energy. Additional details on the form, function, and operation of rotary kilns can be found in Alsop and others (2005) and, in more detail, Duda (1985) and Bhatty and others (2004).

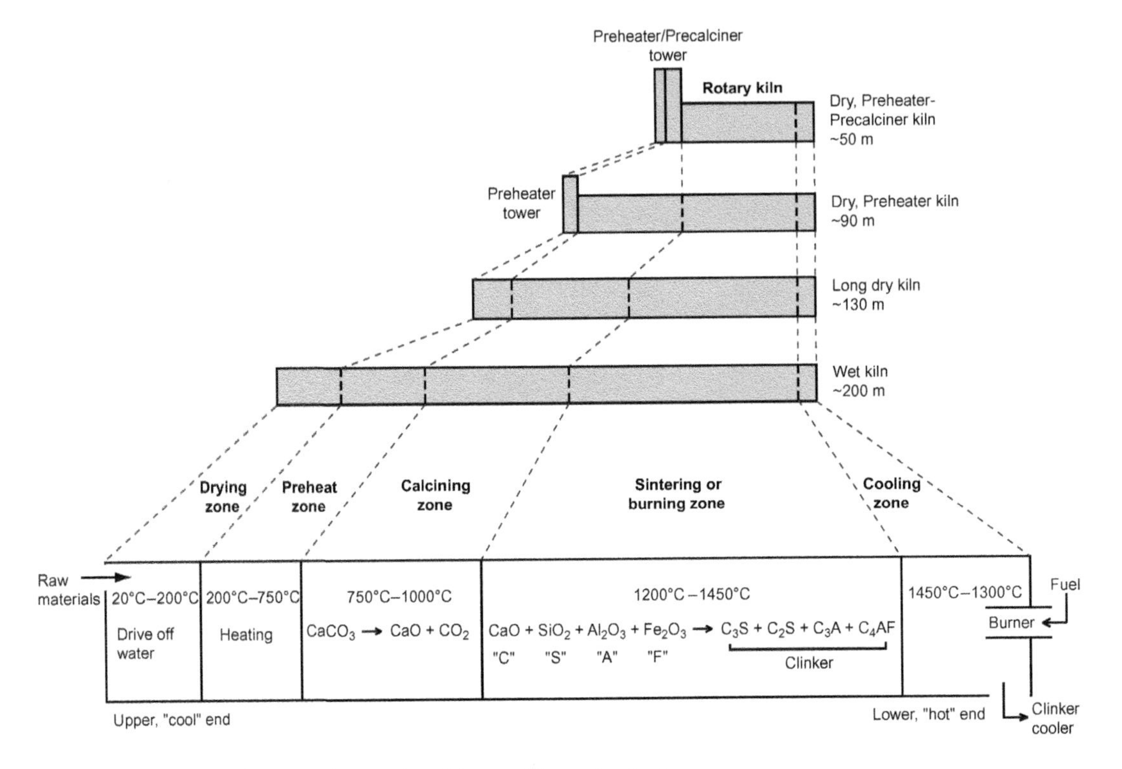

Figure 4. Diagram of functional zones for different kiln technologies.

Process chemistry

Although all of the functions of a kiln line (table 6) in volve a great deal of heat energy, the thermochemical reactions most important to the formation of clinker occur in the calcination and sintering zones. As noted above, there is no sharp transition between these or the other functional zones in the kiln; the terminology refers to the critical reactions occurring dominantly in the particular zone.

Calcination zone reactions

The main oxide in clinker and portland cement is CaO (table 1), and a source of CaO is sought that is abundant, inexpensive, and easily processed to make this oxide available for clinker mineral formation. Limestone and similar rocks are the main raw material sources of CaO (table 5); cement plants are almost invariably located within a few miles of their limestone quarries. The main CaO-bearing mineral in limestone and related rocks is calcite ($CaCO_3$), and calcination simply strips the carbon dioxide from this mineral (or any other carbonate minerals present): $CaCO_3 + \text{heat} \rightarrow CaO + CO_2\uparrow$

If a dolomite [$Ca,Mg(CO_3)_2$] or magnesite ($MgCO_3$) phase is present, calcination via an almost identical equation will yield an MgO component in addition to CaO. This calcination will require somewhat lower temperatures than for calcite, but because the MgO component of cement is kept very small (table 1), the actual calcination temperature for the raw mix overall is not significantly affected. As will be discussed later, the calcination release of carbon dioxide is of great environmental concern.

Within much of the calcination zone temperature range, clay minerals in the raw mix break down into their component oxides or other reactive phases, and other silicate minerals in the mix begin to break down also. There is initial formation of C_2S by the net reaction $2C + S \rightarrow C_2S$. This process continues into the sintering zone.

Sintering zone reactions

A distinction can be made between reactions that straddle the calcination-sintering zone transition (about 900°C to about 1200°C or so), and true sintering (or clinkering) reactions occurring mostly above about 1300°C. The transitional or lower temperature reactions are mainly of two types: a continuation of reactions that began in the calcination zone that involve the thermal decomposition of noncarbonate (mainly silicate) raw materials into their component oxides or other reactive phases, and those that form some of the lower temperature clinker minerals or their immediate precursors from those oxides or phases. The true sintering reactions are mainly those that require high temperatures and which center on the formation of C_3S. For the purpose of the following discussion, the "sintering" reactions will be restricted to those that combine (even at lower temperatures) component oxides into the clinker minerals. These reactions are many and complex, but in terms of an overall oxide balance, the approximate net reaction to form clinker (mineralogical ratios comparable to those in table 3) would be:

$$29C + 8S + 2A + F \rightarrow 6C_3S + 2C_2S + C_3A + C_4AF$$

All of the classic books on cement chemistry have chapters that discuss in detail the sintering reactions in the kiln; the following brief discussion is based on Bogue (1955), Welch (1964), Lea (1970), Taylor (1997), and, especially, the summary by Roy (1985). These references may be consulted for the relevant phase diagrams. It is important to note that the discussions in the chemistry books, as well as that here, tend to focus on pure chemical or mineral phases. In reality, the raw mix and clinker mineral compositions usually contain modest amounts of impurities, and these can affect reaction paths, rates, and temperatures.

Most of the mineral-forming reactions that occur below about 1200–1300°C are in the solid state and tend to be slow and relatively confined to individual raw material particles or to the contacts between adjacent particles. However, in the higher temperature range of the sintering zone there is significant formation of liquid (i.e., partial melting), and this allows oxide combination, and hence clinker mineral formation, to occur much more rapidly and completely. The rate of reaction is important because the clinker raw mix is continuously progressing through the kiln and the residence time in the highest-temperature zone is limited.

Which clinker minerals form, and at what temperatures, depend on the oxides present in the raw mix. The formation and stability of the four clinker minerals (C_3S, C_2S, C_3A, and C_4AF) can be described within the confines of a 4-component (quaternary) system covering the four major component oxides (C, S, A, F). However, it is best to first consider simpler phase systems and then build to the quaternary system, with the goal of understanding how large quantities of C_3S (the dominant mineral in portland cement) can be formed economically—that is, as quickly, and at as low a temperature, as possible.

For a simple binary mix of CaO and SiO_2 (i.e., a C-S system), most or all of the available silica will have been taken up from solid state reactions into C_2S (assuming sufficient CaO) by the time the temperature reaches 1200°C. Initial formation of C_2S begins around 700°C, but most forms in the range of 1100–1200°C. Solid-state binary system formation of the critical C_3S mineral (by the exothermic reaction $C_2S + C \rightarrow C_3S$) starts in the range of 1250–1400°C. However, the reaction is extremely slow at these temperatures and remains so even at 1500°C. Rapid formation of C_3S in the C-S binary system does not occur until a melt forms, and this does not happen below about 2050°C, which is well above the practical material temperatures achievable in a kiln. Fortunately for the portland cement industry, the presence of other oxides (Al_2O_3, Fe_2O_3) dramatically lowers the temperature range at which a melt forms.

For example, in the C-A-S ternary system (important because it accommodates about 90% of the clinker composition), C_3A is a critical phase that lowers the temperature needed to form a melt. This mineral and its aluminate precursors begin to form by solid state reactions at calcination zone temperatures. The three clinker minerals C_3S, C_2S, and C_3A are in stable coexistence with a melt phase at a temperature as low as 1455°C, although there will not be much melt, or much C_3S, at this temperature. Further, 1455°C is still a difficult material temperature to sustain or significantly exceed in a rotary kiln. Thus, as a practical matter, the amount of C_3S that can be made in a C-A-S ternary component system is limited unless higher temperatures can be achieved or fluxes are used.

The occurrence of Fe_2O_3 in clinker raw materials brings the reactions into the C-S-A-F quaternary system, which is highly advantageous to practical clinker production. The key ferrite (C_3AF) phase, which begins to form by solid state reactions at temperatures of about 800°C, acts as a flux to lower the temperature of melt formation. Within the C-A-F ternary subset of the C-S-A-F quaternary system, C_4AF coexists with C_3A plus liquid at the relatively modest temperature range of 1310–1389°C. Likewise, the minerals C_3S, C_2S, and C_3A are in equilibrium with a liquid phase at 1400°C, compared with 1455°C in the pure C-A-S ternary system. In the full C-S-A-F quaternary system, the four minerals C_3S-C_2S-C_3A-C_4AF plus liquid coexist at even lower temperatures. Although the temperature reduction is not all that large with just a few percent Fe_2O_3 in the system, at about 10% Fe_2O_3 the temperature drops significantly. The quaternary eutectic for C_3S-C_2S-C_3A-C_4AF plus liquid is 1338°C, with a modest upwards range of equilibrium point temperatures because of the fact that somewhat different clinker mineral stoichiometries may be present (c.f., the "pure" mineral formulae shown). In any case, the equilibrium temperatures for these four minerals and melt are now in a range readily attainable and sustainable in a kiln, but at the lowest end of this melt temperature range not much liquid is present and so formation of the critical C_3S mineral is still not very rapid. The rate of C_3S formation, and hence the amount of C_3S that will form within a given residence time in the sintering zone of the kiln, increases as the proportion of liquid is increased. For an "average" clinker C-S-A-F composition such as that in table 1, the amount of liquid at about 1340°C is about 20%, and it reaches about 30% (a satisfactory melt component for adequate rate of C_3S formation) at temperatures of 1400–1450°C. Consequently, 1400–1450°C is the approximate targeted maximum temperature range for the sintering zone in the kiln.

In actual clinker manufacture, other compounds that can act as a flux may be present in the raw mix. Among these, MgO and alkalis are the most likely to be naturally present in the raw materials, and fluorine is sometimes added in the form of fluorspar (CaF_2). However, the amount of these materials must be kept low to avoid certain subsequent problems with the operation of the kiln and/or the quality of the finished cement.

Interestingly, in the manufacture of white portland cement, the iron content must be kept very low to avoid iron's coloring effect, and so there forms no (or insufficient) C_4AF to act as flux. In effect, the conditions in a white cement kiln, in the absence of other fluxes, are essentially confined to the C-A-S ternary system. Formation of large percentages of C_3S in line with modern requirements for white cement (table 4) is difficult, and requires that the kilns be operated very hot (sintering temperatures of approximately 1500°C) and the residence times in the sintering zone be extended.

Cooling of clinker in the clinker cooler not only allows for safe subsequent handling, but also stops further changes of the clinker mineral phases and assemblages. The cooling of white clinker must be especially rapid (essentially via water quenching) to minimize any reversion of C_3S to C_2S and to prevent the oxidation of any (slight) iron present to ferric (Fe^{+3}) valence (ferric iron imparts more color than does ferrous iron).

Manufacture of finished cement from clinker

After clinker has been cooled to about 100°C, it is ready to be ground into finished cement in a grinding mill, more commonly referred to as a *finish mill*. At most integrated cement plants, the output or grinding capacity of the finish mill will be at least as large as the sum of the clinker capacity plus the additives to be interground in the finished product. That is, an integrated plant with a clinker capacity of 1 Mt/yr would be expected to have a grinding capacity of at least 1.05 Mt/yr if the usual product is a straight portland cement having a fairly typical 5% gypsum content, but the mill capacity could be significantly higher if the product line involves a lot of ground additives (i.e., to make blended and/or masonry cements). Likewise, finish mill capacity will be higher if the plant was designed to routinely grind supplementary clinker from an outside source. In a few cases, plants report seemingly excess grinding capacities, but include mills that are, in fact, being used to grind granulated blast furnace slag destined to be sold directly to concrete companies as a cement extender (SCM). A few plants have large excess clinker production capacities; these plants routinely transfer out or sell their surplus clinker. A few independent grinding facilities (*grinding plants*) rely on outside sources for all of their clinker.

Generally, separate grinding and/or blending finish mill lines will be maintained at a plant for each of its major product classes (finished portland cements, blended cements, masonry cements, ground slag). In the United States, about 95% of the masonry cement is made directly from a clinker + gypsum + additives feed, with the remainder being made from a (finished) portland cement + additives feed. Additives that commonly require grinding at the mill include gypsum, limestone, granulated blast furnace slag, and natural pozzolans. Additives that generally do not require significant grinding include fly ash, GGBFS, and silica fume, but the finish mill does provide intimate mixing of these with the portland cement base.

Portland and related cements are ground to an extreme fineness (much finer than cosmetic talcum powder). Cement particles average about 10 micrometers in diameter and approximately 85%–95% of cement particles are smaller than 45 micrometers in diameter. This fine particle size helps insure rapid and uniform hydration of the cement and enhances the ease with which it can be mixed into the concrete batch. The fineness of cement is determined using tests that measure the total surface area of a given unit mass of cement powder. Fineness of cement is usually expressed relative to the *Blaine* air-permeability test (ASTM C-204), either in units of m^2/kg or cm^2/g (the latter being more common), but sometimes denoted simply as Blaine, where the reader is expected to know the units based on the number itself. Thus, for example, one might see the fineness of a particular cement casually expressed as either 400 Blaine or 4,000 Blaine; these represent, and would be more rigorously expressed, as 400 m^2/kg (Blaine) or 4,000 cm^2/g (Blaine), respectively. Typically, portland cement in the United States is ground to 3,700–4,000 cm^2/g (Blaine), except that Type III (high early strength) cements are ground much finer (about 5,500 cm^2/g) to speed the hydration of C_3S. As shown in Kosmatka and others (2002, p. 42), white cement is also typically ground very fine (typically about 4,900 cm^2/g (Blaine), but some specifications are ground much finer still). In "older" white cements having low C_3S contents (table 4), this fine grinding helped increase the speed of hydration of the cement, allowing it to meet Type I performances. Modern formulations of white cement have very high C_3S contents and the fine grinding is done mainly to improve the whiteness and brightness of the product. Along with the high heat requirements for white clinker formation, this fine grinding increases the overall energy requirements and hence manufacturing cost of white cement and makes it necessary to sell this product at significantly higher prices (roughly double) than those for gray portland cements of similar structural performance.

Energy requirements for clinker and cement manufacture

Given the huge size of cement kilns, it takes a great deal of fuel (heat energy) to generate and sustain the very high temperatures inside them needed to make clinker and, in the case of wet kilns, to evaporate the water from the raw feed slurry. Likewise, it takes a lot of electricity to crush and grind the raw materials into kiln feed and the clinker into finished cement, as well as to operate the kiln line.

Data on the type and quantity of fuels and electricity consumed by the U.S. cement industry are collected and published annually (Minerals Yearbook) by the USGS. The data are grouped separately by plants operating wet kilns, dry kilns, and those operating both technologies. Data on the heat content of the fuels consumed are also collected by the USGS but have not been routinely published. A summary of recent data for the U.S. industry overall is given in table 7 below, and a compilation for the period 1950–2000 can be found in van Oss and Padovani (2002).

Table 7. Summary of fuel and electricity consumption by the U.S. cement industry in 2000[1,2]

			Fraction of contributed	
Fuel	Quantity	Unit	Heat	Total energy
Coal	10.1	3	67%	60%
Coke, petcoke	1.8	3	14%	13%
Natural gas	338.3	4	3%	3%
Fuel oils	123.7	5	1%	1%
Used tires	0.4	3	3%	3%
Solid wastes	1.0	3	6%	5%
Liquid wastes	929.1	5	6%	5%
Electricity	12.6	6	nil	10%
Average unit consumption of energy:[2]				
Electricity	143.9 kilowatt hours per metric ton of cement.			
Heat	4.7 million Btu[7] per metric ton of clinker.			
Total energy[8]	4.9 million Btu per metric ton cement.			

[1] Source of data: USGS annual survey of U.S. plants.

[2] Fuel and energy consumed reflect the U.S. mix of wet and dry kilns; values likely would differ in countries operating a different mix of technologies.

[3] Million metric tons.

[4] Million cubic meters.

[5] Million liters.

[6] Billion kilowatt-hours.

[7] British thermal units; 1 million Btu = 1.055056 gigajoules.

[8] Includes electricity.

As can be seen from table 7, coal (currently all bituminous) is overwhelmingly the dominant fuel used by the U.S. cement industry; all but a tiny handful of U.S. cement plants burn coal as their primary heat source. However, many U.S. plants routinely burn more than one fuel. For example, when firing up a cold kiln, natural gas or fuel oil is commonly used for the slow, warm-up phase necessary to prevent thermal overstressing of the kiln's refractory brick lining. Once the kiln is sufficiently hot, it will be switched over to coal and/or coke (generally petroleum coke) for production operations. Most U.S. plants are technically capable of burning a variety of fuels, even if they do not routinely do so. Depending on the kiln technology, which may need to be modified, various materials can be burned as alternative or supplementary fuels. An impressive variety of solid and liquid waste materials, including many types of hazardous wastes, can be burned in clinker kilns as supplements or partial replacements to the regular fossil fuels. Some alternative materials used as fuels contribute high unit energy contents (e.g., petroleum coke and used tires), while others are less valued in this respect but are still utilized because the plant is paid to take them. Where waste fuels are incorporated, their contribution to the total heat in the kiln will generally be no more than about 10%–30%. Examples of solid wastes include whole or shredded tires, shredded paper and pulp, spent catalysts, sawdust, scrap wood, rubber residues, shredded packing containers, bone meal, scrap fabrics, dried sewage sludges, oil-contaminated soils, and scrap plastics. Examples of liquid waste fuels include a wide range of spent lubricants and solvents, substandard petroleum refinery products, tars, paints and inks, and miscellaneous chemicals, slurries, and sludges.

Although not shown on table 7, both heat and electricity consumption vary significantly with kiln technology and, for a given technology, tend to be higher for plants operating multiple kilns than for plants with a single kiln of the same overall capacity. Wet kilns consume more fuel on a unit basis than do dry kilns because of the need to evaporate the water in the slurry feed and the much larger size of the wet kilns. In 2000, wet kiln plants in the United States averaged 5.7 million Btu (Mbtu) per ton clinker (fuels only), whereas dry kiln plants averaged 4.4 MBtu/t clinker (USGS data). Data from the Portland Cement Association (PCA) 2000 energy survey of its members (a large subset of the entire U.S. industry) are comparable: 5.8 Mbtu/t clinker for wet plants and 4.2 MBtu/t clinker for dry plants. The PCA's slightly lower values may reflect the exclusion of the white cement plants, which are exceptionally energy-intensive. On a unit clinker basis, fuel consumption tends to be lower in larger capacity kilns and, in the case of dry kilns, decreases with the incorporation of preheaters and precalciners. The PCA 2000 energy survey data illustrate these trends (all data per ton clinker): wet kilns $<$0.5 Mt/yr capacity (6.2 MBtu); wet kilns $\geq$0.5 Mt/yr (5.6 MBtu); dry kilns $<$0.5 Mt/yr (4.9 MBtu); dry kilns $\geq$0.5 Mt/yr (4.1 MBtu); long dry kilns (5.1 MBtu); dry preheater kilns (4.1 MBtu); dry preheater-precalciner kilns (3.8 MBtu). Both the USGS and PCA heat data are based on reported or assigned standard values for the high or gross heat contents of fuels and, for the gaseous and liquid fuels in particular, these will be higher than values for the same fuels reported on a low or net heat basis (as is done in most other countries).

Despite the very much higher temperature requirements for sintering than for calcination, as noted in the kiln functional zone and process chemistry discussions earlier, the actual heat energy input requirements are the opposite. The major heat requirements are for the drying, preheating, and calcination functions, not for sintering. For a dry technology kiln line, of the total theoretical heat inputs to make clinker (i.e., based on reaction thermodynamics, thus ignoring heat losses through the kiln shell and the drying requirements of a wet kiln's feed), about 40% is taken up in the preheating, about 48% in the calcination reaction, and about 12% additional heat is required in the transitional reactions leading to the earliest sintering reactions. The main sintering reaction to make alite ($C_2S + C \rightarrow C_3S$) is exothermic; the sintering process actually yields a net return of heat energy equivalent to almost 9% of the total heat inputs. Thus the total theoretical heat requirements to make clinker are only about 92% of the requirements to preheat and calcine the raw materials. These heat relationships are reflected in the material temperature vs. residence time profiles for a kiln. Figure 5 illustrates this for a typical preheater-equipped dry kiln; very little temperature gain is seen while calcination is ongoing because this reaction is absorbing so much heat energy. For a kiln equipped with a calciner, the 'B' part of the curve (calcination) would be much shorter, as about 80% of the calcination function would overlap the profile of part 'A' (preheating).

In reality, the true heat requirements to make clinker are much higher than the theoretical requirements because of various inefficiencies (totaling about 50%)—mostly heat losses through the kiln shell and, for wet kilns, the enormous heat requirements to evaporate slurry water. On the other hand, most plants will be designed to recover as much heat as possible from the exit gases. For example, hot air from the clinker cooler will be used as combustion air in the kiln. The heat saved by reuse of hot gases can be equivalent to about 50% of the total theoretical heat requirements to make clinker.

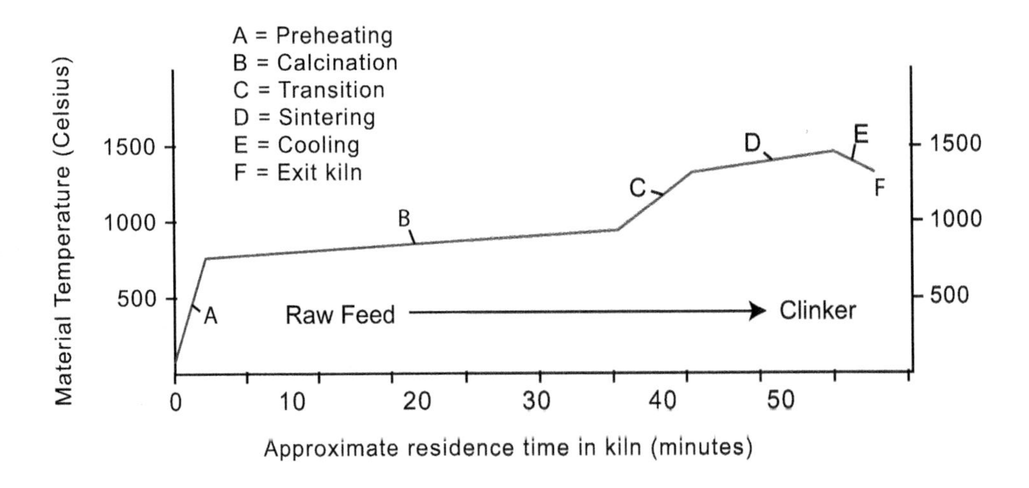

Figure 5. Time-temperature profile for material in a preheater-equipped dry kiln. (After Glasser, 2004)

Surprisingly, the energy (unit fuel or heat) savings noted above for dry kilns is commonly not seen in the plants' electricity consumption, although logic would dictate that a short (hence lighter) dry kiln tube would take less electricity to rotate than a long wet tube. Likewise, there is no inherent reason why the comminution circuits at dry plants should be more efficient than those at wet plants, save that the dry plant category has the majority of the more modern grinding facilities. But for the U.S. industry in 2000, USGS data indicate that wet kiln plants averaged 131 kWh/t cement and dry kilns 148 kWh/t. These values have declined only very slowly and slightly over the years. Comparable data for 2000 from the PCA are 135 kWh/t for wet kilns and 146 kWh/t for dry. The main reason for dry kilns having overall higher unit electrical consumption is that the various fans and blowers associated with preheaters and precalciners are electricity-intensive. However, when examined by type of dry kiln technology, the PCA survey data do not fully support this generalization. For example, long dry kilns averaged about 153 kWh/t cement; preheater kilns about 150 kWh/t cement; and preheater-precalciner kilns about 141 kWh/t cement. Thus, it is evident that other factors are also important determinants of electricity consumption, most probably economies of scale, age of the facility or its upgrade(s), the grinding technology installed, and the type of kiln dust recovery system used. Also, USGS plant survey data (not shown here) reveal that plants that operate multiple kilns (of any technology) have higher unit electricity consumption overall than do plants having similar technology and overall capacity but which have just one kiln. Dedicated clinker-grinding plants typically have electricity consumption rates of 70–80 kWh/t cement. This is about 10%–15% higher than would be expected for a finish mill of similar capacity at an integrated cement plant, and reflects the fact that a grinding plant is a stand-alone facility.

Environmental issues of cement manufacture

Cement manufacturing, as noted earlier, involves two major types of activity: 1) obtaining and preparing raw materials (and fuels), and 2) manufacturing clinker and finished cement. Some of the environmental issues related to these activities will be briefly covered here; the major issue of CO_2 emissions will be covered in some detail. Although the U.S. cement industry is subject to a variety of environmental regulations (most not specific to the industry), it is beyond the scope of this review to describe the regulations and regulatory issues.

Mining of raw materials

The environmental aspects of mining of cement raw materials generally do not attract widespread public attention; certainly not by comparison to the attention accorded to the mining of, for example, metals and coal. The environmental issues for cement raw materials tend to be quite local and are similar to those for aggregates; a review of the latter is provided by Langer (1999). Most raw materials for cement are from surface quarries. A few plants mine underground in cases where, because of stratigraphy, terrain, land acquisition or mine permitting problems, or a need to protect the viewscape, there is a lack of surface reserves. Cement plants generally will not be constructed unless adjacent raw material reserves are sufficient for at least 50 years of operations. Over this time span, plants' limestone quarries can become quite large even though their operational rate will only be on the order of a few thousand tons per day. As with aggregates, the low unit value of limestone and other raw materials precludes high stripping ratios, and thus overburden quantities will be minimal. Likewise, long-distance (more than a few miles) transportation of the major raw materials is generally precluded. Cement raw materials typically are geochemically benign, so issues such as acid mine drainage tend not to apply. Overall, issues with mining of cement raw materials mostly relate to noise, vibrations, and dust from blasting (usually not done daily) and haulage equipment (mostly short-haul transport), as well as aesthetic concerns. Environmental factors related to the procurement of fuels (coal mining, etc.) for cement manufacture are generally not an issue for the cement industry itself. A significant exception is where a cement plant burns waste fuels, which need to be gathered, transported to a location for blending, and then delivered to the cement plant. Especially in cases where hazardous wastes are being used, various aspects of waste fuel handling will likely require environmental permitting.

Manufacture of clinker and finished cement

The major environmental issues associated with cement manufacture are, in fact, associated with the manufacture of clinker rather than with the subsequent intergrinding of clinker and various additives to make cement itself. With clinker manufacture, the environmental issues concern particulate and gaseous emissions from the burning of large quantities of fuels and raw materials—chiefly limestone.

Particulate emissions

All cement plants generate a great deal of fine dust from the kiln line; these dusts will be collectively labeled *cement kiln dust* (CKD). The material in CKD varies among plants and even over time from a single kiln line, but includes particulates representing the raw mix at various stages of burning, particles of clinker, and even particles eroded from the refractory brick and/or monolithic linings of the kiln tube and associated apparatus. In the early days of the portland cement industry, and perhaps still the case at obsolete plants in countries substantially lacking air pollution regulations, CKD was vented to the atmosphere and the resulting dust clouds were no doubt viewed unfavorably by the local communities. Today, however, at all plants in the United States, and at all modern plants worldwide, CKD venting to the atmosphere has been reduced to minute quantities generally invisible to the naked eye. This is because the plants are now equipped with dust scrubbers, either in the form of electrostatic precipitators or filtration baghouses, or both. At many plants, the captured CKD is fed back into the kiln ("recycled"). Likewise, at many kiln lines, CKD-laden exhaust is directly rerouted to the kiln for recycling.

Except to monitor stack emissions (residual venting), CKD generation by cement plants is not routinely measured; indeed, this would be difficult to do for CKD exhaust directly rerouted to the kiln. Consequently, there are only limited data on total CKD generation by cement plants. Some of the available data merely refer to the material captured by the scrubbers. Informal conversations with plant personnel at U.S. plants suggest that typical CKD generation is equivalent to about 15%–20% of the weight of the clinker produced (or about 12–15 Mt/yr at current U.S. clinker output levels).

Apart from environmental concerns, cement plants prefer to recycle (to the kiln) as much CKD as possible, sometimes including material from old CKD piles, because the accumulated CKD represents material that has a substantial value in that its precursors had to be mined, crushed, and burned, and so should not be wasted. However, certain contaminants such as alkalis and some heavy metals tend to concentrate in the CKD, and recycling of CKD to the kiln can thus only be done to the extent that the clinker quality is not compromised. This is a particular issue where local aggregates (for concrete) are susceptible to alkali-silica reactions with the cement. Where not recycled to the kiln, CKD can sometimes be used as a soil conditioner (liming agent), or as a somewhat cementitious material for roadfill, and occasionally as a filler or cementitious extender for finished cement. Where no uses can be found, however, CKD must be landfilled—an increasingly undesirable and costly option. Informal data suggest that about 60%–70% of generated CKD at U.S. plants is currently being recycled to the kilns (7–8 Mt/yr, but only a fraction of this is recorded in the CaO-contributing raw materials in table 4), less than 10% is being used for other purposes, and the remainder is being landfilled..

Gaseous emissions

The principal gaseous emissions from cement plants are nitrogen oxides (NOx), sulfur oxides (SOx), and CO_2. The cement industry is considered to be a significant overall, and large point source of NOx, a modest source of SOx, and a very large point and collective source of CO_2. Nevertheless, for all three, the cement industry's collective emissions are dwarfed by those of thermal powerplants, and by those of motor vehicles for NOx and CO_2.

Nitrogen oxides

The high-temperature combustion of large quantities of fuels, as in cement kilns, can be expected to release NOx in significant amounts, with the nitrogen being derived mostly from the atmosphere and the fuels, but also to a limited degree from the nonfuel raw materials. The formation of NOx in rotary kilns is complex and not fully understood; useful reviews of the subject are given by Haspel (2002), Smart and others (1998), and Young and von Seebach (1998). As noted in these reviews, NOx emissions are dominantly NO (90% or more of the total), with lesser NO_2, and four principal categories or formation mechanisms of NOx are typically identified, namely "thermal" NOx, "fuel" NOx, "feed" NOx, and "prompt" NOx.

Thermal NOx, the dominant type (typically >70% of total), is that formed by direct oxidation of atmospheric nitrogen and forms chiefly by two reactions, both of which are dependent on the dissociation of atmospheric O_2 and N_2:

$$O + N_2 \rightarrow NO + N \quad \text{and} \quad O_2 + N \rightarrow NO + O$$

Thermal NOx begins to form at temperatures above 1200–1500°C, which is well below the burner flame temperature in cement kilns. The formation of thermal NOx increases rapidly with even small increases in temperature when temperatures are in the range of 1370–1870°C, the higher end of which approximates the gas temperatures in the sintering zone of the kiln. Thus, even small shifts in oxygen content of combustion gas in the kiln's sintering zone can have a pronounced influence on the amount of thermal NOx formed.

Fuel NOx refers to NOx formed by the combustion of nitrogen-containing compounds in the fuel. Most fuels (the major exception being natural gas) contain nitrogen in some amount. The oxidation of nitrogen in fuels occurs throughout the entire temperature range of combustion in the kiln line. Based on its nitrogen content, coal has the highest potential to generate fuel NOx and natural gas the least (nil). However, total NOx emissions from kilns burning coal are much lower than from gas-fired kilns, which illustrates the overwhelming importance of thermal NOx in total emissions (natural gas flame temperatures are higher than coal flame temperatures). On the other hand, NOx emissions from calciners are predominantly fuel NOx, as the calciner temperatures can be kept low relative to the sintering zone of the kiln. The formation paths for fuel NOx are very complex, as they are an interplay of oxidizing and reducing reactions (by intermediate nitrogen compounds such as HCN and NH_2^- radicals), and thus vary both with the overall temperature, the amount of oxygen available, and the position (within the flame) of the reaction.

Feed NOx is from the oxidation of nitrogen compounds in the clinker raw materials, and tends to form at relatively low heating temperatures (330–800°C), especially where the rate of heating is slow. Thus feed NOx contributions tend to be greater in wet and long dry kilns than in preheater and precalciner kilns.

Prompt NOx refers to the NO formed in fuel-rich (reducing) flames that is in excess of what would be expected to form by thermal NOx reactions. Prompt NOx appears to be formed by the reaction of CH_2^{-2} and other fuel-derived radicals with atmospheric nitrogen to form cyanide radicals (CN^-) and nitrogen radicals (N^x); the cyanide and nitrogen radicals subsequently oxidize to NO. Prompt NOx is a relatively minor contributor to the total NOx emissions of the kiln line.

As shown by Young and von Seebach (1998), total NOx emissions at cement plants are highly variable over short and long time intervals (minutes to days), and frequent sampling over long periods is required to provide useful data sets for statistical analysis. The majority of emissions (computed as NO_2 and expressed as daily averages) in the long dry kiln example these authors studied in detail were within a range of 0.15%–0.45% of the weight of the clinker produced. Because of the lower fuel requirements and shorter residence times, a preheater-precalciner kiln would be expected to have lower (perhaps by 30%–40%) average overall NOx emissions than a long dry kiln. Johnson (1999) noted EPA 30-day average emissions target guidelines (with NOx-control technology installed) of 0.3% (of the weight of the clinker) for wet kilns, 0.26% for long dry kilns, 0.19% for preheater kilns, and 0.14% for preheater-precalciner kilns.

Approaches to reducing NOx emissions include most general technology upgrades that will reduce fuel consumption or residence times in kiln lines, recycling of CKD, low NOx burners (Johnson, 1999), staged combustion (especially to reduce thermal NOx in precalciners, and via mid-kiln injection of some of the fuel), reduction of excess air, injection of urea or ammonia into the precalciner to reduce oxidizing conditions, switching among major fuels, burning of waste materials (especially whole tires) to create reducing conditions, and, for precalciner kilns, to lower the kiln flame temperatures through water injection (Haspel, 2002). This last would seem to be counterintuitive, but it takes advantage of the fact that, of all the heat energy required to produce clinker, much is consumed in calcination. The subsequent heat requirement to achieve sintering temperatures (about 1450°C) is less than one might expect because some of the sintering reactions (especially that forming C_3S from C_2S and lime) are exothermic. Thus, if most of the calcination is achieved in a precalciner, the actual kiln's flame temperatures can be reduced somewhat from what would be needed in a kiln lacking a precalciner. All emissions reduction strategies benefit from improvements in process controls.

Sulfur oxides

Sulfur from sulfide minerals (mainly pyrite) and from kerogens in both the raw materials (minor) and the fuels yields sulfur oxide (SOx) emissions, almost all of which will be SO_2. Sulfur in fuel oxidizes in the sintering zone of the kiln and in any precalciner apparatus. In contrast, the sulfur in raw materials mostly oxidizes in the preheater apparatus or preheating zone of the kiln. As noted by Schwab and others (1999), much of the SOx evolved in the preheater combines with alkalis to make stable alkali sulfates (e.g., Na_2SO_4), some of which winds up as a buildup or coating in the cooler sections of the kiln line, and some of which becomes resident in the clinker. Scrubbing of SOx by reaction with lime or limestone feed materials (making anhydrite) also occurs during preheating, but the anhydrite is less stable and tends to decompose and rerelease the sulfur (as SOx) as the feed enters the (much hotter) precalciner or calcination zone in the kiln. This SOx is carried with the system air back into the preheating zone and tends to overwhelm the alkali scrubbing capacity of the feed, thus there remains a net evolution of SOx in the exhaust gases. Typical concentrations of SOx in exhaust gases are about 100–200 parts per million, but are highly variable depending on the sulfur content of the fuels and feed materials. Where SOx emissions are in excess of regulatory limits, or where they appear frequently as visible detached plumes, there is pressure on cement companies to install SOx scrubbers. Such scrubbers are readily available, and make use of limestone or lime to form synthetic gypsum.

Carbon dioxide

By far the major environmental issue of concern today related to clinker manufacture is that of carbon dioxide (CO_2) emissions. Interest in calculating CO_2 emissions stems from the global warming debate and the role therein of anthropogenic greenhouse gases. Although the emissions by powerplants that burn fossil fuels and the exhaust from motor vehicles are by far the largest source of anthropogenic CO_2, the cement industry is more or less tied with the iron and steel industry as the largest "industrial" (other than powerplants) emitter of the gas. Overall, the U.S. cement industry emits about 1.4% of total U.S. anthropogenic CO_2 emissions (U.S. Environmental Protection Agency, 2002); in many countries worldwide, the contribution is relatively higher—probably closer to 5%—because of a lower "intensity" of thermal power generation (relative to the overall economy) and less use of motor vehicles.

The emission of CO_2 from clinker manufacture stems from both the calcination of carbonate minerals in the raw feed, and the combustion of fuels. However, many statistical compilations detailing CO_2 emissions by the cement industry and other industrial sources (e.g., U.S. Environmental Protection Agency, 2002) do not directly link the combustion emissions to the specific industries; instead, combustion emissions are only shown as lumped within an all-sources national fuel total. Thus, the industry-specific emissions data may be very incomplete.

Calculation of CO_2 is more properly done on the basis of clinker production data than on data for cement output. This is because a link to cement production assumes that the clinker content of the cement is precisely known. However, with many country-level cement production data, no information is available as to the type(s) of cement produced. As noted earlier, blended cements and masonry cements both contain large fractions of material other than clinker. Two approaches are reasonable using clinker production data as the basis for the CO_2 calculation. For plants calculating their own emissions, it is feasible to calculate CO_2 based on the precise chemical compositions and quantities of the raw materials and fuels consumed. Detailed data like these, however, are generally lacking for the purposes of compiling national or regional emissions totals. For regional totals, the practical approach is to start with clinker production data and work backwards to calculate the CO_2. This approach works well for calcination CO_2, but is more equivocal for fuel combustion emissions.

Carbon dioxide from calcination

For CO_2 generated through calcination, the easiest calculation approach is that advocated in the "good practices" methodology detailed by the Intergovernmental Panel on Climate Change (IPCC) (2000). The IPCC method yields an estimate of emissions good to no better than 5% but generally within 10% based on certain compositional assumptions and typical errors found within reported production and compositional data.

As noted earlier, the calcination reaction for calcium carbonate is:

$$CaCO_3 + \text{heat} \rightarrow CaO + CO_2\uparrow.$$

The basic assumption in the IPCC method is that all of the CaO and CO_2 are derived from $CaCO_3$, but the method advises compensating for cases where it is known that a significant amount of CaO is being contributed from non-carbonate sources, such as ferrous slags. It does not matter if all of the $CaCO_3$ is within limestone. Given the range of CaO values in typical clinkers (60%–67%), the clustering of most values in the range of 64%–66%, and imprecisions in the compositional control in manufacture and chemical analysis of clinker, a default composition of 65% CaO (e.g., per table 1) for clinker is a satisfactory assumption in the absence of more specific data.

In the calcination equation given above, the CaO fraction is 56.03% of the original weight of the $CaCO_3$, and the CO_2 fraction is 43.97%. Accordingly, the amount (X) of $CaCO_3$ required to yield 1 t clinker containing 0.65 t CaO (i.e. 65%) would be:

$$X = 0.65\ \text{t}/0.5603 = 1.1601\ \text{t (unrounded)}.$$

This weight of $CaCO_3$ yields CO_2 in the amount of:

$$1.1601 \text{ t} \times 0.4397 = 0.5101 \text{ t (unrounded)}; = 0.51 \text{ t (rounded)}.$$

This amount (0.51 t CO_2 per ton of clinker) is the IPCC default emissions factor for calcination CO_2 and, again, assumes that 100% of the CaO is from $CaCO_3$. For comparison, 1 t of clinker of 60% CaO content would back-calculate to 0.47 t (rounded) of CO_2, and a 67% CaO clinker would calculate to 0.53 t of CO_2.

The IPCC method (2000, but currently is being updated) offers little guidance for cases where carbonates other than $CaCO_3$ are present in clinker feeds, but this turns out to be a relatively insignificant problem. Examples of such carbonates would include dolomite $CaMg(CO_3)_2$, magnesite ($MgCO_3$), siderite ($FeCO_3$), rhodochrosite ($MnCO_3$), and various solid solutions among these. The effects of these carbonates on CO_2 emissions can be easily calculated by the same procedure as for $CaCO_3$ above.

For example, MgO is commonly present in small amounts in clinker raw materials—it is present in many limestone feeds where it typically forms a minor dolomitic phase, even though pure dolomite would not commonly be used as a kiln feed, and magnesite even less so. And MgO is also common in many non-carbonate clinker feeds, such as the silicate minerals in shales and slags. If the assumption is made, however unrealistic, that 100% of the MgO comes from a carbonate phase, then it can be shown that, for a clinker of 65% CaO, the default calcination CO_2 emissions factor would become, in unrounded terms:

$$[0.5101 + M(0.011)] \text{ t } CO_2 \text{ per ton clinker}$$

where the 0.5101 is the emissions factor (see above) for a pure $CaCO_3$ system, and M is the percentage of MgO in the clinker. This modest M(0.011) component would be a maximum MgO contribution to CO_2. For the small amount of MgO in clinker (e.g., 2% in table 1; the amount in portland cement is limited to a maximum of 6%, per ASTM standard C-150, and few U.S. portland cements would approach this limit), it may be argued that the Mg-carbonate contribution to calcination is small enough to be ignored because it would be subsumed within the 5%–10% overall error range of the IPCC methodology. The effect of iron carbonates is smaller still: for a clinker of 65% CaO, the combined emissions factor, if iron is assumed to be 100% from carbonate (very unlikely), becomes just:

$$[0.5101 + F(0.0055)] \text{ t } CO_2 \text{ per ton clinker (unrounded terms)}$$

where F is the percentage Fe_2O_3 in the clinker. Thus, for the modest amount of iron in clinker (e.g., 3% Fe_2O_3 in table 1), an iron-carbonate contribution to CO_2 will be trivial. Accordingly, for regional clinker data, it is reasonable to look just at $CaCO_3$ and to ignore the effects of the other carbonates. For plant-level reporting, sufficient compositional data may be available to warrant including the other carbonates, if present.

A remaining issue in the calculation of calcination CO_2 emissions is the component represented by "lost" CKD. This component is very difficult to quantify. As noted earlier, total generation of CKD is large at all cement plants. In most countries today, and at all modern plants, essentially all of the CKD that is generated is either directly rerouted to the kiln's raw material feed stream or is captured by electrostatic precipitators or filtration baghouses. Captured CKD can be recycled to the kiln (the preferred use for it), used for other purposes, or landfilled. The CO_2 emissions associated with the CKD rerouted or recycled to the kiln become part of the clinker emissions calculation. All the other CKD, however, is "lost" to the CO_2 calculation because the dust is not part of the clinker production tonnage. How much CO_2 this lost CKD represents depends upon the weight of the lost CKD, the degree to which the CKD represents an original carbonate (i.e., CO_2-bearing) feed, and the degree to which this carbonate material was calcined. Data on all these factors are generally poor or lacking, but within very broad constraints, the IPCC recommended a "best practice" default addition of 2% to the calcination CO_2 calculated for the clinker itself for plants or regions where it is believed that significant amounts of CKD are not being recycled to the kilns.

Carbon dioxide from fuel combustion

Few data exist on the CO_2 emissions from fuel combustion by the cement industry because most major compendia (e.g., EPA, IPCC) choose the expedient of combining emissions from fuel combustion from all sectors of the economy rather than attempt to show (or survey) each of the myriad emissions sources (fuel consumers) separately. Nevertheless, the contribution from fuel combustion is required to have a reasonably complete picture of CO_2 emissions by the cement industry. It is generally assumed that carbon monoxide released by cement plants (a relatively small amount because of the high combustion efficiencies of cement kilns) is ultimately oxidized to CO_2.

It is more difficult, and much less precise, to calculate CO_2 emissions from the combustion of fuels than emissions from calcination because of uncertainties in the reported quantities and identification of fuels (particularly differentiating among the many types of waste fuels), and the different reported (or assigned) heat "contents" (the heat energy released from burning) of the fuels. Data on the heat contents are required because they are integral components of published fuel carbon factors. Instead of tables showing how much carbon (from which CO_2 is readily calculated) is in a ton of a given fuel, carbon factor data instead are presented in terms like "million tons of carbon equivalent per quadrillion Btu" for the fuel in question. Thus, the energy content of the fuel is needed to isolate the carbon content. Where heat contents must be assigned in the absence of (correctly) reported data, a complication is faced in that published standard heat contents commonly show significant ranges for individual fuels. Few published data exist for the heat or carbon contents of waste fuels; further, waste fuels commonly are consumed in diluted or impure forms. Finally, because the reporting units commonly are in terms with large exponents (e.g., quadrillion Btu and petajoules, both of which involve 10^{15}) and are rounded, these carbon-isolation calculations are very prone to propagation of rounding errors. Overall, calculated combustion emissions, whether presented rounded or not, should be viewed as being good to within a range of 5%–10% at best, and probably a 15%–20% error range would be safer.

With the foregoing in mind, the 2000 fuel data in table 6 yield U.S. average unit emissions of about 0.43 t CO_2/t clinker (van Oss and Padovani, 2003). Although earlier years' fuel consumption data are not shown here, annual average unit emissions calculated for them show a gradual decline over the years from about 0.63 t CO_2/t clinker in 1950. The decline reflects an increased reliance on dry kiln technology and parallels a decline in unit energy consumption. The U.S. industry in 2000 had about 76% of its clinker output from dry plants (not all of which operated modern, preheater-precalciner technology, however), with the remainder from wet plants. Countries with a high proportion of modern (i.e., dry) cement plants likely have unit emissions somewhat lower than the U.S. average.

Overall carbon dioxide emissions

Calcination emissions (about 0.51 t per ton of clinker—more or less worldwide), and combustion emissions (current U.S. level about 0.43 t per ton of clinker) noted above, sum to a total of about 0.94 t CO_2/t clinker. Given the likely imprecision of the combustion emissions calculation, and changes with time in fuel use, the combined calcination and combustion emissions are better rounded to about 0.9 or even 1 t CO_2/ t clinker. This is a good average for first order emissions estimates for most countries or regions.

Not included in this average is an estimate of the CO_2 released in generating the electricity purchased by the cement plant. Its exclusion is the norm, and is reasonable because the outside electricity would customarily be assigned to electric utilities in national emissions inventories. The amount of electricity CO_2 would depend on the fuels used to generate the electricity and would vary nationally and regionally. But, overall, for the U.S. cement industry and electrical grid, the purchased electricity consumption of integrated cement plants would equate to an additional 7%–8% of current total (calcination plus combustion) emissions. In contrast, electricity cogeneration at cement plants (currently, very rare in the U.S. industry) utilizes waste heat from the kiln line and should not directly add to the CO_2 output.

Instead of relating CO_2 emissions to clinker, some studies casually quote a 1:1 mass ratio for CO_2 and (portland) cement; the rationale given for the cement linkage is that cement production data are more readily available than clinker data for most regions. Given that a straight (clinker + gypsum) portland cement will have a clinker ratio or clinker factor of about 93%–97% (95% being a useful average for estimation purposes), this generalization is reasonable for cases where the cement production data actually refer to straight portland cement. Unfortunately, this is not a safe assumption for USGS and most other country-level tabulations of cement production data. These data sets normally are for total hydraulic cement, which, although likely dominated by portland cement, may include for some countries significant quantities of blended cements, portland-limestone cements, and masonry cements, and perhaps even misassigned cementitious admixtures (SCM or pozzolans) as yet uncombined within finished cement. Blended and masonry cements have lower (commonly much lower) clinker factors than straight portland cements, and the admixtures (other than CKD) have zero clinker factors. For masonry cements that incorporate lime as the main additive, there is a separate CO_2 legacy related to lime manufacture (generally done at a different facility), but it will be somewhat lower for lime than for the equivalent weight of clinker because of the lower temperatures (and hence fuel consumption) to achieve calcination for lime (as a product) than that to achieve clinker formation. For countries believed to have significant output of these other cements, an overall clinker ratio of 75%–85% may be a better approximation, the IPCC "good practices" suggestion is to use a 75% ratio for these countries.

Strategies to reduce carbon dioxide emissions

There are four main strategies to reduce CO_2 emissions by the cement industry. The first is to switch to lower carbon fuels, such as from coal to natural gas. Many plants are equipped to switch among fuels, but a fuel switch is not always desirable. One issue is that a switch could lead to fuel cost/availability problems. This is a particular problem with switching to natural gas—a low carbon fuel. More importantly, a fuel switch can adversely affect the performance of the kiln because fuels vary in heat contents and in the heat-transfer and shape characteristics of the flames they generate. A fuel switch may counter efforts to reduce NOx; as noted earlier; natural gas, because it burns with a hotter flame, generates more thermal NOx than relatively high-nitrogen content coal. An alternative fuel switching strategy is to burn a measure of waste fuels, as these may have lower carbon contents, or the plant might receive some form of carbon "credit" for them because of the reduced consumption of standard fuels.

The second major strategy is to upgrade the kiln line to be more fuel-efficient. There are many options for this, such as by the installation of more efficient burners and improved process control systems, and innumerable minor "tweakings" that cumulatively make the plant more efficient. But for major fuel reductions, a plant generally needs to go through a major upgrade, such as converting from wet (or older dry) kiln technology to modern preheater-precalciner systems. Major technological conversions are expensive (tens of millions of dollars or more) and may thus not be economical for an old plant lacking long-term (say, 50 years) reserves of raw materials, or that is located in a small market. Fuel-reduction strategies essentially target combustion CO_2.

The third reduction strategy is to target calcination CO_2 emissions by using raw materials that will contribute part of the CaO needed to make clinker from a source other than $CaCO_3$. Many plants already get minor amounts of CaO from various silicate minerals in the feed, but this type of contribution can be increased by incorporating feeds such as slags and fly ash or bottom ash. The key consideration in such a CaO-source substitution is to make sure the alternative source does not require significantly more heat (and hence fuel) to process. A highly promising CaO source has proven to be steel slag. This material had been tried at various time in the past and, although chemically suitable, had been viewed unfavorably because it was difficult (hence costly) to grind. A key discovery regarding steel slag was made by the cement company Texas Industries, Inc. as a result of a program to find a more beneficial use (than as coarse aggregate) for slag produced by an adjacent electric arc furnace steel plant owned by a subsidiary company. The discovery, patented under the name CemStar, was that the slag did not require fine crushing or grinding. With just coarse crushing (to about 2–2.5 cm diameter), the material proved to be easily incorporated by a cement kiln. The slag's mineralogy already contained C_2S and/or compounds (including iron) that made C_2S at low temperatures, and because the slag melted easily (just 1260–1300°C), it provided a relatively low-temperature melt environment for the C_2S to combine exothermically with lime (from calcination) to rapidly form the critical C_3S clinker mineral. Using CemStar, additional clinker is produced in roughly a 1:1 ratio to the slag added, the unit fuel consumption is reduced, as are the unit calcination CO_2 emissions. Typical slag additions with CemStar are as a 3%–10% substitution for limestone (or, more properly, a kiln's throughput capacity can be increased by these percentages by using CemStar). A review of the process is provided by Perkins (2000).

The fourth strategy to reduce CO_2 emissions is to reduce the clinker content of finished cement through the use of SCM additives (i.e., make more blended cements) or admixtures; on a societal level, this strategy would also mean encouraging concrete companies to increasingly use SCM as a partial substitute for portland cement. At both levels, increased SCM use can only proceed to the degree that construction codes allow it. A similar substitution is that of ground limestone or similar "inert" material to finished straight portland cement, in an amount of 1%–5%. This limestone substitution is common in Europe and, at the low range shown has been proposed and accepted by ASTM as a modification of the standard for portland cement (ASTM C-150); higher substitution levels are common in Europe. In all cases with substitution, the emphasis is not on reducing absolute clinker production but the clinker fraction of finished cement. This allows increased production of finished cement without adding a commensurate amount of clinker production capacity.

Environmental benefits of cement manufacture

In the increasingly popular industrial ecology paradigm, it is desirable to have industries and industrial processes that are interconnected and interdependent, particularly in the context of having no, or greatly reduced, net wastes by the entire multi-industry complex. In other words, it is desirable to have industries that consume other industries' waste products (the CemStar process, mentioned earlier, is a good example of this). Given that cement and concrete are produced and consumed in huge quantities worldwide, the manufacture of cement to meet this demand ensures that enormous quantities of raw materials and fuels will also be required. The manufacture of cement (actually clinker) has virtue in its inefficiencies of high requisite temperatures and long residence times for clinker formation. These allow for the complete destruction or conversion into clinker and/or heat of virtually anything that enters the kiln, including a wide variety and large quantities of waste fuels and other waste materials, some of them classified as hazardous. This use of wastes saves on standard fossil fuels and more costly disposal or storage strategies for the wastes. It is in this production of a valuable product from the destruction of other industries' wastes that the cement industry is seen as an ideal driver in existing and future industrial ecosystems (van Oss and Padovani, 2003; Vigon, 2002).

References Cited in Part 1

Alsop, P.A., Chen, H., Chin-Fatt, A.L., Jackura, A.J., McCabe, M.I., and Tseng, H.H., 2005, The cement plant operations handbook, 4th ed.: International Cement Review, Tradeship Publications Ltd. UK, 257 p.

Bentz, D.P., 1997, Three-dimensional computer simulation of portland cement hydration and microstructure development: J. American Ceramics Society, v. 80, No. 1, p. 3–21.

Bhatty, J.I., Miller, F.M., and Kosmatka, S.H., (eds), 2004, Innovations in portland cement manufacturing: Skokie, IL, Portland Cement Association, 1367 p.

Bogue, R.H., 1955, The chemistry of portland cement, 2nd. ed.: New York, Reinhold Publishing Corp., 793 p.

Campbell, D.H., 1999, Microscopical examination and interpretation of portland cement clinker, 2nd ed.: Skokie, IL, Portland Cement Association, 202 p.

Duda, W.H., 1985, Cement-data-book, 3rd ed.: Weisbaden, Bauverlag GmbH, vol. 1., 636 p.

Glasser, F.P., 2004, Advances in cement clinkering: in Bhatty, J.I., Miller, F.M., and Kosmatka, S.H., (eds), 2004, Innovations in portland cement manufacturing: Skokie, IL, Portland Cement Association, p. 331–368.

Haspel, David, 2002, Lowering NOx for less: International Cement Review, January, p. 63-66.

Intergovernmental Panel on Climate Change, 2000, Good practice guidance and uncertainty management in greenhouse gas inventories: Intergovernmental Panel on Climate Change, p. 3.1–3.18.

Johnson, S.A., 1999, Low NOx burners-what are the options?: World Cement, vol. 30, no. 10, p. 82–86.

Kemp, B.J., 1994, Concrete—over three thousand years in the making: International Cement Review, February, p. 52–53.

Klemm, W.A., 2004, Cement manufacturing—A historical perspective: in Bhatty, J.I., Miller, F.M., and Kosmatka, S.H., (eds), 2004, Innovations in portland cement manufacturing: Skokie, IL, Portland Cement Association, p. 1–35.

Kosmatka, S.H., and Panarese, W.C., 1988, Design and control of concrete mixtures, 13th ed.: Skokie, IL Portland Cement Association, 205 p.

Kosmatka, S.H., Kerkoff, B., and Panarese, W.C., 2002, Design and control of concrete mixtures. 14th ed.: Skokie, IL, Portland Cement Association, 358 p.

Lan, Wang, 1998, Advances in Chinese VSK technology: World Cement, vol. 29, no.12, p. 44–48.

Langer, W.H., 1999, Environmental impacts of mining natural aggregates: Proc. 35th Forum on the Geology of Industrial Minerals—The Intermountain West Forum, Utah Geol. Survey Misc. Pub. 01-2; p. 127–137.

Lea, F.M., 1970, The chemistry of cement and concrete, 3rd ed.: New York, Chemical Publishing Co., 727 p.

Leming, M.L., 1996, Alkali-silica reactivity—mechanisms and management: Mining Engineering, v. 48, no. 12, p. 61–64.

Lesley, R.W., 1924, History of the portland cement industry in the United States: Chicago, International Trade Press, Inc., 330 p.

Moir, G. K., 2003, Gaining acceptance: International Cement Review, March, p. 67–70.

Perkins, David, 2000, Increased production and lower emissions: World Cement, vol. 31, no. 12, p. 57–59.

Portland Cement Association, 2000, U.S. and Canadian labor-energy input survey: Skokie, IL, Portland Cement Assoc., 48 p.

Roy, D.M., 1985, Portland cement—Constitution and processing, Part II—Cement constitution and kiln reactions: in Roy, D.M., ed., Instructional modules in cement science: J. Materials Education, p. 73–92.

Schmidt, M., Middendorf, B., Vellmer, C., and Geisenhansluekke, C., 2004, Blended cements, in Bhatty, J.I., Miller, F.M., and Kosmatka, S.H., (eds), 2004, Innovations in portland cement manufacturing: Skokie, IL, Portland Cement Association, p. 1107–1148.

Schwab, J., Wilber, K., and Riley, J., 1999, And SO2 can you: International Cement Review, January, p. 54-55.

Smart, J.P., Mullinger, P.J., and Jenkins, B.G., 1998, Combustion, heat transfer and NOx: World Cement, vol. 29, no. 12, p. 14–25.

Smeaton, J., 1791, A narrative of the building and a description of the construction of the Eddystone lighthouse with stone: London, H. Hughes. 198 p.

Stanley, C., 1999, Where it all began: World Cement, v. 30, no. 12, p. 20–24.

Taylor, H.F.W., 1997, Cement chemistry, 2nd ed.: London, Thomas Telford, 459 p.

U.S. Environmental Protection Agency, 2002, Inventory of U.S. greenhouse gas emissions and sinks, 1990-2000: U.S. Environmental Protection Agency, p. ES-3 and 3.2–3.7.

van Oss, H.G., and Padovani, A.C., 2002, Cement and the environment; Part I—Chemistry and technology: J. of Industrial Ecology, vol. 6, no. 1, p. 89–105.

van Oss, H.G., and Padovani, A.C., 2003, Cement and the environment; Part II—Environmental challenges and opportunities: J. of Industrial Ecology, vol. 7, no. 1, p. 93–126.

Vicat, L.J., 1837, A practical and scientific treatise on calcareous mortars and cements, artificial and natural: trans. J.T. Smith, London, John Weale, 302 p.

Vigon, Bruce, 2002, Industrial ecology in the cement industry: World Business Council for Sustainable Development, Substudy 9: Toward a sustainable cement industry, Battelle, 103 p.

Welch, J.H., 1964, Phase equilibria and high-temperature chemistry in the CaO-Al2O3-SiO2 and related systems: in Taylor, H.F.W., ed., The chemistry of cements, London, Academic Press, v. 1, p. 49–88.

Wilcox, Simon, 1995, From the mists of time...: International Cement Review, July, p. 73–75.

Young, J.F., 1985, Hydration of portland cement: in Roy, D.M., ed., Instructional modules in cement science: J. Materials Education, p. 5–21.

Young, G.L., and von Seebach, Michael, 1998, NOx variability and control from portland cement kilns: Proc. 34th International Cement Seminar, Salt Lake City, UT, p. 13–44.

Glossary of Terms

Brief definitions are provided below of technical terms and abbreviations found in this report as well as some other terms found in the external literature on cement and concrete. For many of the terms, fuller definitions or additional information can be found in the text of the report. Words in italics are defined elsewhere in the glossary.

A 1) Cement chemistry shorthand for alumina (Al_2O_3). 2) As a capitalized suffix (e.g., Type IA portland cement) it denotes the addition to the cement of an *air-entraining* agent.

AAR Alkali-aggregate reactivity. Adverse reactions within concrete between certain aggregates and the alkali hydroxides in cement. The most common type of AAR is alkali silica reactivity (*ASR*).

AASHTO American Association of State Highway Transportation Officials. An alternative to ASTM for setting of standards; however, many cement- related AASHTO standards are similar or even identical to those of ASTM.

Accelerator An agent (admixture) added to concrete to speed *setting* and hardening, and/or to speed *hydration*, and/or to speed strength development; c.f. *retarder*.

Additive Material intermixed with hydraulic cement to form a different finished cement product.

Admixture Ingredient (other than cement, water, and aggregates) added to a concrete mix.

Aggregates Particulate materials such as sand, gravel, crushed stone, and crushed *slag*, used in construction.

Air-entraining agent Chemical agent added to cement or concrete that causes the formation of tiny bubbles in the resulting concrete.

Alite A cement mineral, generally equated to *C_3S* but usually somewhat impure.

Alumina Aluminum oxide (Al_2O_3; or A in cement chemistry shorthand). 1) As a solid material, its major use is in the production of aluminum metal, but it also has refractory and chemical applications (including as a secondary raw material or *sweetener* in clinker manufacture). 2) term pertaining to the aluminum oxide content of a material.

Aluminate a) Casual term for the cement mineral *C_3A*; b) referring to C_3A or similar phases containing aluminum oxide in cement chemical reactions.

Aluminous cement Hydraulic cement based on clinker made from a mix of *limestone* and *bauxite*. Used for certain high temperature and rapid-setting applications.

Anhydrite Anhydrous calcium sulfate ($CaSO_4$ or, in cement chemistry shorthand, $C\bar{S}$). A mineral sometimes interground with portland cement clinker to control setting times in portland cement; in this role it partially substitutes for *gypsum*.

Aragonite A mineral composed of calcium carbonate ($CaCO_3$).

ASR Alkali-silica reactions or reactivity. Undesirable reactions in concrete between disordered silica in some aggregates and alkali hydroxides in the cement.

ASTM American Society for Testing and Materials; organization has now been renamed ASTM International. Sets standards for testing and performance of construction and other materials.

Bauxite An earthy material consisting of a mix of iron and aluminum oxides, hydroxides and silicates. It is processed into *alumina* for subsequent reduction to aluminum metal and for various chemical and refractory applications. Bauxite can be a supplementary raw material for portland cement clinker production, and is a major raw material for *aluminous cement* production.

Belite A cement or clinker mineral, generally equated to C_2S but usually somewhat impure.

Blended cement A hydraulic cement made of a mixture of portland cement (or clinker plus gypsum) plus *pozzolans* or other *SCM.*

Blending plant An independent (of a *portland* cement company) facility that purchases cement and then blends it with other materials to make a different type of cement, typically blended cements (by addition of *SCM*), *colored* cements (by addition of pigment), or *masonry* cements (by addition of crushed limestone or other materials). Blending plants are considered to be *final customers.*

Burnability An informal term, generally encountered in the context of choosing raw materials for clinker manufacture, pertaining to the relative amount of heat energy required to break down the specific raw material into its component oxides. High burnability, then, refers to a material that breaks down easily, requiring relatively little heat input.

Burning An imprecise term referring variously to a) the combustion of fuels in the cement plant; b) the extreme heating and *thermochemical* decomposition of raw materials into their component oxides; c) the hottest part (zone) of the kiln where the actual clinker minerals are formed; a.k.a. *sintering* or *clinkering.*

C 1) Cement chemistry shorthand for calcium oxide (CaO); 2) conventional chemical notation for carbon.

C_2S Cement chemistry shorthand for calcium disilicate, one of the four principal minerals in portland cement clinker. Sometimes referred to as *belite.*

C_3A Cement chemistry shorthand for tricalcium aluminate, one of the four principal minerals in portland cement clinker. Commonly referred to as *aluminate* (as in aluminate content or phase).

C_3S Cement chemistry shorthand for calcium trisilicate, the dominant of the four principal minerals in modern portland cement clinker. Sometimes referred to as *alite.*

C_4AF Cement chemistry shorthand for tetracalcium aluminoferrite, one of the four principal minerals in portland cement clinker. Strictly, C_4AF is the mean compositional value of a solid solution between C_6A_2F and C_6AF_2. Commonly referred to as *ferrite* (as in ferrite content or phase).

C-S-H Cement chemistry shorthand for calcium silicate hydrate. A colloidal gel made up of a family of related calcium silicate hydrates (e.g., $C_3S_2H_3$; $C_3S_2H_4$) formed, chiefly, from the hydration of the cement minerals C_3S and C_2S. In the older literature, the simplest of the C-S-H formulations is sometimes called *tobermorite*. C-S-H is the dominant contributor of strength to concrete.

Calcination 1) The heat-induced removal, or loss, of chemically-bound volatiles, usually other than water. 2) In cement and lime manufacture, it involves the thermal decomposition of calcite and other carbonate minerals to a metallic oxide (mainly CaO) plus carbon dioxide.

Calciner See *precalciner*.

Calcite A mineral composed of calcium carbonate ($CaCO_3$); the dominant mineral in limestone and hence the most common single mineral raw material for portland cement manufacture.

Carbonate 1) Refers to a mineral containing the carbonate radical CO_3^{-2}; 2) the act of *carbonation*.

Carbonation The re-formation of carbonate minerals through the absorption of carbon dioxide by metallic oxides (e.g., carbonation of *lime* yields *calcite*).

Cement 1) A binding agent. In construction, this agent is a powder to which water is added and which develops binding properties either through hydration of the component minerals in the cement (*hydraulic* cement) or through *carbonation* (e.g., *lime* mortars). 2) Informal term for *cement paste*.

Cement chemistry shorthand The use of single letters to denote the most common oxides in cement chemistry; e.g., C for CaO. *See individual letters.*

Cement paste A mix of hydraulic cement plus sufficient water to ensure full hydration of the cement minerals. Cement paste contributes virtually all of the strength to concrete and mortars.

Cement rock 1) An impure limestone containing a complete set of oxides, and in the correct proportions, to make clinker or cement with little or no addition of other raw materials; 2) less commonly, a term used in a quarry to distinguish the limestone that is mined for cement manufacture from other rock that is unsuitable for cement; 3) less commonly, a term denoting the major limestone feed to the kiln as opposed to other limestone used less frequently (perhaps as a *sweetener*).

C.i.f. Cost, insurance, and freight. A common value indicator and is inclusive of the base cost (see *customs value*; *f.o.b.*) of a shipment, plus insurance and freight to an agreed-upon destination (typically a port or land terminal). It does not include costs of unloading the material and other port or terminal fees, or any import duties or tariffs. For U.S. trade data, the c.i.f. valuation is generally based on the location (port) of official entry into the country.

Ciment fondu A type of *aluminous cement.*

CKD Cement kiln dust; casually refers to all dust generated in the *kiln* or *pyroprocessing line.*

Clinker An intermediate product of hydraulic cement manufacture. Clinker is produced in a kiln and consists of semifused nodules that contain a controlled and intimate mix of clinker (or cement) *minerals. Portland cement* clinker consists, chiefly, of the four minerals C_3S, C_2S, C_3A, and C_4AF. Clinker is finely ground to make finished cement; in the case of portland cement, the clinker is interground with a small amount of *gypsum* and/or *anhydrite.*

Clinkering The *thermochemical* formation of the actual clinker minerals, especially to those reactions occurring above about 1300°C; also the zone in the kiln where this occurs; a.k.a. *sintering* or *burning.*

Clinker ton A unit of measure used at some plants to directly relate clinker production to the potential output of portland cement: 1 clinker ton of clinker is sufficient to make 1 ton of portland cement. The actual weight or mass of a clinker ton, therefore, is dependent on the particular plant's recipe for portland cement; that is, its ratio of clinker to other ingredients in the cement.

Colored cement A cement to which pigments have been added. Excludes *white* cement (although white and colored cements have a common tariff code); see also *gray cement.*

Comminution Reduction of particle size by crushing and/or grinding.

Composite cement *Blended cement.*

Concrete A proportioned mix of *hydraulic* cement, water, fine and coarse *aggregates*, and sometimes additives, that hardens to a tough, rocklike material used for construction.

Curing The process of maintaining the moisture content of concrete to allow full hydration of the component hydraulic cement minerals and hence the development of full strength.

Customs value In trade data, it is the base value or price of the merchandise being imported and generally equates to the *f.o.b.* value. It excludes onward shipping and insurance costs (see *c.i.f*), ship unloading and other destination terminal costs, and import duties and tariffs.

Dead burned Refers to a material, generally a metallic oxide, that has been heated to a point where it is no longer chemically reactive (relative to its lower temperature reactivity). For example, dead burned magnesia or *periclase* (*M* in cement chemistry shorthand) no longer readily *carbonates* to *magnesite*, whereas reactive magnesia readily does so.

Dehydration Removal or loss of chemically or structurally bound water; c.f. *drying*.

Dolomite A mineral composed of calcium magnesium carbonate $CaMg(CO_3)_2$. Also a sedimentary rock composed primarily of this mineral.

Drying Removal of water other than that which is chemically and structurally bound; c.f. *dehydration*.

Dry kiln (plant) Refers to a kiln for which the raw materials are crushed, ground, proportioned, and fed into the kiln line in a dry state; c.f. *wet kiln*.

Ettringite A mineral (formula $C_6A\bar{S}_3H_{32}$ in cement chemistry shorthand) formed by the hydration of the cement mineral C_3A in the presence of excess sulfate.

Extender As in 'cement *extender*'; term used in some countries to denote cementitious *admixtures* or *SCM*.

F 1) Cement chemistry shorthand for ferric oxide (Fe_2O_3); 2) conventional chemical notation for the element fluorine.

F.a.s. Free alongside ship. The base value or price of merchandise delivered to an agreed-upon port or terminal of debarkation (e.g., export) and placed alongside the ship (or train). It excludes the cost of loading the ship and all other onward costs (see *customs value*, *c.i.f.*).

Ferrite Casual term for the cement mineral C_4AF. The formula represents the mean value of a solid solution with end members C_6A_2F and C_6AF_2.

Final customer A term of convenience used in the cement industry to denote a purchaser of cement other than a rival cement company or a sister plant or terminal owned by the selling company. As a practical reporting matter, most final customers are concrete companies, construction contractors, or building material suppliers, and are not individual citizens or companies owning the location where the concrete or mortar was actually put into place. *Blending plants* independent of the originating or rival cement companies are included as final customers. In the case of *swaps*, the final customer is that which paid the originating company for the cement.

Finished cement A cement ready for sale, i.e., which needs no further processing. The product(s) of a cement plant's *finish mill*.

Finish mill The section of a cement plant where *clinker* and other ingredients are finely ground and combined into *finished cement*.

Flux 1) A material that reduces the temperature and/or energy input requirements of a chemical reaction or physical change (such as melting). 2) In *clinker* manufacture, a material that lowers the temperature and energy requirements of the clinker-forming (especially the *sintering*) reactions by promotion of the development of a liquid phase. Casually synonymous with *mineralizer*.

Fly ash Fine grained glassy *silicate* particles released through the burning of coal in power plants and recovered by scrubbers. Some varieties of fly ash are useful as *pozzolans* or *SCM* and others can be used as raw material for clinker manufacture and as fine-grained construction *aggregates*.

F.o.b. Free on board. The base value of merchandise at an agreed-upon location from which the material will be sent to the customer, or at which the customer will pick up the merchandise. It includes the cost of loading the onward carrier (truck or train or ship). It is thus comparable to the *f.a.s.* price plus loading costs.

GGBFS Ground granulated blast furnace slag. A form of blast furnace *slag* produced by quenching molten slag in a water stream to form sand-sized grains of glass. When finely ground, this material is a *latent cement*, although it is generally included as a *pozzolan* or *SCM*. Increasingly, GGBFS is being sold under the imprecise term *slag cement*. See also *pelletized* slag.

GHG Greenhouse gas(es). A gas (e.g., carbon dioxide) that causes retention of heat in the atmosphere; usually cited in the context of the global warming debate.

Granulated slag A form of blast furnace *slag* that when quenched by a water stream forms sand-sized grains of silicate glass. When very finely ground (*GGBFS*), this material is an *SCM*. Can also be used as a *grinding aid* in the *finish mill*.

Gray cement Cement other than *white* or *colored* varieties. Generally synonymous with gray *portland* cement, but would include other, similar use cements (e.g., *blended cements*). It may or may not include gray *masonry* cement.

Grinding aid A material added in the *finish mill* to aid in the grinding of clinker into *finished cement*.

Grinding circuit The parts of a cement plant where grinding of raw materials is done (*raw mill*) or where clinker is ground into finished cement (*finish mill*).

Grinding plant A stand-alone cement manufacturing facility that grinds clinker that was made at another (usually foreign) location; c.f. *integrated plant*.

Gypsum Calcium sulfate dihydrate ($CaSO_4{\cdot}2H_2O$ or in shorthand $C\bar{S}H_2$); it is a mineral component of portland cement and its function is to control setting time.

H 1) Water (H_2O) in cement chemistry shorthand notation; 2) hydrogen in conventional chemical notation.

Hydrate(d) Refers to a mineral or compound formed from another mineral or compound that has undergone hydration, e.g., *CSH*; *hydrated lime*.

Hydrated lime The compound $Ca(OH)_2$; also called slaked lime. In solid form, sometimes called *portlandite*.

Hydration Chemical combination of water with another compound. Hydration of cement minerals (to form new minerals called *hydrates*) is the key reaction in the hardening and development of strength in concrete. 2) Absorption of structural water into a crystal lattice.

Hydraulic (cement) Refers to a cement's ability to set and harden under, or with excess, water through the hydration of the cement's constituent chemical compounds or minerals.

Integrated plant An informal term used to describe a cement plant that produces both clinker and finished cement; c.f. *grinding plant.*

K 1) Cement chemistry shorthand for potassium oxide K_2O. 2) Potassium in conventional chemical notation.

Kiln The heating apparatus in a cement plant in which *clinker* is manufactured. Unless otherwise specified, may be assumed to refer to a *rotary kiln.*

Kiln line A.k.a. *pyroprocessing line*. The part of the cement plant that manufactures clinker; comprises the kiln itself plus any preheaters and precalciners, plus the clinker cooler apparatus.

Latent cement An infrequently encountered term referring to material having some cementitious character but whose hydraulic cementitious properties are significantly enhanced when interacted with free hydrated lime. The term is most commonly associated with *GGBFS.*

Lime A general term for: 1) the compound CaO (denoted C in cement chemistry shorthand), also called quicklime; 2) the hydrated compound $Ca(OH)_2$ (denoted CH in shorthand) that is more properly termed *hydrated lime* or slaked lime. Solid hydrated lime is sometimes called *portlandite.* 3) The foregoing plus high magnesian or dolomitic forms, e.g., CaO·MgO or $(Ca,Mg)(OH)_2$; 4) *hydraulic* lime, which contains hydraulic silicates.

Limestone 1) A sedimentary rock composed primarily of calcium carbonate (generally as the mineral *calcite*). Limestone is generally the main raw material for cement manufacture. 2) Locally, any rock (e.g., limestone, cement rock, marble) composed primarily of calcium carbonate and used by the plant as its primary raw material in cement manufacture.

M Cement chemistry shorthand for magnesia MgO.

Magnesite A mineral composed of magnesium carbonate $MgCO_3$.

Masonry 1) Refers to construction using natural or manufactured blocks (e.g., bricks, dimension stone, cinderblock), either shaped or unshaped; 2) see *masonry cement.*

Masonry cement A general term for cements used as the binder in mortars. Commonly consists of a mix of portland cement plus plasticizing agents such as lime or ground limestone. Loosely, the term includes true masonry cements, portland-lime cements, plastic cements, and cements for *stucco.*

Mill net value The total or unit value of cement (or clinker) sold to *final customers f.o.b.* the plant, regardless of whether the cement was, in fact, sold from an associated terminal. It thus excludes all shipping costs from the plant to the terminal. It includes any packaging charges, but excludes any discounts. For sales from an independent or stand-alone terminal (usually an import terminal) reporting separately from the plant, the comparable valuation is the *terminal net* value.

Mineral 1) In geology, a naturally occurring inorganic material (or synthetic version thereof) having a defined chemistry and crystal lattice and which has defined physical and chemical properties. 2) In the literature on cement, *mineral* has its geologic meaning but also refers to various synthetic solid phases in clinker or cement that may or may not occur in nature.

Mineralizer Casually synonymous with *flux*, but more properly refers to an addition to the *raw mix* that both promotes the development of a liquid phase and promotes the formation of specific *clinker minerals,* especially *alite.*

Monosulfate Informal term for certain compounds that form during the hydration of the cement mineral C_3A when sulfate is not present in excess. The presence of monosulfate makes the concrete susceptible to later *sulfate attack.*

Mortar 1) The binder in masonry construction. Generally a proportioned mix of masonry (or similar) cement, water, and fine aggregates. 2) In the oldest historical literature, sometimes used synonymously with *cement.*

N 1) Sodium oxide (Na_2O) in cement chemistry shorthand. 2) Nitrogen in conventional chemical notation.

Ordinary As in *ordinary portland cement (OPC)*: a designation used in some countries for *straight* portland cement for general purpose use; OPC is generally comparable to an ASTM C-150 Type I portland cement.

PCA Portland Cement Association. The principal U.S. private organization representing the cement industry.

Pelletized slag Blast furnace slag cooled by quenching with water to generate copious steam and hence a vesicular texture in the slag; a form of expanded slag. Most commonly used for lightweight aggregate, it can also be used as an *SCM* if very finely ground.

Periclase A mineral composed of magnesia MgO.

Plaster 1) Short term for *plaster of Paris*, which is ground calcium sulfate hemihydrate $CaSO_4 \cdot \frac{1}{2}H_2O$ and which hydrates to *gypsum.* 2) A plastic material used to coat and/or decorate walls and similar surfaces (see *stucco*).

Portland cement The most common hydraulic cement. A proportioned and finely interground mixture of portland cement *clinker* and a small amount of calcium sulfate (generally as *gypsum*). In practice, minor amounts of other additives may also be incorporated. Strictly, the term in the United States is limited to the Types I through V varieties (and their air-entrained variants) as defined in ASTM C-150; these types are also collectively called *straight* portland cement. Apart from the straight varieties, "portland cement" when used loosely (a common industry practice) can also include a number of similar hydraulic cements, including *blended* cements, that are based on portland cement clinker plus gypsum.

Portlandite A mineral composed of hydrated lime $Ca(OH)_2$ (denoted CH in shorthand).

Pozzolan(ic) 1) A natural or synthetic *silicate* material that develops *hydraulic* cementitious properties when interacted with hydrated lime. Pozzolans and similar materials are commonly lumped under the term *SCM*. 2) Used loosely, the term is synonymous with *SCM*.

Pozzolana A pozzolanic volcanic ash or tuff.

Precalciner A kiln line apparatus, usually combined with a *preheater*, in which partial to almost complete calcination of carbonate minerals is achieved ahead of the kiln itself, and which makes use of a separate heat source. A precalciner reduces fuel consumption in the kiln, and allows the kiln to be shorter, as the kiln no longer has to perform the full calcination function.

Preheater An apparatus used to heat the *raw mix* before it reaches the dry kiln itself. In modern dry kilns, the preheater is commonly combined with a *precalciner*. Preheaters make use of hot exit gases from the kiln as their heat source.

Premix A bagged product containing proportioned dry ingredients for concrete (i.e., sand, gravel, and cement powder); the customer adds the required amount of water (for cement hydration) at the jobsite.

Pyroprocessing Chemical transformation using intense heat from a flame. In cement, it refers to the manufacture of clinker, which is achieved in a kiln utilizing the flame from an internal burner tube. The pyroprocessing circuit or line is also called the *kiln line* of a cement plant.

Quartz A common mineral having the formula SiO_2. A major source of additional silica in clinker manufacture.

Quicklime The compound CaO; see *lime*.

Raw mill The part of the cement plant in which the raw materials are crushed, ground, and proportioned to form the feed for the kiln.

Raw mix/meal/feed The crushed, ground, proportioned, and thoroughly mixed raw material-feed to the *kiln line*.

Ready-mix(ed) Also spelled as one word (readymix). Common type of *concrete* in which all the ingredients, including water, are preproportioned at the concrete plant and placed into the rotatable drum of a mixing truck. The concrete is then thoroughly mixed via drum rotation while the truck is in transit from the concrete plant to the jobsite. Data on ready-mixed concrete generally include similar concrete made at semimobile batch plants positioned near the jobsite. Batch plants are used where the jobsite is too far from the main concrete plant to allow convenient delivery of ready-mixed concrete by truck.

Retarder An agent (*admixture*) added to a concrete mix to delay *setting* and hardening; c.f. *accelerator*.

Rotary kiln A kiln consisting of a gently inclined, rotating steel tube lined with refractory brick. The kiln is fed with raw materials at its upper end and heated by flame from, mainly, the lower end, which is also the exit end for the product (clinker); c.f., *vertical shaft kiln*.

S 1) Cement chemistry shorthand for *silica* (SiO_2); 2) conventional chemical notation for sulfur.

$\bar{S}$ Cement chemistry shorthand for the sulfite (sulfur trioxide) radical (SO_3^-).

SCM Supplementary cementitious material(s). Materials that can be incorporated within blended cements or in concrete mixes as partial substitutes for portland cement. Common examples are *GGBFS*, *fly ash*, *silica fume*, and *pozzolana*. Casually synonymous with *pozzolan*.

Semidry kiln (plant) A plant in which an initially wet (slurry) raw material feed is dried before pyroprocessing in a dry kiln line.

Set or setting Hydration-induced stiffening of cement paste or concrete. Initial set is the loss of fluidity and plasticity of the material; final set is the development of a certain degree of hardness. Concrete is difficult to work once setting has commenced.

Silica 1) Silicon dioxide, SiO_2; denoted S in cement chemistry shorthand. 2) Pertaining to the silicon dioxide content of a material.

Silica fume Ultrafine particles of disordered *silica* formed as a byproduct of the manufacture of silicon metal, silicon carbide, and silicon alloys (e.g., ferrosilicon). It is used as a *pozzolan* or *SCM*.

Silicate Refers to minerals or compounds whose formulae include silica as a component oxide.

Sintering In clinker manufacture, refers to the process of, or the *thermochemical* reactions, forming the actual clinker minerals, especially those reactions occurring above about 1300°C. The sintering zone of a kiln is that part of the kiln where the sintering reactions occur; it is the highest temperature zone of the kiln. A.k.a. *clinkering* or *burning*.

Slag A silicate melt produced during metal smelting and which essentially is the residuum of the fluxing agents used and the impurities from the metal ores and fuels or reductants. The term also applies to the silicate material after it has cooled to a solid. In the general context of cement and concrete, *slag* (unmodified) refers to iron or steel (furnace) slag. More specifically, as a cementitious component of finished cement or concrete admixture, *slag* refers to the *granulated* variety used either unground as a grinding aid in the finish mill or ground (*GGBFS*) as an *SCM*. As a raw material for clinker manufacture, *slag* generally refers to steel furnace slag. As an aggregate in concrete, *slag* generally refers to air-cooled blast furnace slag.

Slag cement 1) Properly, an ASTM C-595 *blended* cement (Type S), defined as having ≥70% *GGBFS*. 2) Increasingly on the U.S. market, the term *slag cement* is used for a 100% *GGBFS* product that is sold as an *SCM*.

Slaked lime *Hydrated lime*; see also *lime*. Also refers to a liquid solution containing hydrated lime.

Slurry 1) A suspension of insoluble particles in a liquid (generally water) which overall still flows like a liquid. 2) Denoting the raw material feed to a wet kiln.

Straight Refers to portland cement defined in the strict sense (i.e., straight portland cements are those within ASTM standard C-150), as opposed to the general grouping "portland cement" which may also include a variety of other, similar, cements that are based on portland cement clinker.

Stucco A mix of portland cement (or sometimes lime), plasticizers, fine aggregates, and water that will adhere to a steep surface and retain imposed surface impressions and textures, and which is used for coating walls and other surfaces. Also called portland cement *plaster*.

Sulfate attack Deleterious expansion of concrete caused by reaction of certain hydrated *monosulfate* phases in the cement with sulfate-bearing groundwater or soils. The reaction re-forms *ettringite* (a higher-volume phase).

Sweetener An informal term used for a clinker raw material, generally of high purity, that is added to the *raw mix* to rectify a small deficiency in one or more oxides. For example, silica sand is a common sweetener to boost silica.

Terminal net value The total or unit value of cement or clinker sold to *final customers* at a terminal, including any packaging costs and charges for loading the onward conveyance vehicle (typically a train or truck), but excluding onward delivery charges and customer discounts. For import terminals, the terminal net would be comparable to the *c.i.f.* value, plus all terminal charges and markups. Terminal net is analogous to *mill net* for a plant.

Thermochemical Refers to chemical reactions induced by high heat (as in the making of clinker in a kiln).

Tobermorite A somewhat discredited name for the gel phase resulting from the hydration of the cement minerals C_3S and C_2S. The formula of tobermorite, in shorthand, is $C_3S_2H_3$ (sometimes denoted $C_3S_2H_4$). Because C_3S and C_2S actually hydrate to form a whole family of compounds of related formulation, the more general name 'calcium silicate hydrate' or *C-S-H* is preferred to tobermorite.

Transfer-in In USGS cement reporting, the receipt by cement plant X, or its terminals, of material from sister (same company) cement plant Y or its terminals. The term includes cement from Y that was directly delivered to a final customer of X, where X was paid by the customer for the cement. See also *transfer-out.*

Transfer-out In USGS cement reporting, a shipment of material from cement plant X or its associated terminal(s) to sister (same company) cement plant Y, Y's terminals, or Y's final customers (where the customer pays Y). The transfer-out transfers "ownership" of the cement from X to Y. A shipment from a cement plant to its own terminal is not a transfer-out, nor is any shipment to rival (different company) cement companies. See also *transfer in.*

Vertical shaft kiln A vertical, cylindrical, or chimney-type kiln, heated from the bottom, which is fed either with a batch or continuous charge consisting of an intimate mix of fuel and raw materials. Generally considered obsolete for cement manufacture.

VSK *Vertical shaft kiln.*

Wet kiln (plant) Refers to a kiln that takes its crushed and ground raw material feed as a wet (aqueous) slurry; compare with *dry kiln.*

White cement A cement made from white *clinker*, and is based upon raw materials having very low contents of iron (oxides) or other transition elements to avoid the coloring effects of these elements. Unless otherwise specified (e.g., white *masonry* cement), white cement generally is confined to white *portland* cement. White cement is used to make white *concrete* and *mortar*, and serves as a base for *colored* cements, and is generally much more expensive than equivalent-performance *gray* cement varieties.

Links to Other Sources of Information on Cement and Concrete

AASHTO: *http://www.transportation.org*

American Coal Ash Association: *http://www.acaa-usa.org*

American Concrete Institute: *http://www.aci-int.org*

American Society of Civil Engineers (ASCE): *http://www.asce.org*

ASCE National Concrete Canoe Competition: *http://www.asce.org/inside/nccc2006*

ASTM International: *http://www.astm.org*

Cement Kiln Recycling Coalition: *http://www.ckrc.org*

European Cement Association (Cembureau): *http://www.cembureau.be*

The Fly Ash Resource Center: *http://www.rmajko.com/flyash.html*

National Institute of Standards and Technology (NIST): *http://www.nist.gov*
NIST (Virtual Concrete): *http://ciks.cbt.nist.gov/vcctl*

National Ready Mixed Concrete Association: *http://www.nrmca.org*

National Slag Association: *http://www.nationalslagassoc.org*

Portland Cement Association: *http://www.cement.org*

Silica Fume Association: *http://www.silicafume.org*

Slag Cement Association: *http://www.slagcement.org*

U.S. Environmental Protection Agency (EPA): *http://www.epa.gov*
EPA: Coal Combustion Products Partnership: *http://www.epa.gov/epaoswer/osw/conserve/c2p2/*

U.S. Geological Survey (Home page): *http://www.usgs.gov*

U.S. Geological Survey (Minerals information): *http://minerals.usgs.gov/minerals*

World Business Council for Sustainable Development —Cement: *http://www.wbcsdcement.org*

Part 2: Issues Related to Cement Industry Canvasses and Data Interpretation

The U.S. Government has been collecting data on the domestic cement industry for more than a century, mostly through canvasses sent directly to the producers. Annual production and sales data extend back to 1879, and some decadal summations go back to the middle of the century. Monthly sales data have been collected since about the mid-1960s. Collection and reporting of cement data was conducted by the USGS through the 1923 data year, by the U.S. Bureau of Mines (USBM) for the data years 1924–94 (and most of 1995 for monthly data), and again by the USGS for the data years 1995 onwards (monthly data were 1996 onwards). However, for simplicity, reference hereafter to USGS canvasses and publications on cement will include those by the now-closed USBM. Annual data compilations and commentaries have been published as chapters on cement in the USGS Minerals Yearbooks, whereas the monthly data have been published as individual Mineral Industry Surveys reports. In recent years, hard copy distribution of cement reports has been supplemented by electronic dissemination of reports, most recently via the Internet.

The purpose of Part 2 is to better familiarize readers with the USGS periodic canvasses of the U.S. industry, and to discuss some of the issues associated with the physical collection of the data, their completeness and accuracy, and their interpretation. The discussion will make better sense if the reader has at hand a copy of a monthly MIS and an annual report, both of which are available on the Web at URL: *http://minerals.usgs.gov/minerals*. Although we believe that the U.S. cement industry data published by the USGS are the best, most complete, and sometimes the only data available, the data are not perfect. It is hoped that the following discussion will clarify the limitations of the cement data beyond what is possible with footnotes to the actual data tables or with explanations in the periodic reports' texts. Further, although there will be no mention of specific companies, plants, or personnel (except on a fictitious example basis), it is hoped that the discussion will help respondents to the USGS cement industry canvasses complete their data reporting.

The canvasses sent out by the USGS to the cement industry are filled out on a voluntary basis and the high response rate over the years is quite remarkable given the highly proprietary nature of the information requested. It also is a testament to the usefulness of the data to the industry. The USGS reports are neutral sources of consumption and production data that support the accurate evaluation of cement market conditions by interested parties in the cement and concrete industries, academe, and Government agencies.

Protection of proprietary data

Data received through the USGS canvasses are checked for accuracy and are aggregated into non-proprietary tabulations of State or district totals for public dissemination. Protection of proprietary data is done primarily through two tests. The first is the Rule of Three (ROT) test, wherein unless a specified region has three or more companies active (not just plants), the regional total must be withheld (symbolized in tables as W/) or combined with other regions until enough companies are present. Where ROT is not an issue, the data are then examined relative to the Dominant Company (DC) test. The DC test is failed if any one company in a region accounts for 75% or more of the activity, or any two combined account for 90% or more of the activity. A DC test failure requires that the regional total be withheld or combined. Exceptions to the ROT and DC rules are where the USGS has written permission from all of the critical companies in the region to publish the regional total. Thus, for example, in a region having just two producing companies (fails the ROT test), both companies would need to provide written permission to print the regional total. The USGS questionnaires include a question asking (Yes or No) whether the data in the form may be revealed in a nonproprietary way. With rare exception, this question is either left unanswered or is answered 'NO.' If left blank, a 'NO' answer is assumed.

Types of cement canvasses

Currently, the USGS sends out two types of cement surveys. The first is the D16 form, which is a monthly canvass of, primarily, sales of cement to final customers. Some D16 forms are filled out on a specific plant (including its distribution terminals) or independent terminal basis, and others are filled out on a consolidated basis where a single form covers the activities of more than one plant or terminal. Currently, approximately 100 forms are sent out monthly, and the response rate is generally 100%, if not always on a timely basis. This survey misses a few, mostly small, importers that have yet to agree to take part in the survey. The second canvass (D15) is the annual questionnaire, which covers a range of activities of plants or terminals, and is sent to individual plants or independent (mostly import) terminals. Currently, approximately 140 annual forms are sent out, and although the response rate is generally high, the timeliness of the responses is a common problem.

Because both the monthly and annual canvasses are official Government forms, substantive changes to the forms themselves are made only with concurrence of the U.S. Office of Management and Budget, and generally only after negotiation with the industry as to the need for the changes. However, minor changes, such as in rewording of instructions for better clarity, may be made at the discretion of the USGS. Substantive changes have been fairly infrequent.

Reporting units and accuracy

Currently, all of the USGS cement data are collected in nonmetric units, as these are believed to be more familiar to the majority of U.S. cement industry personnel (hence fewer reporting errors), and because they remain the units of domestic cement commerce. More than 80% of the U.S. cement industry is now foreign-owned and is becoming increasingly consolidated and dependent on centralized metric unit bookkeeping. Thus, the industry and the PCA may eventually prefer the USGS to switch to metric unit canvasses, and may even start selling cement in metric units. The USGS cement data have been published in metric units for several years and the ASTM standards to which the industry adheres are in metric units.

In recognition of inherent inaccuracies in determining the weight of shipments and production for most mineral commodities, the USGS rounds most commodity data to 3 significant digits. Given a perfectly accurate original datum having more than 3 significant digits, rounding to 3 significant digits maintains the integrity of the original number to within 0.5%. For most commodity measurements, 3 significant digit reporting will in fact provide more precision than is warranted by the accuracy of the original data. Nevertheless, the U.S. cement industry likes to track cement sales on the basis of individual tons, and has indicated a strong preference that the USGS not round its published cement data except where required for brevity (e.g., tables showing units in thousands). Despite the industry's tracking their activities on the basis of individual tons, it is not uncommon for the USGS to receive revisions to data, some quite large. While it is the current policy of the USGS to accommodate the cement industry's preference for unrounded data, the USGS believes that the unrounded cement data are not accurate to more than 3 significant figures. Exceptions to reporting unrounded data are where estimates have been incorporated; unless qualified by a footnote, such estimated data will have been rounded to no more than 3 significant digits.

USGS Form 9-4039-M
Ind. (rev. 4/7/04)

OMB Control No. 1028-0062
Approval expires: 4/30/07

D16

UNITED STATES
DEPARTMENT OF THE INTERIOR
U.S. GEOLOGICAL SURVEY
986 NATIONAL CENTER
RESTON, VIRGINIA 20192

PORTLAND, BLENDED, AND MASONRY CEMENT

INDIVIDUAL COMPANY DATA - PROPRIETARY

Unless authorization is granted in the section above the signature, the data furnished in this report will be treated in confidence by the Department of the Interior, except that they may be disclosed to Federal defense agencies, or to the Congress upon official request for appropriate purposes. Unless objection is made in writing to the USGS, the information furnished in this report may be disclosed to the respondent's State Geological Survey (or similar State Agency) if the State has appropriate safeguards to prevent disclosing company proprietary data.

(Please correct if name or address has changed.)

FACSIMILE NUMBER
1-800-543-0661

Public reporting burden for this voluntary collection of information is estimated to average 30 MINUTES per response. A Federal agency may not conduct, sponsor, or require a person to respond to a collection of information unless it displays a valid OMB control number. Comments regarding this collection of information should be directed to: U.S. Geological Survey, Statistics and Information Systems Section, 988 National Center, Reston, VA 20192. **Please do not mail survey forms to this address.**

Collection of nonfuel minerals information is authorized by Public Law 96-479 and the Defense Production Act. This information is used to support executive policy decisions pertaining to emergency preparedness, national defense, and analyses for minerals legislation and industrial trends. The USGS relies on your voluntary and timely response to assure that its information is complete and accurate. Please complete and return the form in the enclosed envelope or fax to the above toll-free number **BY THE 10TH OF THE MONTH** following the report month. Use zero (0) when appropriate. Do not report decimals or fractions. Additional forms are available upon request.

You may report on an individual plant basis or on a consolidated basis. For each month, complete Question A and all of page 2.

If you have any questions concerning completion of this form, please contact the Mineral Commodities Data Unit, U.S. Geological Survey, 985 National Center, Reston, VA 20192, Telephone (703) 648-7960.

Name and title of person to be contacted regarding this report	Tel. area code	No.	Ext.
Address No. Street	City	State	ZIP Code

A. This report is *(check one)* **(a) ☐ Plant report (b) ☐ Consolidated report**

If you checked (a): Is the plant covered in this report the one appearing in the address imprint? ☐ Yes ☐ No

If you checked (b): Number of plants covered ____________

COMPLETE ONLY ON THE JANUARY REPORT OR IF A CHANGE OCCURS IN PLANT STATUS OR COVERAGE.

1. If you checked A (a), give

Name of plant ____________

Location of plant State ____________ County ____________ Nearest city ____________

2. If you checked A (b), complete the following item for each plant covered by the consolidated report. If necessary, continue on a supplemental sheet.

* Name of plant ____________

Location of plant State ____________ County ____________ Nearest city ____________

* Name of plant ____________

Location of plant State ____________ County ____________ Nearest city ____________

* Name of plant ____________

Location of plant State ____________ County ____________ Nearest city ____________

* Name of plant ____________

Location of plant State ____________ County ____________ Nearest city ____________

3. Status of the reporting establishment(s)

Operated as a subsidiary ☐ Yes ☐ No

If yes, give name and address of controlling company ____________

OVER

B. Disposition of finished portland, blended, and prepared masonry (incl. portland-lime and plastic) cement during month, by destination. Report the quantity of cement shipped or used by this reporting unit. Cement shipped includes all cement billed, whether produced by this unit or purchased, and should be reported by State of destination. Blended cements are those within ASTM standards C 595 and 1157. Include all cement used directly by this reporting unit except that used to prepare masonry cement. Include all imported cement or cement produced from imported clinker. **Shipments are only those to final customers (including masonry cement blending plants) and exclude inter- and intra- (cement) company transfers, and shipments to terminals. Report to the nearest short ton.**

Destination (1)	Code	Portland cement (short tons) (2)	Blended cement (short tons) (3)	Masonry cement (short tons) (4)
Alabama	010			
Alaska	020			
Arizona	040			
Arkansas	050			
N. California	061			
S. California	062			
Colorado	080			
Connecticut	090			
Delaware	100			
District of Columbia	110			
Florida	120			
Georgia	130			
Hawaii	150			
Idaho	160			
Illinois *(ex. Chicago)*	170			
Chicago Metro	176			
Indiana	180			
Iowa	190			
Kansas	200			
Kentucky	210			
Louisiana	220			
Maine	230			
Maryland	240			
Massachusetts	250			
Michigan	260			
Minnesota	270			
Mississippi	280			
Missouri	290			
Montana	300			
Nebraska	310			
Nevada	320			
New Hampshire	330			
New Jersey	340			
New Mexico	350			
E. New York	364			
W. New York	365			
New York Metro	366			
North Carolina	370			
North Dakota	380			
Ohio	390			
Oklahoma	400			
Oregon	410			
E. Pennsylvania	424			
W. Pennsylvania	425			
Rhode Island	440			
South Carolina	450			
South Dakota	460			
Tennessee	470			
N. Texas	481			
S. Texas	482			
Utah	490			
Vermont	500			
Virginia	510			
Washington	530			
West Virginia	540			
Wisconsin	550			
Wyoming	560			
Puerto Rico	720			
U.S. possessions and territories	810			
Foreign countries	850			
TOTAL	899			

C. Disposition of finished portland, blended, and prepared masonry (incl. portland-lime and plastic) cement, during month, by origin. Report the total quantities of cement noted in SECTION B (LINE 899) by origin of finished cement. Indicate the State where domestic finished cement, including that made from imported clinker, was produced (ground) and the country of origin for imported finished cement. The totals in SECTION C (LINE 899) should equal the totals in SECTION B (LINE 899).

State or Country of origin (1)	Code	Portland cement (short tons) (2)	Blended cement (short tons) (3)	Masonry cement (short tons) (4)
Domestic:				
Foreign:				
TOTAL	899			

D. Production of clinker during month. Report the domestic production of clinker by State of origin. Exclude clinker imports. **The clinker total (LINE 999) will be independent of cement totals (LINE 899).**

State of origin (1)	Code	Clinker production (short tons) (5)
	999	

May tabulations be published which could indirectly reveal the data reported above? ☐ (1) Yes ☐ (2) No

Signature	Title	Date

Printed on Recycled Paper

Monthly canvass and data

The D16 monthly canvass collects data on the disposition of cement sales and the production of clinker. The data gathered via this canvass, together with trade data from the U.S. Census Bureau, are published in the monthly Mineral Industry Surveys (MIS) for cement. The D16 form is a 2-page document, the first page of which, however, has only the address and identification codes for the company or facility in question, and solicits news of changes as to which facilities are being covered by the form.

Sales destination data

Page 2 of the D16 canvass is divided into 3 sections. The first section tracks the cement sales tonnages to final customers by the State of *destination* (location of the customer). The sales are broken out into three cement types: portland cement, blended cement, and masonry cement, and include sales of imported cement and cement made domestically from imported clinker. A few States are subdivided (north vs. south or east vs. west), and two metropolitan areas (Chicago and New York) are broken out, reflecting sufficient market activity to warrant the additional detail. The State sales data are considered to represent the consumption of cement in that State. These data are published in the monthly MIS publications for cement as the tables 2 and 3 series; and (summed for the year) as tables 9 and 10 in the annual reports. In these tables, data for all States are revealed separately. Because a great deal of cement is transported across State lines, and in highly variable quantities, the State destination or consumption totals are generally not subject to proprietary protections.

The destination data are primarily used by analysts interested in regional and temporal comparisons of consumption levels, and by cement and concrete companies seeking to determine their individual market shares in a given State. Although not a problem with market share analyses, general State-level consumption analyses are complicated by the fact that the term *final customer* is taken to mean a concrete company or similar customer (e.g., a building contractor), and, the destination State could simply be that of the customer's headquarters or an address printed on the order invoice instead of the actual concrete plant or cement storage facility owned by the customer. Importantly, the final customer is not the person whose concrete driveway is being redone. A sale to a ready-mixed concrete company in eastern Pennsylvania could easily involve delivery of cement to a concrete batch plant along a highway in New Jersey, or the delivery may have been to a concrete plant in eastern Pennsylvania, but the concrete was then transferred by truck to a jobsite in western New Jersey. Thus, the State destinations data, although defined as the State consumption levels, may actually include some ultimate consumption in adjacent States.

Cement origins data

The second data section of the D16 form breaks out the total sales of each type of cement to final customers in terms of the State(s) and/or countries of origin. Origin is defined as the location where the finished cement was manufactured (ground from clinker). These data are a proxy for monthly production by State, but do not truly equate to this because the cement sales can include material from stockpiles. Because of the link to actual production sites, the cement origins data are subject to ROT and DC proprietary protections, and many States thus require grouping into districts for the totals to be shown. A number of States are not listed at all because they lack cement plants. Origins data are published in the table 1 series in the monthly reports, but appear in the annual report (table 9) only as overall sums showing total domestic vs. foreign origins. Origins data (as a proxy for production) are mainly used to look at production capacity utilization levels in given regions, by comparing the data with grinding capacity data for individual plants published by the PCA. For example, high capacity utilization rates in a given area combined with even higher consumption levels suggest opportunities for a company to add to its existing production or cement distribution (terminals) capacity in a given area.

A significant problem has come to light in recent years with the data on cement origins. As the industry has consolidated, and as cement sales have been by means of an increasingly complex network of plants and terminals (perhaps involving intra- and intercompany transfers), cement companies are finding it increasingly difficult to identify where the cement that they sold in a given month was ground. Instead, the origins being reported increasingly just represent the point of last possession by the company; such data are of little use in the analysis of production capacity utilization.

Breakout of blended cement

Prior to January 1998, the D16 canvass collected data on the disposition (destinations and origins) only of portland and masonry cement, and blended cement data were included with portland cement. Following discussions with the industry, the D16 form and the resulting published tabulations were altered, starting with the January 1998 report, to show blended cements separately from portland cement. The blended cement category (published as table 2b) was created to better track what was perceived to be a rapidly growing market for this material. However, the origins data for blended cement continue to be published as totals combined with portland cement (table 1a). This is because the origins are supposed to refer to the plant location where the cement (i.e., the clinker) was ground and not, as is the case for some blended cement, the location of a terminal where the blending (SCM addition) took place.

Table 3 in the monthly report shows combined portland and blended cement consumption for the given month over a 5-year interval; the combination of the two cement types reflects the fact that they both feed essentially the same concrete markets. In any multiyear comparison of consumption levels (monthly table 2 series) for "portland" cement, it is important to compare pre-1998 portland cement with 1998 and later data for portland cement plus blended cement.

Clinker production

The third section of the D16 canvass also was added in January 1998 at the request of the industry, and captures State-level production of clinker during the month. This addition followed a negotiated decision to define cement origin as being the location where the clinker was ground into finished cement, not where the clinker was manufactured. Clinker production data are subject to ROT and DC proprietary protection. The data are published in the table 4 series of the monthly MIS publication, and the State groupings are usually the same as for the cement origin data.

Treatment of trade data collected by the D16 canvass

Although the D16 form captures the sales and country origins of imported cement, these import data are kept completely proprietary, except that they are summed for the entire industry to a single line total (as "foreign" origin) in the monthly MIS table 1 series. The D16 canvass includes many multiplant/terminal consolidated forms, so much of the country-specific import data recorded therein cannot be linked to a specific cement plant or terminal. And the canvass does not ask for the name of the port handling the imports. Instead, the USGS publishes nonproprietary import data collected by U.S. Customs and made available by the U.S. Census Bureau. These data (monthly MIS tables 5–7), however, do not show who the importers were, or where the cement was consumed. There is thus a complete break between the published trade data (which show import tonnages by country of origin and Customs District of entry) and the State-level consumption data collected through the D16 canvass.

Issues concerning the D16 monthly canvass

The major issues or problems concerning the monthly surveys concern timeliness, completeness, and accuracy. There is an inherent assumption to the survey—one based on general industry and PCA agreement—that companies will report completely and accurately; they will not knowingly misreport their data (to attempt to influence the market or confound their competitors).

Timeliness

Currently, although the USGS is getting a 100% response rate, or very close to it, on the monthly canvass, many responses are slow to arrive. In accordance with an informed agreement between the PCA and the USGS, companies are requested to return their D16 forms (by mail, fax, or Email) so that they will arrive at the USGS within 10 days following the end of the data month in question. But, in fact, many of the responses arrive later than this—sometimes more than 30 days later—and sometimes reminder notices to the nonrespondents must be sent by the USGS. With this response pattern and the time requirements for data entry, computer processing, and data checking, the USGS is currently able to meet its guidelines to 1) electronically disseminate the preliminary monthly tables within 45 days after the end of the data month, and 2) to issue the full report electronically within 60 days. To meet the day 45 preliminary data target release date, all major company data must be in hand by day 43; data still missing from small companies may be obtainable at the last minute during the data checking on day 44 and incorporated. It is important to note that, in fact, the USGS time requirements to process and release the monthly data are only about 5 working days; thus, if the industry met its day 10 day reporting target, the USGS could advance its monthly data releases by as much as 30 working days.

Completeness

The current response rate is essentially 100% of the facilities or companies canvassed. However, the D16 canvass currently misses some of the independent importers of cement. Most of these importers are directly tied to specific concrete companies that are using the material to supply their own cement requirements (rather than selling cement into the open market). These concrete companies already know their own markets, so have little incentive to contribute to the USGS voluntary canvass. Most of the missing import volumes are relatively small, but can still be important in the local market because they represent material not being purchased from domestic producers.

Accuracy

The cement industry tries very hard to supply accurate monthly (and annual) data, both by agreement with the PCA and the USGS, and because their own market analysts are major users of the USGS reports. The main accuracy issue has long been the elimination of double-counting of cement sales. As noted earlier, respondents are only supposed to report, in terms of destinations, the cement sales to *final customers*, not transfers-out of cement to sister cement plants and/or terminals nor sales of cement to rival cement companies. Thus, if cement company A were to report sales to cement company B as if to a final customer, then this cement would be double counted when company B properly reported the sale of the same cement to a true final customer. This mischaracterization of the buyer is fairly common and can lead to large errors if not caught and corrected. For example, in a recent year, a correction for double counting among three companies led to a downward revision of that year's cement consumption in Florida of approximately 500,000 tons.

Situations particularly prone to double counting are those where a cement company arranges for a rival cement company to supply cement to one of the first company's final customers. Such an arrangement could stem from unforeseen production shortfalls at the first cement company, the customer's requirement for a cement type only made by the rival cement company, or a location of the customer much closer to the rival cement company. And the arrangement could be bidirectional (i.e., a swap). The problem is to determine which cement company should report the sale to the USGS. The answer basically is that the reporting to the USGS should be by the cement company that receives payment from the final customer.

Consider an example involving a swap arrangement between "Peach" Cement Co. in Georgia and "Orange" Cement Co. in Florida. For geographical convenience, Peach arranges for Orange to supply 10,000 t of portland cement to a Peach customer in Florida. The Peach customer pays Peach for the cement and so Peach should report (to the USGS) a sale of 10,000 t of portland cement into Florida. In turn, Peach agrees to supply 12,000 tons to an Orange customer in Georgia. The Orange customer pays Orange for the cement received from Peach, so Orange should report a sale of 12,000 tons into Georgia. These reporting "vectors" are regardless of the fact that neither cement company ever had possession of the cement sold into the other's State. And the reporting vectors are unaffected by whether or not the sales arrangement between Peach and Orange involved an exchange of money between the two cement companies or whether it was based simply on supplying comparable tonnages or values of cement to each other's customers.

Miscellaneous reporting problems with the D16 form

a) *Identification of cement type:* Another type of reporting error on the D16 canvass is where a company may report sales of certain types of cement in the wrong cement category. Examples of these errors include inclusion of plastic cements with portland cement (the correct assignation would be with masonry cement), and putting masonry cements into the blended cement category or vice versa. These errors, particularly where consistently made, are difficult to detect. Also difficult to detect is where a company may omit its sales of white or colored cements—these are supposed to be included with the reporting of gray cement.

b) *Assignment to State:* A very common error (both by cement company respondents and the USGS in data entry) is where cement sales are attributed the wrong State; the tonnage is usually misreported in the State immediately above or below the correct State on the form. Where the tonnage involved is large, or the mislocation obvious (such as a southern California plant reporting sales into Maine), the error is quickly spotted on the computer printouts and corrected. But it is easy to miss location errors where the shipments are small and typically erratic. The company may actually sell the occasional truckload of cement into a very distant State, but the inclusion or lack thereof in a given month's canvass could instead be an error. Similarly, a sudden large increase in sales into a certain State could be an error or could represent a short-term, but large, sales contract.

c) *Assignment to State subdivision:* In some USGS published cement tables, four States (CA, NY, PA, and TX) are subdivided (eastern-western or northern-southern) and two metropolitan areas are split out (metropolitan Chicago from IL, and metropolitan New York from NY). Cement shipments are sometimes reported into the wrong State division and this error, unless breaking a pattern, can be very difficult to identify. Accordingly, consumption levels of State subsets must be viewed as being potentially less accurate than that for the total State. For the divided States, and for the Chicago and New York metropolitan areas, the USGS publishes a list of the counties included in, or defining the critical boundaries between, the State subsets (see table 2 in the annual report).

d) *Clinker production:* With clinker production, the most common reporting error stems from the fact that clinker production is generally not routinely weighed. Instead, the amount is calculated by the plant based on the amounts and ratios of raw materials consumed by the kiln. Plants will periodically conduct audits of the calculation, where clinker production for a time will actually be weighed and/or the actual weight of clinker in a storage facility of known volume or capacity will be determined. Such audits commonly reveal a need to revise the monthly data. This revision may be apportioned over several months or the entire tonnage correction accumulated over several months between audits may be credited to the current or next month's production figure. The latter adjustment gives unwarranted spikes in the data.

Another clinker issue is that of reporting units (clinker tons instead of short tons); this will be discussed under the annual canvass section below.

Revisions to monthly data

Companies are encouraged to promptly submit revisions to data should they or the USGS discover errors in past reporting. Revisions are accepted back to January of the preceding year, but no farther. The USGS may choose to ignore very small revisions. It is evident that some companies are not submitting revisions for many of their "small" (single tons to a few hundreds of tons) monthly errors, except where such are part of a long string of errors (wherein, the tons in question have cumulatively become "large"). In the monthly MIS tables, revised data will be indicated by a footnote next to the new number or sometimes alongside column or row headers. The revision indicator footnote will appear only in the publication issue that first shows the revised number; subsequent issues of the report will not indicate the presence of a revision. It is important in the analysis of time-series data to always work backwards in time so as to catch any revisions to data.

The monthly Census Bureau trade data sometimes contain errors; these will be discussed under the annual form section below.

Annual canvass and data

The D15 annual canvass and the data derived from it differ from the monthly (D16) canvass in some important aspects. Unlike the monthly canvass, all of the annual canvass responses represent individual plants (or plant complexes, including their distribution terminals) or independent terminals; that is, there is very little consolidated reporting. Except for the facility location itself, no regional information is gathered; that is, data are not collected on the destination of sales. The annual canvasses focus on the characterization of the total cement tonnages sold, the fuels and raw materials consumed by the plants, and the performances and capacities of individual kilns. The D15 canvass consists of a 4-page questionnaire plus a sheet of detailed instructions.

Page 1 information

The first page of the D15 questionnaire has a preprinted name and address for the specific plant, plus identification codes, but asks that the respondent supply the county and nearest city information. A few, but important, summary questions follow that ask for the production of clinker, clinker purchases, the beginning and yearend clinker stockpiles, and the total plant grinding capacity. In Section 4, a mass balance is set out for portland cement (loosely defined and including blended cement) activity in terms of beginning stockpiles, inflows, outflows, and ending stockpiles. The inflows are split among cement production, transfers-in, purchases from other domestic cement companies, and imports. The outflows consist of shipments to final domestic customers, transfers-out, sales to other cement companies, exports, cement transferred to production of masonry cement, and cement consumed by the plant for miscellaneous purposes.

Subtraction of the outflows from the inflows yields an entry for yearend final stockpiles, or a combination of an inventory adjustment and the actual stockpiles.

Page 2 information

The second page is all based on the tonnage reported on page 1 (section 4 line 240) for the total sales of portland cement to final domestic customers, and is divided into sections 5 through 7. Section 5 asks for the sales to be broken out by tonnage in terms of the types of cement involved. Currently, 15 varieties of portland cement are identified, including Types I–V portland cement and 5 general types of blended cements. Gray portland sales are reported in a separate column from those of white portland cement. *Mill net values* are requested on the total sales of gray portland cement and the total sales of white portland cement. The mill net values are the ex-factory or f.o.b. factory values, inclusive of any bagging charges, but exclusive of any onward transportation costs (to customer or terminal) and discounts. The value may be reported as total dollars or as average dollars per ton—the former is preferred. If the reporting facility is an independent terminal, then the value sought is a *terminal net value*, which is the c.i.f. cost of the cement, plus all terminal unloading and storage charges, any bagging charges or other value added, and the normal terminal markup. Onward shipping costs and discounts are excluded.

Section 6 on page 2 asks for the total sales to final customers to be apportioned among about 15 types of final customers (e.g., ready-mixed concrete producers, brick and block manufacturers, road pavers, etc.). Section 7 asks for the total sales to be apportioned among various methods of transportation (rail, truck, boat or barge) to final customers and/or terminals, subdivided between bulk shipments and bag shipments.

Page 3 information

Page 3 also has three sections (8–10). Section 8 is a mass balance among the inflows, outflows, and stockpiles for masonry cement (including portland-lime and plastic cements), much like the balance for portland cement in section 4 on page 1. Section 9 requests information on the quantities and total heat contents of the individual fuels burned by the kiln line, and electricity consumed by the entire plant. Section 10 asks for information on the performance and technological specifications of the plant's kiln lines.

USGS Form 9-4041-A
Ind. (rev. 4/13/04)

OMB Control No. 1028-0062
Approval expires: 4/30/07

D15

UNITED STATES
DEPARTMENT OF THE INTERIOR
U.S. GEOLOGICAL SURVEY
986 NATIONAL CENTER
RESTON, VIRGINIA 20192

PORTLAND AND MASONRY CEMENT

INDIVIDUAL COMPANY DATA - PROPRIETARY

Unless authorization is granted in the section above the signature, the data furnished in this report will be treated in confidence by the Department of the Interior, except that they may be disclosed to Federal defense agencies, or to the Congress upon official request for appropriate purposes. Unless objection is made in writing to the USGS, the information furnished in this report may be disclosed to the respondent's State Geological Survey (or similar State Agency) if the State has appropriate safeguards to prevent disclosing company proprietary data.

FACSIMILE NUMBER
1-800-543-0661

(Please correct if name or address has changed.)

Name of Plant ____________
County ____________ State ______
Nearest city or town ____________

Public reporting burden for this voluntary collection of information is estimated to average 5 HOURS per response. A Federal agency may not conduct or sponsor, and a person is not required to respond to a collection of information unless it displays a valid OMB control number. Comments regarding this collection of information should be directed to: U.S. Geological Survey, Statistics and Information Systems Section, 988 National Center, Reston, VA 20192. **Please do not mail survey forms to this address.**

Collection of nonfuel minerals information is authorized by Public Law 96-479 and the Defense Production Act. This information is used to support executive policy decisions pertaining to emergency preparedness, national defense, and analyses for minerals legislation and industrial trends. The USGS relies on your voluntary and timely response to assure that its information is complete and accurate. Please complete and return a separate form for each plant in the enclosed envelope or fax to the above toll-free number before March 1. REPORT FIGURES TO THE NEAREST SHORT TON UNLESS OTHERWISE SPECIFIED. *(Please check appropriate boxes, where applicable.)* Use zero (0) when appropriate. Please do not make entries in shaded areas. Additional forms are available upon request.

1. Nature of plant operations covered by this report.

(1) ☐ Grinding mill, kiln, and distribution facilities (2) ☐ Grinding mill only (3) ☐ Distribution facility only

If you checked: (1) Complete ALL Sections. (2) Complete ALL Sections EXCEPT 2 and 10.
(3) Complete ONLY Sections 4 through 8 (EXCLUDE Columns 1-4 of Section 7), and 11 (if applicable).

2. Actual Clinker Production ____________ short tons **Clinker Stockpiles** ____________ short tons *(beginning of year)* ____________ short tons *(end of year)*

3. Actual Grinding Capacity ____________ short tons
(Based on the fineness necessary to grind your normal product mix and making allowances for normal maintenance downtime.)

3A. Total Clinker Purchases and/or transfers in ____________ short tons

FINISHED PORTLAND AND BLENDED CEMENT

** *(DO NOT INCLUDE MASONRY, PLASTIC, AND PORTLAND LIME CEMENTS - SEE SECTION 8.)* **

4. Production, receipts, shipments, and stocks of finished portland and blended cement *(in short tons)*.

Item (1) ***(Include all terminals controlled by reporting facility)***	Code	Quantity (2)
Stocks at beginning of year (include stocks at distribution terminals)	210	
Produced at this plant from clinker (1) ☐ Gray (2) ☐ White (3) ☐ Colored	221	
PLUS		
Transfers-in (**domestic** plant-to-plant within company)	230	
Purchases from other domestic cement producers	231	
Imports from outside the United States and Puerto Rico	232	
TOTAL (CODES 210 through 232)	239	
LESS		
Shipments to **final domestic customers** (directly from plant and/or from distribution terminals)	240	
Invoiced shipments to other domestic cement producers and/or cement suppliers	241	
Transfers-out (**domestic** plant-to-plant within company)	242	
Exports out of the United States and Puerto Rico	243	
Portland cement used to produce masonry, PC lime, and/or plastic cement *(EXCLUDE any masonry, PC lime, and/or plastic cement produced directly from clinker)*	244	
Consumption at facility, other than that in CODE 244	245	
TOTAL (CODES 240 through 245)	249	
Inventory adjustments (+/-)	250	
End of year inventory (CODE 239 - 249 +/- 250)	260	

OVER

DO NOT INCLUDE MASONRY, PORTLAND LIME, AND PLASTIC CEMENT IN SECTIONS 5-7

5. Domestic shipments, by type, of finished portland and blended cement. *(in short tons)*
Mill or terminal net value equals the value of shipments to domestic customers, f.o.b. this reporting facility (i.e., the actual value of shipments less all discounts, allowances, freight charges to customers, and freight charges from domestic producing facilities to distribution facilities.)

Portland and Blended Cement Types *(include air-entrained versions)* (1)	Code	Gray		White	
		Quantity (2)	Mill Net Value (3)	Quantity (4)	Mill Net Value (5)
Types I, II, general use and moderate heat	310				
Type III, high-early-strength	320				
Type IV, low-heat	330				
Type V, high-sulfate-resistance	340				
Block	345				
Oil-well *(API Spec. 10)*	350				
Portland-natural pozzolan blended	360				
Portland-blast furnace slag blended	361				
Portland-fly ash blended	362				
Portland-silica fume blended	363				
Other blended portland *(exclude PC Lime)*	365				
Expansive (ASTM C845)	370				
Regulated fast setting	375				
Waterproof	380				
Other *(please specify)*					
TOTAL (equals CODE 240, SECTION 4)	399		$		$

6. Domestic shipments of finished portland and blended cement by customer type. *(in short tons)*

Type of customer or user (1)	Code	Quantity (2)
Ready-mixed concrete	870	
Building material dealers	810	
Concrete product manufacturers *(please specify below)*:	820	
Brick/Block	821	
Precast/Prestress	822	
Pipe	823	
Other concrete products	824	
Contractors *(please specify below)*:	830	
Airport	831	
Road paving	832	
Soil cement	833	
Other	834	
Federal, State, and local government agencies	840	
Oil well drilling	850	
Mining *(other than oil)*	860	
Waste stabilization	880	
Other *(please specify)*		
TOTAL (equals CODE 240, SECTION 4)	899	

7. Shipments of finished portland and blended cement to final domestic customer by mode of transportation. *(in short tons)*
Total shipments in SECTION B, CODE 649 should equal CODE 240, SECTION 4.

Mode of transportation *(includes pick-up by customer)*	Code	A. Shipments to terminal as first destination		B. Shipments to final customer *(COLUMNS 3-6 should equal LINE 240, SECTION 4)*			
		Plant to terminal		Plant to final customer		Terminal to final customer	
		Bulk (1)	Bag or package (2)	Bulk (3)	Bag or package (4)	Bulk (5)	Bag or package (6)
Rail	610						
Truck	620						
Barge or Ship	630						
Other *(please specify)*							
	649						

PREPARED MASONRY, PORTLAND LIME, AND PLASTIC CEMENT

Report the quantity of prepared masonry, portland lime and/or plastic cement actually produced, received, and shipped. Mill or terminal net value equals the value of shipments to domestic customers, f.o.b. this reporting facility (i.e., the actual value of shipments less all discounts, allowances, freight charges to customers, and freight charges from domestic producing facilities to distribution facilities.)

8. Production, receipts, shipments, and stocks of prepared masonry, portland lime, and/or plastic cement. *(in short tons)*

(400) Specify in percent, the quantity produced as: Gray________% White________% Colored________%

Item (1) *(Include all terminals controlled by reporting facility)*	Code	Quantity (2)	Mill Net Value (3)
Stocks at beginning of year *(include stocks at distribution terminals)*	410		
Produced at this plant: From clinker	421		
From portland and blended cement	422		
PLUS			
Transfers-in *(**domestic** plant-to-plant within company)*	430		
Purchases from other domestic producers	431		
Imports from outside the United States and Puerto Rico	432		
TOTAL (CODES 410 through 432)	439		
LESS			
Shipments to final domestic customer *(from plant and/or from distribution terminal)*	440		$
Invoiced shipments to other domestic cement producers and/or cement suppliers	441		
Transfers-out *(**domestic** plant-to-plant within company)*	442		
Exports out of the United States and Puerto Rico	443		
Consumption at facility	444		
TOTAL (CODES 440 through 444)	449		
Inventory adjustments (+/-)	450		
End of year inventory (**CODE 439 - 449 +/- 450**)	460		

PRODUCTION FACILITY INFORMATION

9. Fuel and energy used at this plant for all operations during the year. *(exclude fuels used at distribution facilities)*

Item (1)	Code	Unit of measure (2)	Quantity (3)	**Million** BTU's (4)	Fuel used for: Kiln	Power plant	Dryer	Other
Natural gas	901	Thousand cubic feet (MCF)			☐	☐	☐	☐
Fuel oil	902	**Thousand** gallons			☐	☐	☐	☐
Coal Bituminous	903	**Thousand** short tons			☐	☐	☐	☐
Anthracite	904	**Thousand** short tons			☐	☐	☐	☐
Coke *(from coal)*	905	**Thousand** short tons			☐	☐	☐	☐
Petroleum coke	906	**Thousand** short tons			☐	☐	☐	☐
Waste fuel: Tires	921	Thousand short tons			Specify type of waste fuel:			
Other solid	922	Thousand short tons						
Liquid	923	**Thousand** gallons						
Electric energy Generated	916	**Thousand** kilowatt hours						
Purchased	917	**Thousand** kilowatt hours						

10. Kiln data for the report year.

Please list each kiln separately. Exclude data for idle kilns that cannot be restarted in less than six months.

Code	Number of days NOT in production: Total (COL. 2 + 3) (1)	Routine maintenance downtime ONLY (2)	Other days NOT in production (3)	Total length of each kiln *(feet)* (4)	Internal diameter of lined kiln *(feet)*: Upper end (5)	Lower end (6)	Maximum output for each kiln per 24 hour day *(short tons)* (7)	Pollution control equipment *(number)*: Glass bag house (8)	Electrostatic precipitator (9)	Pre-heater (10)	Pre-calciner (11)	Type of process: Wet (12)	Dry (13)
701										☐	☐	☐	☐
702										☐	☐	☐	☐
703										☐	☐	☐	☐
704										☐	☐	☐	☐
705										☐	☐	☐	☐

11. RAW MATERIALS CONSUMED IN THE PRODUCTION OF CLINKER AND FINISHED PORTLAND AND MASONRY CEMENT.

Report quantity of each material consumed during the year at this facility. Check whether the material was mined (M) or purchased (P) by you. The materials (quantities) consumed in the production of finished cement exclude those consumed for the production of clinker. Report quantities in short tons, if possible; otherwise please specify the units used.

Raw material (1)	Code	M	P	QUANTITY TO MAKE	
				Clinker (1)	Finished cement (2)
Clinker (purchased):					
Imported (foreign)..	533				
Domestic.............	534				
Calcareous:					
Cement rock.........	501				
Limestone...........	502				
Marl.....................	503				
Oyster, seashells...	504				
Chalk...................	505				
Marble.................	506				
Aragonite............	507				
Coral....................	508				
Cement kiln dust...	561				
Lime....................	593				
Other *(specify)* ________					
Aluminous:					
Shale and schist....	511				
Clay *(all kinds)*......	512				
Staurolite............	513				
Bauxite...............	514				
Aluminum dross....	515				
Alumina..............	518				
Other *(specify)* ________					
Ferrous:					
Iron ore *(all types)*.	528				
Pyrite cinder.........	529				
Mill scale..............	530				
Other *(specify)* ________					

Raw material (1)	Code	M	P	QUANTITY TO MAKE	
				Clinker (1)	Finished cement (2)
Siliceous and pozzolanic:					
Sandstone...........	521				
Silica, other sand..	522				
Quartzite............	523				
Calcium silicates...	524				
Pumice, tuff.........	516				
Pozzolana...........	563				
Other rock pozzolans..........	577				
Other igneous rocks................	571				
Rice husk ash.......	583				
Silica fume..........	584				
Other micro-crystalline silica..	592				
Granulated blast furnace slag.......	586				
Other blast furnace slag.......	587				
Steel furnace slag.......	589				
Other slag *(specify)* ______	591				
Fly ash................	536				
Other ash *(specify)* ______	598				
Other pozzolans and cementitious additives *(specify)* ________	520				
Other:					
Gypsum...............	538				
Anhydrite............	539				
Fluorspar.............					
Other *(specify)* ________	540				

12. List any noteworthy changes in quarry or mill equipment made at this plant during the year

____

____

____

13. List any expansions/modernizations to plant/quarry during reporting year ____

____

Please use additional sheet if needed

Name of person to be contacted regarding this report	Tel. area code	No.	Ext.

Address No. Street	City	State	ZIP Code

May tabulations be published which could indirectly reveal the data reported above?

Value data ☐ (1) Yes ☐ (2) No Other (including quantity) data ☐ (1) Yes ☐ (2) No

Signature	Title	Date

Printed on Recycled Paper

Page 4 information

Section 11 is mainly a list of nonfuel raw materials, grouped by major oxide contribution, wherein the plant's actual consumption (in tons) of each is split between materials burned in the kiln to make clinker and those introduced afterwards to the finish mill to make finished cement. Consumption of outside clinker, split between domestic and foreign manufacture, is also requested.

Data from the D15 annual questionnaire are collated and published by State or district for most items, as national totals for some others (e.g., raw materials consumption, types of portland cement sold), and split between wet and dry technology plants for still others (fuels and electricity consumption). The State-level presentations are subject to ROT and DC proprietary data protection and, thus, many of the State totals require the aggregation of individual data within districts. The data are published in the USGS annual MIS report for cement. This report becomes the cement chapter in the Minerals Yearbook (MYB). These reports also contain U.S. annual trade data supplied by the U.S. Census Bureau, presented in somewhat more detail than in the monthly MIS. The annual report, as with the monthly reports, is available electronically at: *http://minerals.usgs.gov/minerals*.

Issues concerning the D15 annual canvass

As with the monthly survey, the annual canvass is subject to timeliness, accuracy, and completeness issues.

Timeliness

The D15 canvass is mailed out by the USGS in early January following the data year in question, with a requested return due date of March 1. Typically, only 10%–20% of the forms have been returned to the USGS by the requested return date, perhaps 50% are in by the end of April, and about 80% by the end of May. Companies that have not responded by the end of March receive reminder telephone calls from the USGS. Generally, the remaining forms come in June and July, with the last few coming in either August or not at all.

The timeliness problem of the annual canvass reflects the fact that the industry relies less on the annual report than on the monthly reports. In the case of the monthly surveys, both the respondents and users of the data typically are the marketing divisions of the plants or companies and the data are used in monthly economic analyses critical to the business of the company. In contrast, the annual reports are seen as historical documents. Although useful to a company seeking long-term trend data for the purpose of evaluating plans to build new or added capacity or doing other long-term analyses, the corporate need for the annual data appears to be less pressing than for the monthly data.

Another factor in the lateness of annual responses is that cement sales activities and data are increasingly being centralized within companies, which means that the actual production site personnel are commonly unfamiliar with data pertaining to cement sales. That is, the plan's job is mostly just to make cement and ship it to a terminal or make it available for pickup at the plant. Accounting now tends to be done at central or regional sales offices. Thus, plant personnel may find it difficult to fill out sections 4–8 of the D15 canvass, as these sections pertain to sales, and may thus postpone filling out the form. Apart from causing general delays to the publication of the annual report, late responses make more difficult the resolution of any errors found on individual forms. A particularly frustrating problem is where respondents have relied heavily on the previous year's form for guidance in filling out the current form; this leads to errors being perpetuated.

Where, after perhaps repeated telephone requests, data still have not been reported to the USGS, estimates for the missing data will be made, using monthly data and previous years' annual reporting for guidance. The estimated data, while generally a small fraction of national totals, can be significant for specific regions of the country.

Completeness issues

The D15 annual form is sent to all of the currently operational integrated production facilities in the United States and Puerto Rico, and as many grinding plants and independent distribution and import facilities of which the USGS is aware and which agree to participate in the survey. Currently, several independent import terminals are not part of the survey.

For the facilities canvassed, response rates have generally been high (covering commonly 90% or more of the facilities and the total U.S. tonnages). The missing import terminals are a small part of the overall tonnage nationally, but can be important locally.

Even for forms returned to the USGS, it is common for certain questions to have been skipped by the respondents. Where the missing data cannot be obtained by follow up inquiry, they are estimated.

Accuracy issues

Discussion of accuracy will be grouped by topic.

Sales and production tonnage issues

In a typical year, U.S. total cement sales to final customers reported monthly (and summed for the year in tables 9 and 10 in the annual report) may significantly exceed the shipments to final domestic customers plus exports reported on the annual forms. The difference has been as high as 7 million metric tons (Mt) in some recent years (i.e., about 6%–8% of sales). To locate and reduce the discrepancies, all annual forms are now compared with the monthly data and an attempt is made to resolve at least the large discrepancies (say, more than 1% of a plant's total sales). A common source of discrepancies is when a plant's annual form fails to include associated or other terminals that were included on its monthly forms. In such a case, a possible solution is for the company to submit a separate annual form for the missing terminal(s). Another common problem is when a plant's shipments are allocated improperly among sales to final customers, sales to other cement companies, and transfers-out to sister cement plants. The third most common problem is where, contrary to instructions, masonry cement is included in the portland and blended cement data. A careful comparison of annual to monthly sales data has succeeded in reducing the sales tonnage discrepancy significantly; it was just 4 Mt or about 3.8% of total sales for portland cement in 2000, less than 0.2 Mt in 2001 and 2002, about 1.7 Mt in 2003, and less than 0.1 Mt in 2004. However, as a matter of practicality, it is generally not possible for a given plant to resolve most differences in portland cement sales of about 20,000 t or less for the year. This approximate threshold is typically less than about 3% of the given plant's total sales. In most cases, the sales discrepancies have been found to be the result of errors in the annual form.

In contrast, clinker production data—the other data directly comparable between the monthly and annual forms—tend to show only small (< 1%) discrepancies or none at all. The errors in the clinker data have generally been with the monthly reports, and then only with a few months' entries. The errors stem from the fact that plants do not routinely weigh their clinker output; instead they calculate it based on raw materials consumed and then do periodic audits to check their calculations. The USGS is not always informed of the revisions resulting from the audits. Another problem, of unknown extent but which was revealed by one plant's recent survey, was that of reporting clinker in "clinker tons" rather than short tons. A "clinker ton" is a unit of convenience for a cement plant and is defined as that weight of clinker that will yield 1 short ton of portland cement. The weight of a clinker ton varies plant-to-plant; in this particular instance, the clinker ton was 1,915 lbs.

Although monthly origins of sales data are proxies for actual cement production, they are, in fact, rather poor checks against annual production data. This is because the origins data can include stockpiled material and may erroneously reflect a location of last possession instead of production. Also, as with clinker, plants do not routinely weigh their cement output. Cement sales, in contrast, are weighed very precisely.

Another data check done with the annual forms is to see if the clinker production (adjusted for clinker purchases, transfers-in, changes to stockpiles, and for consumption of pozzolans and gypsum) is sufficient to make the finished cement quantities reported.

The reported grinding (or cement) capacity is checked to see if it is adequate for the cement production shown. A problem sometimes encountered is where the grinding capacity appears to be too small. In some cases, this is because it has been misreported (perhaps does not account for a mill upgrade). In some other cases, the low number represents the correct capacity to grind clinker, rather than the capacity needed to grind clinker plus the additives to make the finished cement.

Regionalized production and related data for portland and masonry cement are given in tables 3 and 4, respectively, of the annual report. For portland cement, the reported grinding capacity also is shown, as is a capacity utilization percentage (production/capacity). The capacity utilization data shown in table 3 of the annual report are entirely based on portland plus blended cement output and are thus conservative (ideally, utilization would be compared to the production of total hydraulic cement).

Some regions show an apparently abnormal (usually low) cement or clinker capacity utilization percentage. In most cases, this is an artifact of the USGS policy of including all capacity that was in operation during the year, regardless of for how long. Thus, a new plant that came on-stream late in the year, or an old plant that was closed early in the year, will have its full capacity included in the regional total, but will have little offsetting actual cement or clinker production for the year.

Other data related to sales tonnages

Types of cement sold

Where major differences in total sales to final customers have been resolved, company annual data as to the types of portland cement sold are generally quite accurate in terms of tonnages, except that they may indicate an erroneous type of blended cement (e.g., fly ash blended cements reported as natural pozzolan blended cements), and sometimes are inconsistent as to how hybrid portland cements are reported (e.g., will a Type II/V hybrid be registered on the entry line for (combined) Types I, II or for Type V?).

Mill net values of sales

A far more common problem is where the mill net value of the total portland cement sold has been omitted or is in error. Most omitted values can be obtained through follow-up inquiries, but some cannot, either because the company finds value data to be too sensitive to release or where the canvass form is not returned. For the remaining missing data (typically 10%–15% of the total forms), estimates are made using values from plants that did provide data and which are in the same market area. For plants reporting values in the form of an average value per ton, a total value (rounded) is calculated by the USGS. The mill net values reported on some canvass forms appear to be too low, and likely reflect the omission of bagging and pallet charges (bagged cement is invariably more expensive per ton than bulk cement). Occasionally, a very low unit value will prove to reflect the erroneous reporting of a production cost (this, normally, would be less than the sales price).

In the annual report, separate tabulations (tables 12–14 in reports through 2003, and 11–13 for 2004) show regionally reported mill net values for portland cement (gray and white combined) and masonry cement. Gray cement sales dominate the total value for portland cement, but the unit value of white cement is much higher. For the national average values, a separate tabulation provides the split out unit value averages for gray portland cement, white portland cement, total portland cement, masonry cement, and total cement. The white cement average shown is likely somewhat too high because it includes both primary sales and resales (with markups).

Although the regional average value data are generally presented in the annual report unrounded, the annual report cautions readers that the data contain estimated components and should not be taken too literally; regional or temporal value differences of less than $0.50– $1.00/t are probably not significant.

Sales by customer type

Portland and blended cement sales are broken out by the major construction activities of the customer (customer type) rather than the actual uses of the cement sold; the former classification is believed to be easier for cement companies to track.

Although a noticeable improvement has been seen in recent years, a number of facilities still fail to provide customer-type breakouts of sales. In some cases, the respondents claim not to track this information at all; this is rather surprising for a company overall but may be reasonable for production personnel isolated at the plants themselves. The USGS encourages respondents that lack adequate data to put in their best estimates of the customer types; it is virtually certain that company-provided estimates are preferable to the USGS estimates that would have to be made otherwise. Estimates by the USGS and, it appears, some by the companies, favor the major customer-type categories over the minor ones. Another difficulty is that a company may categorize its customers in ways that do not match the breakouts on the USGS form. In addition, some of the D15 breakout categories are overlapping. A perennial example of this overlap is characterizing cement sales to ready-mixed concrete producers (the largest category) that are also engaged in road-paving (one of the subcategories of "contractors"). Also, the general categories "Concrete products" and "contractors" each include a subcategory called "other" that is intended to capture miscellaneous customers, but instead is used by a few respondents as a catch-all to register undifferentiated sales within the more general category.

Sales by mode of transportation

The final section of the D16 form related to portland cement sales is fairly complicated but seeks to apportion shipments to final customers by method of transportation (rail, truck, boat), form of sale (bulk vs. bag or packages), and location of transfer to the customer (plant vs. terminal). The section asks for the quantity in tons of cement destined for final customers that were transferred during the year from the plant to terminals. This transfer does not require that the cement be fully sold during the year (i.e., some can remain in stockpiles at yearend). Transfers from plant to terminal exclude imported cement, but sales to customers from the terminals include imported material. Thus, there can be a fundamental disconnect in tonnage between domestic loading of terminals and total sales from terminals.

Another problem with this section that has surfaced recently is confusion regarding the split between "bulk" and "container" shipments The form's instructions explain that jumbo bags (a.k.a. "supersacs") are to be registered as "containers" despite their common use for bulk shipments to terminals unequipped for true bulk cement handling. But the term "container" also is an example of how word meanings can change over time. Through about 1970, cement shipment data were still being reported in barrels (1 barrel = 376 lbs), even though no cement had actually been shipped in a barrel in the United States for many decades. Instead, cement was increasingly being shipped in bulk form or, for the non-bulk deliveries, in various (usually 94-lb) bags, sacks, or packages. For brevity, the word "container" was adopted to describe all these non-bulk shipments, and proved satisfactory for many years. However, "container" now is sometimes confused with the large rectangular wood or steel-sided general cargo containers carried on the decks of general freighters, by "ro-ro" ships, and as full loads for tractor trailers and individual railcars. Little, if any, cement gets sold or carried in these rigid containers. So, rather than erroneously register non-bulk sales as (rigid) container sales, some annual respondents have instead erroneously included their bag or package shipments in the "bulk" category. It is unknown how long this error has been occurring, but it has led to a wording change from "container" to "bag or package" on the D15 form.

Regional presentation of data

Cement and clinker production data in the annual report are presented as regional tabulations showing the locations of manufacture. However, with the exceptions of table 9 (and 10 in reports prior to 2004), it is important to note that the annual reports' State and district breakouts of cement sales represent the location of the reporting facility, not the location of consumption. For example, data shown for ". . . cement shipped by producers . . ." in eastern Pennsylvania represent the sales tonnages reported by the producers located in eastern Pennsylvania, including their sales into adjoining States, but the data exclude sales into eastern Pennsylvania from producers outside eastern Pennsylvania. The true consumption of cement in eastern Pennsylvania (and the other States, shown individually) is given in table 9. Individual State production in tons shown in table 9 will not match those in the other tables. The totals for the United States overall, however, should match among all tables, but commonly do not because of reporting errors already noted. A problem with the regional representation of sales occurs if a plant erroneously reports all of its cement shipments as sales to final customers. Some of this material could have been transferred out to sister cement plants in other States.

Fuel and electricity data

Data on fuels burned by the cement industry are of increasing interest for environmental studies, chiefly those regarding CO_2 emissions. In the D15 canvass, plant-level fuel consumption is broken out by fuel type, both in terms of the quantity of fuel and the heat energy realized for each. The fuel types specified are coal (bituminous vs. anthracite), metallurgical coke, petroleum coke, natural gas, fuel oil, and three types of waste fuels: tires, "other solid wastes," and liquid wastes.

A number of problems have been encountered with the reporting of fuel data. The most common is with the reporting units. The D15 form currently specifies nonmetric units of "thousand short tons" for the solid fuels, "thousand cubic feet (MCF)" for natural gas, and "thousand gallons" for the liquid fuels. If other units are, in fact, used, the respondent is asked to specify them precisely. The problem is that respondents commonly report fuel data in full rather than thousand-units. This becomes obvious for solid fuels: no single plant burns 67,525 thousand short tons of coal; the entry should have been 67.525 (or 68) thousand tons. The same error is far less obvious, however, for the liquid and gaseous fuels, as they may be burned in relatively small amounts to warm up a cold kiln or in large quantities as a major fuel or co-fuel during regular kiln operations. Natural gas is especially difficult to pin down, as plants track this fuel from the gas company invoices, and these are generally denoted in MCF (thousand cubic feet), which many respondents misinterpret to mean "million cubic feet." The gas may even be reported as some very large number (actual cubic feet) annotated "MCF" or worse, "MMCF". Since the addition by the USGS of the "MCF" qualifier to the units specification on the form, it is believed that the incidence of errors has decreased significantly.

A check on the fuel units is possible when the company also provides, as requested, the heat (in million Btu, high or gross heat value basis) realized for each fuel. Where Btu data are provided, a Btu/ton fuel ratio can be calculated and compared with standard gross heat values to check for order of magnitude errors in the fuel reporting units. Unfortunately, heat data are commonly omitted or are sometimes reported in amounts that make no sense regardless of the fuel reporting unit. Where omitted, standard heat values can be applied to see if the overall plant's heat consumption appears reasonable, but this is a crude check at best. It is increasingly common for plants to report heat energy on a low or net heat basis instead of the high or gross heat basis requested. A few plants erroneously apportion heat values among the fuels based on the total unit heat consumption per ton of clinker and the mass ratios among the fuels. Because the quality of the heat data provided are so poor, the USGS has not routinely published heat data. However, corrections to the Btu data are being increasingly sought and the quality of the collected data is improving.

Another problem is that the fuel splitout on the D15 form is not very comprehensive. No space is offered for gasoline, for example, although it is known that some plants' consumption (generally very small) is included, for lack of a better place, with fuel oil. And "fuel oil" does not distinguish between distillate fuel oils (like diesel) and residual fuel oils; their heat values are somewhat different. Likewise, no provision is made for liquid petroleum gas. For many years, the D15 form did not distinguish between metallurgical coke (i.e., devolatilized coal) and the completely unrelated petroleum coke (petcoke). Petcoke is by far the more common fuel but is sometimes erroneously put on the metallurgical coke line. Finally, only rarely do respondents identify the waste fuels burned (other than as tires vs. other solid wastes vs. liquid wastes); it is difficult to evaluate or assign Btu values for unidentified waste fuels, or for that matter, liquid wastes identified merely as mix of "spent oils, solvents, alcohols, inks, lubricants, etc . . ."

Electricity data are commonly reported in total kilowatt-hours instead of the "thousand kilowatt-hours" requested; this error is easily spotted and corrected. Occasionally, a plant will record purchased electricity as having been cogenerated. These errors usually are spotted. For many years, electricity consumption was reported on a per-ton portland cement basis instead of, more properly, on a total cement basis. Although U.S. masonry cement output is equivalent to only about 5% of the portland cement production, most is made directly from clinker rather than from portland cement, and so imposes a significant electricity burden within the grinding mill.

Raw materials data

Instead of just requesting the total tonnages of various raw materials consumed, the D15 form now asks that the respondent indicate whether the particular raw material was used to make clinker or was added in the finish mill to make finished cement. This cement vs. clinker distinction caused some confusion when it was introduced for the 1998 canvass, but the industry appears now to have become accustomed to the split. A few materials are, perhaps by habit, still reported in the wrong usage columns, but most of these errors are easily spotted. A few materials are habitually underreported. For example, the CKD space is generally left blank, despite the fact that, at many plants, CKD is returned to the kiln. The problem is that this flow is not routinely measured, and the plant thus has very incomplete or no data on it. Material reported as limestone one year might become marble the next year, or cement rock. The data for gypsum will include natural gypsum and may include synthetic gypsum (a product of flue gas desulfurization at thermal power plants) or the synthetic gypsum may be put in the space for "Other" raw materials (but usually identified) or, unfortunately, it may be put in the space for anhydrite.

The raw materials data are published in table 6 of the annual report. A grand total tonnage is shown at the bottom of the table but it is misleading because, until recently, it included the actual tons of imported (foreign) clinker consumed. Although imported clinker can be viewed as a raw material (because the cement produced from it is counted as U.S. production), it is better if the imported clinker is first converted to an approximate weight of raw materials used to make that clinker. This conversion commenced with the 2002 report.

Kiln data

Data are collected separately for each "operational" kiln at a plant. Operational kilns are those that had clinker production during the year as well as some idle kilns. An idle kiln is deemed operational if it can be restarted, with all operating permits in hand, within a period of 6 months or less; very few U.S. cement plants report idle kilns at present. The data collected for each kiln include its length and internal diameter, its daily (24-hour) clinker output capacity, its technology (wet, dry, dry with preheater, dry with preheater-precalciner, and type of dust control system), and the number and characterization of downtime days.

The characterization of downtime is between that for routine maintenance (previously called "scheduled" downtime) and all other downtime combined. The key datum here is the days for routine maintenance (M); this is used in conjunction with the reported daily capacity (C) to calculate an apparent annual capacity (AC) by the formula:

$$AC \text{ (tons/year)} = (365 - M) \text{ days/year} \times C \text{ (tons/day)}$$

The "working year" component (365 – M) becomes (366 – M) for leap years.

A serious problem derives from how "routine" or "scheduled" maintenance is defined at individual plants, USGS instructions notwithstanding. It is supposed to cover all the planned outages for routine maintenance, including and generally concurrent with the annual or semiannual task of rebricking the kiln, and it is the amount of time that this work is expected or planned to take. Any extensions to this expected work period (delay in arrival of parts, condition of equipment worse than expected, etc.), all breakdowns, all shutdowns for poor market conditions, shutdowns for major upgrades (in excess of routine maintenance done at the same time) are to be entered as "other" downtime. Most kilns need 10–30 days annually for routine maintenance, taken in one or two scheduled kiln outages. However, some plants do not schedule a set period for routine maintenance; they wait for the first problem requiring a shutdown to occur, and then do the routine work as well as the needed repair and "plan" on taking as much time as it takes. Some plants include all of the time for upgrades, arguing that this work was certainly planned and scheduled. For these plants, excessively long routine maintenance periods are shown, and this leads to a calculated apparent annual capacity that is too small, and an annual capacity utilization (clinker production/AC) that is too high; not uncommonly > 100%.

The daily capacities reported for the kilns are supposed to represent a realistic maximum 24-hour sustained output, which may well exceed the conservative ratings provided by the kiln manufacturers. Daily capacities should change (year-to-year) if the equipment has been upgraded, a bottleneck has been removed, or the raw mix changed in some way so that the throughput increases. A potential problem of unknown magnitude is when plants report daily clinker capacities in "clinker tons" rather than short tons. As noted earlier, a clinker ton is a measure of convenience and represents the amount of clinker in 1 short ton of portland cement.

Regionalized kiln performance and clinker production data are published as (currently) table 5 in the annual report. A frequent question is why the published regional daily capacities and average routine maintenance downtimes do not yield the regional annual capacities shown, using the footnoted formula for apparent annual capacity. The answer is that the regional daily and apparent annual capacities are the sums of the actual (not average) individual daily and annual capacities, respectively, for each kiln, whereas the average downtime figure is an average of all the kilns.

Trade data

Trade data, other than single-line totals for cement imports and exports, are not collected on the annual canvass. Instead, as with the monthly reports, nonproprietary U.S. Census Bureau data are utilized for the annual report. Currently, seven trade tables (17–22 through the 2003 report; 16–21 for 2004) are published, all showing quantities (tons) and values. Table 17 lists U.S. exports of hydraulic cement and clinker (combined), by country of destination, and is the only export table in the report. Table 18 shows a summary of combined hydraulic cement and clinker imports, by country of origin. Table 19 breaks out the table 18 data by country of origin and the customs district ("port") of entry into the United States (including San Juan, Puerto Rico). Table 20 is a subset of table 18 showing just the imports of gray portland cement, by country of origin. Table 21 does the same thing for white portland cement, and table 22 shows the subset for imports of clinker.

Values shown on the trade tables are "free alongside ship" (f.a.s.) for exports, and both "customs" and "cost, insurance, freight" (c.i.f.) values for imports.

Removing the portland cement imports (tables 20–21) and clinker imports (table 22) from table 18 leaves a small residuum of imports representing other forms of hydraulic cement. Some of these imports calculate to very high unit values and likely represent either aluminous cement or some highly specialized material.

As earlier noted, both the monthly and annual canvasses miss the activities of a few, mostly small, importers, but the activities of these importers are included in the trade tables. Except for antidumping tariffs still imposed on imports from Mexico and Japan, cement imports into the United States are not subject to import tariffs so there is no incentive to smuggle the material or misreport its tariff code to U.S. Customs. Nevertheless, apparent errors in the monthly or annual trade tables are occasionally discovered by, or brought to the attention of, the USGS. Some of the data on white cement imports calculate to unit values ($/t) that are unrealistically low for white cement. In at least some of these cases, the USGS has determined that the low unit values actually represent gray portland cement (or even gray clinker) imports that the importer had mistakenly invoiced under the white cement tariff code. This mistake is understandable if one considers the abbreviated explanatory wordings offered in some older code tabulations. For example, under the Harmonized Tariff Schedule of the United States (HTS), item HTS code 2523.21.000 is commonly described just as "portland cement" and the next code 2523.29.00 is described as "portland cement, other than white and colored." It is only by reading the second code that one realizes that the first code is actually for white (and/or colored) cement; more recent tabulations now have code 2523.21.00 reading "Portland cement, white or colored." The second code is the correct one for gray portland cement.

Another issue with official trade data is that they do not include relatively low-valued (<$2,000 customs valuation on the shipment) entries for commodities lacking import tariffs or are otherwise deemed innocuous. This omission has been noticeable for clinker imports from Canada that come in by truck into the Seattle, WA (and possibly the Detroit, MI, and Milwaukee, WI) Customs Districts. At least some finished cement entering by truck from Canada appears also to be missing from the official data, but the magnitude of such has been hard to determine, as the tonnages are likely small compared with large waterborne or train-conveyed bulk cement imports from that country.

Where trade data errors appear to be significant, the USGS informs the Census Bureau of the problem, but does not unilaterally issue corrections.

APPENDIX

II

Concrete Pavement Technology Update

OUTLINE

CONCRETE PAVEMENT
CPTP
TECHNOLOGY PROGRAM

Concrete Pavement Technology Update

CPTP has sponsored precast concrete pavement demonstrations in Missouri and other States (see page 288).

The Concrete Pavement Technology Program
CPTP is an integrated, national effort to improve the long-term performance and cost-effectiveness of concrete pavements by implementing improved methods of design, construction, and rehabilitation and new technology. Visit www.fhwa.dot.gov/pavement/concrete for more information.

About CPTP Updates
The CPTP Update is one facet of CPTP's technology transfer and implementation effort. Updates present new products and research findings that emerge from CPTP studies. Place your name on the mailing list with a call (202-347-6944), fax (202-347-6938), or e-mail (sstamey@woodwardcom.com).

In This Update

U.S.Department of Transportation
Federal Highway Administration

Concrete Overlays—An Established Technology With New Applications

The need for optimizing preservation and rehabilitation strategies used to maintain the Nation's highway pavements has never been greater. Concrete overlays have a long history of use to preserve and rehabilitate concrete and asphalt pavements, and many of the practices are well established. However, of recent origin are techniques that use thinner concrete overlays with shorter joint spacing. Field experience over more than 15 years with the thinner concrete overlays under a range of traffic and site conditions has demonstrated their viability as a cost-effective solution to extend the service life of deteriorated asphalt and concrete pavements.

The Federal Highway Administration (FHWA) has initiated several activities to support technology transfer related to concrete overlays. These activities include reviews, on a regional or statewide basis, of current applications of concrete overlays, identification of gaps in technology, and assistance in developing a program—jointly with State departments of transportation (DOTs) and industry—for technology transfer and demonstration projects. FHWA is assisting with organization of meetings at the State and regional levels to help coordinate concrete overlay technology transfer activities.

Overview of Concrete Overlays

Concrete overlays offer a broad range of applications for preserving and rehabilitating asphalt, concrete, and composite pavements. Concrete overlays can be designed for a range of traffic loading to provide long performance lives, 15 to 40+ years, to meet specific needs. Well-designed and well-constructed concrete overlays require low maintenance and can have low life cycle costs. Applications include the following:

- Over existing asphalt pavements
  - Bonded overlay of asphalt pavements
  - Unbonded (directly placed) overlay of asphalt pavements
- Over existing concrete pavements
  - Bonded overlay of concrete pavements
  - Unbonded (separated) overlay of concrete pavements
- Over existing composite pavements
  - Bonded overlay of composite pavements
  - Unbonded (directly placed) overlay of composite pavements

Bonded overlays are typically thin, 2 to 6 in. (50 to 150 mm) in thickness. When bonded to a milled asphalt surface, the overlay panels are typically 6 by 6 ft (1.8 by 1.8 m) or less in dimension.

The *Guide to Concrete Overlay Solutions* (National Concrete Pavement Technology Center, January 2007) is available at www.cptechcenter.org/publications/overlays/. Print copies are available from ACPA (800-868-7633).

BONDED CONCRETE OVERLAYS

Over asphalt

Over concrete

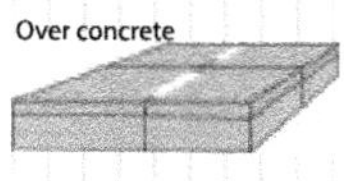

Over composite

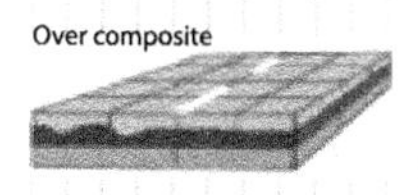

UNBONDED CONCRETE OVERLAYS

Over asphalt

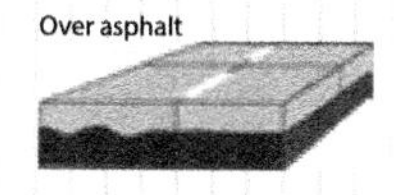

Over concrete

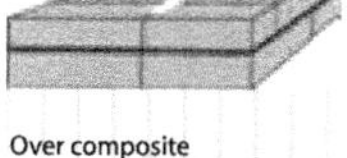

Over composite

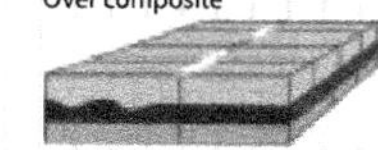

Construction of an unbonded concrete overlay on concrete pavement, I-44, Missouri.

When bonded to a prepared concrete surface, the overlay jointing pattern matches the jointing pattern of the existing jointed concrete pavement. In the case of continuously reinforced concrete pavement, transverse jointing is not provided.

Unbonded overlays are of two types:

- *Conventionally thick overlays,* 6 in. (150 mm) or thicker, are full-width and have transverse joint spacing of 12 to 15 ft (3.7 to 4.6 m).
- *Thinner overlays* are 4 to 6 in. (100 to 150 mm) thick, and the overlay panels are typically 6 by 6 ft (1.8 by 1.8 m) or less in dimensions.

Irrespective of thickness, unbonded overlays are always placed over an asphalt concrete surface—whether an asphalt pavement or an asphalt interlayer/resurfacing placed over a concrete pavement.

Regional Meetings

A keystone activity of FHWA's concrete overlay program is meeting with highway agency senior management to assist with reviewing their agency's application of concrete overlays, identifying gaps in technology, and developing a program, jointly with FHWA, highway agencies, and industry, for technology transfer activities. In November 2007, a meeting was held in Springfield, Virginia, among FHWA; senior management from highway agencies in Delaware, District of Columbia, Maryland, Virginia, and West Virginia; and the American Concrete Pavement Association (ACPA) Virginia Chapter. The mid-Atlantic experience with concrete overlays was reviewed, and the successful experiences of Colorado and Michigan with concrete overlays were discussed. Meetings were also held with the Louisiana DOT staff in Baton Rouge in April 2008 and the South Dakota DOT staff in Pierre in July 2008, where the North Dakota DOT staff participated as well. Local concrete overlay experience was reviewed and assistance discussed.

Concrete Overlays Guide

With support from FHWA, the National Concrete Pavement Technology Center (NCPTC) at Iowa State University developed a best practices *Guide to Concrete Overlay Solutions*. Prepared by a joint industry/State DOT Task Force on Concrete Overlays, the guide presents an overview of concrete overlay systems for resurfacing or rehabilitating pavements and includes detailed guidelines for overlay use:

- Evaluating existing pavements to determine whether they are good candidates for concrete overlays.
- Selecting the appropriate overlay system for a specific pavement condition.
- Managing concrete overlay construction work zones under traffic.
- Accelerating construction of concrete overlays when appropriate.

The guide is available from NCPTC and ACPA (see sidebar).

Best Practices Workshop

Under FHWA's Concrete Pavement Technology Program (CPTP), a 6-hour workshop based on the *Guide to Concrete Overlay Solutions* has been developed and is available to highway agencies. The concrete overlay workshop covers the following topics:

- Pavement fundamentals for overlay applications
- Existing pavement evaluation
- Technical discussion of concrete overlay design, construction, and materials
- Special considerations—geometrics, transitions, shoulders
- Concrete overlay case studies
- Maintenance of traffic
- Overlay performance and life cycle cost considerations
- FHWA/NCPTC Concrete Overlay Field Applications Program

The overlay workshop has been presented to State DOT staff from Pennsylvania, Louisana, North and South Dakota, and Washington State.

Field Applications Program

To advance the use of concrete overlays as cost-effective solutions for a wide variety of pavement conditions, FHWA and NCPTC

Precast Pavement Technology—Moving Forward Fast

Precast pavement technology is a new and innovative construction method that can be used to meet the need for rapid pavement repair and construction. Precast pavement systems are fabricated or assembled off-site, transported to the project site, and installed on a prepared foundation (existing pavement or re-graded foundation). The system components require minimal field curing time to achieve strength before opening to traffic. These systems are primarily used for rapid repair, rehabilitation, and reconstruction of asphalt and portland cement concrete (PCC) pavements in high-volume-traffic roadways. The precast technology can be used for intermittent repairs or full-scale, continuous rehabilitation.

In *intermittent repair* of PCC pavement, isolated full-depth repairs at joints and cracks or full-panel replacements are conducted using precast concrete slab panels. The repairs are typically full-lane width. The process is similar for full-depth repairs and full-panel replacement. Key features of this application are slab panel seating and load transfer at joints.

In *continuous applications,* full-scale, project-level rehabilitation (resurfacing) or reconstruction of asphalt concrete and PCC pavements is performed using precast concrete panels.

This article provides a summary of current initiatives related to precast pavement technology.

FHWA CPTP Initiative. Recognizing the need for effective, rapid rehabilitation methods, the Federal Highway Administration (FHWA), through its Concrete Pavement Technology Program (CPTP), and the Texas Department of Transportation (TxDOT), sponsored a study during the late 1990s that investigated the feasibility of using precast concrete for pavement rehabilitation. At the conclusion of the study, performed by the Center for Transportation Research (CTR) at The University of Texas at Austin, a concept for precast concrete pavement was developed. In March 2002, using this innovative concept, TxDOT completed the first pilot project that incorporated the use of post-tensioned precast concrete pavement along a frontage road near Georgetown, Texas. Since then, FHWA has actively promoted the concept of precast pavement systems to State departments of transportation (DOTs), and demonstration projects have been constructed in California, Missouri, and Iowa to develop field experience with this technology.

Demonstration of the precast post-tensioned concrete pavement system in Sikeston, Missouri.

FHWA's CPTP also sponsored the development of precast pavement technology for full-depth repair of concrete pavements. This work was conducted at Michigan State University and has resulted in several field trials of this technology in Michigan and Ontario, Canada.

For information, contact Sam Tyson, FHWA (sam.tyson@dot.gov).

FHWA Highways for Life Program Initiatives. The purpose of the Highways for LIFE (HfL) program is to accelerate the adoption of innovations and new technologies, thereby improving safety and highway quality while reducing congestion caused by construction. Since about 2006, the HfL program has identified precast concrete pavement technology as a high-priority, innovative technology to produce major benefits for the Nation's highways. In May 2008, the HfL program sponsored a web-based conference that attracted a large audience. The conference provided an update on precast concrete pavement technology and alerted participants to the HfL program's initiative to consider support for funding of several demonstration projects over next few years.

For information, contact Byron Lord, FHWA (byron.lord@dot.gov); www.fhwa.dot.gov/hfl.

A specification clearinghouse and other information on the use of precast concrete pavement is available at http://www.aashtotig.org/?siteid=57&pageid=1826.

Industry Initiatives. Parallel to FHWA's efforts, several organizations in the United States initiated independent development activities to refine precast concrete pavement technologies. These technologies have certain proprietary features and require licensing for product use. Privately developed technologies include the following:

1. The Fort Miller Super Slab system. For information, contact Peter J. Smith (psmith@fmgroup.com) or Michael Quaid (mquaid@fmgroup.com).

2. The Uretek Stitch-in-Time system. For information, contact Mike Vinton (mike.vinton@uretekusa.com).

3. The Kwik Slab system. For information, contact Malcolm Yee (info@kwikslab.com).

Since about 2001, the Fort Miller system has been used on several production projects (continuous and intermittent) for repair and rehabilitation applications. In continuous application, this system simulates conventional jointed plain concrete pavement sections. The Uretek system has also been widely used, according to the developer for intermittent repairs. The Kwik Slab system has been used on a limited basis in Hawaii. This system simulates long jointed reinforced concrete pavement sections. In addition to the proprietary precast concrete pavement systems, generic systems have also been used and are under development. The Port Authority of New York and New Jersey (PANY/NJ) installed generically developed precast concrete pavement test sections at La Guardia International Airport in New York to investigate the feasibility of rapid rehabilitation of a primary taxiway.

Intermittent repairs along a section of I-295 in New Jersey using the Fort Miller Super-Slab system (May 2008).

Highway Agency Initiatives. In the last few years, several agencies have developed specifications that allow the use of precast concrete pavement systems. These agencies include the departments of transportation (DOTs) of New York State, Minnesota, Michigan, and Virginia, as well as the Ontario Ministry of Transport and PANY/NJ. Also, several agencies have installed test sections to demonstrate the feasibility of the precast concrete pavement systems. Accelerated testing of the Fort Miller precast concrete pavement system performed in California indicates that the system tested is capable of long-life service.

For information on agency initiatives, contact the following:

- New York State DOT: Mike Brinkman (mbrinkman@dot.state.ny.us).
- New Jersey DOT: Robert Sauber (rsauber@dot.state.nj.us).
- Caltrans: Tom Pyle (Tom_Pyle@dot.ca.gov).
- New York Thruway: Tom Gemmitti (tom_gemmitti@thruway.state.ny.us)
- Illinois Tollway: Steve Gillen (sgillen@getitpass.com).
- PANY/NJ: Scott Murrell (smurrell@panynj.gov).

AASHTO Technology Implementation Group Activities. Recognizing the increasing interest in precast concrete pavement technologies, AASHTO established a Technology Implementation Group (TIG) during 2006 to support technology transfer activities related to precast concrete pavements. The mission of this AASHTO TIG is to promote the use of precast concrete panels for paving, pavement rehabilitation, and pavement repairs to transportation agencies and owners nationwide and to present an unbiased representation to the transportation community on the technical and economic aspects of the precast paving systems currently available in the market place. In June 2008, the AASHTO TIG completed work on the following documents:

Generic Specification for Precast Concrete Pavement System Approval.

Guidance and Considerations for the Design of Precast Concrete Pavement Systems.

Generic Specification for Fabricating and Constructing Precast Concrete Pavement Systems.

The AASHTO TIG, in cooperation with the FHWA HfL Program, is planning to conduct several technology outreach activities. Roadshows are planned in Delaware, New

Jersey, and Illinois (Illinois Tollway Authority), and at other locations during late 2008 and in 2009.

For information, contact Tim LaCoss (timothy.lacoss@fhwa.dot.gov).

Strategic Highway Research Program 2 (SHRP 2) Project R05, Modular Pavement Technology. The objective of SHRP 2 activities is to achieve highway renewal that is performed rapidly, causes minimum traffic disruption, and produces long-lived facilities. A related objective is to achieve such renewal not just on isolated, high-profile projects, but consistently.

The focus of Project R05 is to develop tools that public agencies can use for the design, construction, installation, maintenance, and evaluation of modular pavement systems. By necessity, the primary focus of this study will be precast concrete pavements. Project funding was established at $1 million. Phase I of the study, to be completed by December 2008, includes a review of modular pavement systems, review of highway agency and industry experience, and identification of successful strategies, promising technologies, and future needs related to modular pavement systems.

For information, contact James Bryant (jbryant@nas.edu; www.trb.org/shrp2/SHRPII_ETG.asp).

Technical Society Activities. Recognizing the high level of interest in precast concrete pavement technologies and to support their members' need for technical information, the following technical organizations have formed task forces to develop technical information on precast concrete pavement technologies:

American Concrete Institute—ACI's Committee 325 established the Subcommittee on Precast Concrete Pavements. The subcommittee is developing a document summarizing current technologies and providing case studies. For information, contact Shiraz Tayabji, Fugro Consultants, Inc. (stayabji@Fugro.com; 410-997-9020).

Precast/Prestressed Concrete Institute—PCI has established a Pavement Committee to develop interest in pavement applications by the precast industry and to develop guidelines for use of precast concrete pavements. For information, contact David Merritt, the Transtec Group (dmerritt@thetranstecgroup.com).

National Precast Concrete Association—NPCA has established a Pavement Committee to develop interest in pavement applications by the precast industry and to develop guidelines for use of precast concrete pavements. For information, contact Peter Smith, The Fort Miller Group (psmith@fmgroup.com).

Developments Outside the U.S.A. Recently, several European countries have started to investigate the use of precast concrete and other precast concrete systems for rapid repair and rehabilitation of pavements. The Dutch have developed the ModieSlab system. The Japanese have used precast concrete slab systems for high-speed slab track applications, tunnel roadways, and airports.

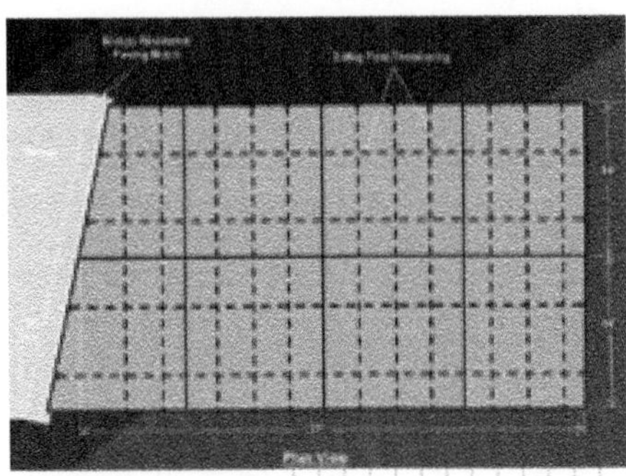

Plan for the precast slab system approach for a Highway 60 bridge near Sheldon, Iowa.

Below, the integrally tied approach using precast post-tensioned concrete slabs on the Sheldon bridge.

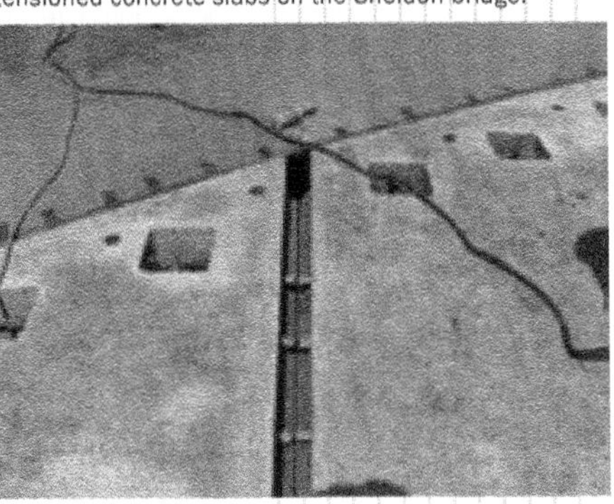

Summary

Precast concrete pavement technologies have come of age and are ready to be implemented. There is still room for innovations to optimize system components and to reduce costs. Current systems have higher initial costs compared to conventional procedures. However, the rapid process (fewer traffic control issues and shorter lane closure times) and better pavement durability can easily offset current higher initial costs. Properly engineered and well-constructed precast concrete pavement systems are capable of providing long service life.

Article prepared by Shiraz Tayabji, FHWA CPTP Implementation Team (stayabji@Fugro.com; 410-997-9020).

MEPDG Development Continues

The MEPDG documentation and the latest version of the MEPDG software are available at http://www.trb.org/mepdg.

Pavement design has come a long way and is preparing to take a new turn—a turn to the future with the development and eventual implementation of a new pavement design procedure. Referred to as the *Mechanistic-Empirical Pavement Design Guide* (MEPDG), the procedure analyzes basic pavement system responses (stresses, strains, and deflections) and relates them to pavement performance in terms of common distress types (e.g., cracking, faulting). This process allows engineers to design for allowable levels of distress over the selected design period.

Development of the MEPDG was sponsored by the National Cooperative Highway Research Program (NCHRP) under Project 1-37A. Initiated in 1996, the project developed a comprehensive approach to the design of pavement structures, with formal documentation produced in 2004 and ongoing refinements to the accompanying software program culminating in the release of version 1.0 in 2007.

Primary benefits of the new design guide are the improved reliability of the resultant designs and the ability to handle any and all combinations of materials, traffic, and climatic conditions. This capability is a major improvement over the widely used 1986/1993 *AASHTO Design Guide for Pavement Structures*, which was based on the limited materials, short test period, single climatic zone, and low traffic levels associated with the conduct of the AASHO Road Test from 1958 to 1960.

Strictly speaking, the MEPDG is not a design tool, but rather a pavement analysis tool. In other words, an initial design is first selected, along with definitions of material properties, climatic conditions, and traffic loadings. The response of that initial design to the climatic and traffic loading conditions is determined, and related to the development of pavement damage and distress. If the projected distress is within tolerable limits, the design is considered adequate; if not, the pavement design is modified and another iteration is performed to determine its suitability. This process is illustrated below, left.

Since the release of the research version in 2004, a number of state highway agencies have been conducting investigatory and feasibility studies on the MEPDG, with an eye towards implementing the procedure. A separate report providing recommended calibration procedures for the MEPDG will be available from NCHRP soon, which should help agencies in their implementation efforts.

During the summer of 2007, the American Association of State Highway and Transportation Officials (AASHTO) Highway Subcommittee on Design and Highway Subcommittee on Materials approved the MEPDG, and in late 2007 the AASHTO Standing Committee on Highways voted to adopt the MEPDG as an interim specification. Much of the work now is focusing on enhancement and refinement of the current version of the MEPDG software and turning it into an AASHTOWare® software program, DARWin ME.

The overall timeline for DARWin ME calls for the development of the program to begin in late 2008 or early 2009 and to be completed in 15 to 18 months. Following the release of DARWin ME, the current version of DARWin will be retired within 12 months.

Article prepared by Kurt Smith, CPTP Implementation Team (ksmith@pavementsolutions.com).

The MEDPG allows engineers to predict how a pavement system with given material properties might respond to climatic conditions and traffic loadings over time.

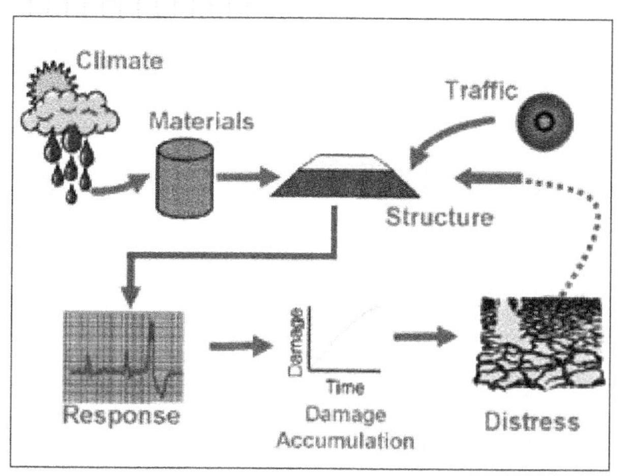

Ongoing Surface Characteristics Research Activities

On March 31, 2008, the American Concrete Pavement Association (ACPA) Noise and Surface Characteristics Task Force was updated on ongoing research activities by Iowa State University (ISU), the Federal Highway Administration (FHWA), Purdue University, and ACPA.

Paul Wiegand, ISU, reported the results of on-board sound intensity (OBSI) testing of 500 unique surface textures and 1,200 unique pavement test sections. Noise emissions ranged from 98 to 109 dBA. Results indicated that sub-100-dBA projects can be obtained using any of the conventional tining or diamond grinding treatments. The causes of excessive noise levels were identified as a too-smooth or too-bumpy surface texture, presence of an ill-designed repeating pattern, and impulse loading effects from joints. ISU will produce a report on texturing guidelines.

Mark Swanlund, FHWA, highlighted some activities of the Pavement Surface Characteristics Program, which covers smoothness, friction, tire–pavement noise, and vehicle splash/spray:

- Working toward an improved functional performance indicator, particularly important as more distress data will be available in the Highway Performance Monitoring System, that may be a better indicator of functional performance than the International Roughness Index alone.
- Updating the 1980 Technical Advisory on skid crash reduction programs.
- Continuing the High Friction Surface Demonstration project. Plans are for three to five States to identify up to 10 locations in each where a high-friction surface can be applied, before-and-after crash data monitored, and effectiveness evaluated. Friction testing equipment (fixed slip or variable slip) other than the locked wheel skid trailer may be recommended. FHWA is conducting an equipment loan program featuring the Dynamic Friction Tester, the Circular Track Meter, and the Grip Tester.
- Enhancing the Traffic Noise Model, which will evaluate the impact of pavement type on tire–pavement noise emissions. An Expert Task Group developed provisional standards for OBSI measurement that were adopted by AASHTO. Projects are also underway to develop quieter pavement technology for both portland cement concrete and hot-mix asphalt surfaces.
- Developing a model to predict the likelihood of splash and spray conditions on existing pavement considering factors like cross-slope, longitudinal profile grade, texture, and rainfall (intensity, duration, and frequency). FHWA's review of methods for quantifying splash and spray found none to be particularly well suited for use with existing roads.

Robert Bernhard, Ph.D., provided an overview of the three-phase research effort on diamond grinding, sponsored by ACPA, at Purdue. One of the more interesting findings is that the results tend to be tire dependent. The grooving experiment involved including an acoustic medium in the groove, and this appeared promising. Innovative textures of various geometric patterns (circles, waffles, or variations) are being evaluated.

Larry Scofield, ACPA, briefed the Task Force on several topics. Current noise and friction testing results on the Next Generation Concrete Surface were presented; three test sections have been placed, and four more potential locations are being pursued. ACPA is collecting friction data to compare longitudinal- and transverse-tined concrete surfaces. In addition, the use of longitudinal grooving to affect roadway surface drainage is being evaluated. ACPA is developing an interactive sound demonstration system that will provide audio comparisons of roadways with different surfaces and textures, to be available soon.

It is only recently that significant research has been devoted to functional performance of concrete pavements, which is heavily influenced by pavement surface characteristics. Perhaps the most challenging task will be to relate those characteristics to the prediction of highway crashes. The results of this research will help bring safe, quiet, durable, and economical pavements to the traveling public.

Two of the 500 unique surface textures tested for on-board sound intensity at Iowa State University. Above, diamond grinding; below, longitudinal tining.

More information on surface characteristics research is available at these sites:

http://www.cptechcenter.org/projects/detail.cfm?projectID=1106856762

http://www.cproadmap.org/research/probstatement.cfm?psID=1646

http://www.tcpsc.com/LittleBookQuieterPavements.pdf

Article prepared by Roger Larson (rlarson@pavementsolutions.com), CPTP Implementation Team.

FHWA, South Dakota, and consultant staff visit a candidate concrete overlay site.

For more information:

Field Applications Program:

- Sam Tyson, FHWA (sam.tyson@dot.gov)
- Dale Harrington, NCPTC (dharrington@snyder-associates.com)

CPTP Concrete Overlay Workshop:

- Sam Tyson, FHWA (sam.tyson@dot.gov)
- Shiraz Tayabji (stayabji@aol.com)

Article prepared by Shiraz Tayabji, CPTP Implementation Team.

are implementing the Field Applications Program. The overall objective of this 2-year program is to increase awareness and knowledge related to concrete overlay applications among State DOTs, cities, counties, contractors, and engineering consultants by demonstrating and documenting various concrete overlay applications. At least six States from diverse regions will participate in the selection, design, and construction of concrete overlay demonstration projects.

The Field Applications Program includes the following components:

- A workshop based on the *Guide to Concrete Overlay Solutions* through CPTP.
- Site visits to select candidate concrete overlay demonstration projects.
- Meetings with an expert team managed by NCPTC for guidance on design and construction of the projects.
- Access to Iowa State University's mobile concrete testing laboratory to document quality control at the project sites.
- Documentation of the design and construction processes.
- Development of recommendations for updating concrete overlay practices.

The Louisiana and South Dakota DOTs have agreed to participate in the FHWA/NCPTC Concrete Overlays Field Applications Program, have hosted the Best Practices workshops, and are in the process of identifying candidate projects.

CPTP Update – New Products and Recent Activities

Products and Reports

Several CPTP reports and related products have been completed recently. These include:

- Cost-Effectiveness of Sealing Transverse Contraction Joints in Concrete Pavements (CPTP Task 9). This report reviews the history of transverse joint construction and sealing in concrete pavements, summarizes the results of a field survey program in which distress and deflection data were collected from 117 test sections, and presents the findings of a detailed analysis of that data.
- Inertial Profile Data for PCC Pavement Performance Evaluation (CPTP Task 63). The ProVal software program (which allows users to view and analyze pavement profiles in many different ways) was employed to help minimize the harmful effects of thermal curling and moisture warping in concrete pavement slabs. Work conducted under this CPTP task contributed to the development of ProVal 2.0.
- Computer-Based Guidelines for Job-Specific Optimization of Paving Concrete (CPTP Task 64). A beta-version software program called COMPASS has been developed to assist engineers in determining the ideal paving mixture for a particular project (see page 293).

A complete listing of CPTP-related reports and products is available at www.fhwa.dot.gov/pavement/pub_listing.cfm.

ETG Plans Next Steps

The CPTP Engineering Expert Task Group met on November 6, 2007, in conjunction with a CPTP-sponsored international conference. The group heard updates on CPTP and related projects and discussed priorities for CPTP activities during the coming months. The following action items were identified:

- Conduct an aggressive marketing effort to promote the CPTP Best Practices workshops.
- Develop a TechBrief on joint sealing following approval of final report.
- Emphasize the need for more concrete pavement training at all levels.
- Update and emphasize the CPTP Product Catalog.
- Consider a TechBrief on COMPASS after the final report is available.
- Organize a conference in spring 2009 focused on effective strategies for concrete pavement maintenance and rehabilitation.

In addition, Bill Farnbach summarized activities at Caltrans related to concrete pavement with a focus on "green" concrete, environmental impact, and long-life pavements.

COMPASS—
Concrete Mixture Performance Analysis System Software

With the emphasis on accelerated construction and long-lasting pavements coupled with a wider variety of materials options, concrete mixture optimization has become more challenging than ever. Mixture design requires the consideration of a wide array of aggregate sources, cement sources and types, chemical admixtures, supplementary cementitious materials, and recycled materials. The designer must also consider the interaction of ingredients within the mixture and how a given environment may affect the construction and the long-term performance of the pavement. "What was needed was a mixture optimization tool that could simplify the approach to the mix design and proportioning process based on job-specific conditions," said Peter Kopac of the Federal Highway Administration's (FHWA) Office of Infrastructure Research and Development. To meet that need, FHWA developed the new COMPASS software.

The Windows-based system consists of two main parts, a knowledge base (expert system) and four computer modules. The optimization process is illustrated below.

The knowledge base is a compilation of information on concrete properties, testing methods, material characteristics, and material compatibilities with one another and with the environment. The information is interactively accessed, filtered, and logically presented to the user.

Four computer modules accept user defined job-specific inputs and perform the analysis:

1. Mix Expert
2. Gradation
3. Proportioning
4. Optimization

Each module has the ability to be used independently, or the user can perform a comprehensive analysis across modules 1 to 4 building upon each consecutive module's analysis. The flow for the comprehensive analysis is as follows:

1. Identify the performance criteria relative to site-specific conditions (Module 1–Performance Criteria).
2. Select materials that best suit the conditions (Module 1–Materials Selection).
3. Determine optimal aggregate gradations for maximum packing density and workability (Module 2–Gradation).
4. Determine preliminary mixture proportions (Module 3–Proportioning).
5. Optimize the concrete mixture proportions based on job-specific criteria (Module 4–Optimization).

The inputs to the first module are cross-referenced with the information in the knowledge base to help guide the user in selecting the performance criteria and materials to meet those criteria. This information is used to optimize the paving mixture for the environment at the pavement project. Modules 2 to 4 include analytical subroutines to optimize materials proportions based on job-specific criteria. Yet, to simplify and reduce training requirements, a number of input options are fixed.

The COMPASS system has undergone extensive peer review and field application testing, and the draft final report and a user's manual have been completed and are undergoing FHWA review.

For more information, contact Peter Kopac at FHWA (peter.kopac@fhwa.dot.gov) or J. Mauricio Ruiz at Transtec (mauricio@thetranstecgroup.com).

To download the beta version of COMPASS, visit http://www.pccmix.com.

Resources:

"Computer-Based Guidelines for Job-Specific Optimization of Paving Concrete, Final Report," The Transtec Group, Inc., Austin, TX, February 2008.

Focus. Federal Highway Administration, FHWA-HRT-08-008, Washington, DC, November 2007.

Article prepared by Ken McGhee, CPTP Implementation Team (kvmcghee3@aol.com).

Optimization Process Employing a Knowledge Base and Computerization

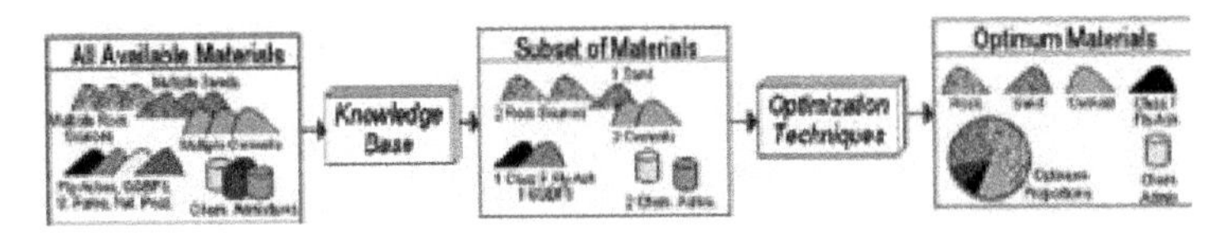

Concrete Pavement Research Roadmap Tracks Underway

Over the past few years the FHWA, in cooperation with Iowa State University (ISU) and the American Concrete Pavement Association (ACPA), has developed the Concrete Pavement Road Map, which outlines a collaborative approach to strategic concrete pavement research and technology transfer for the future. The development process relied heavily on input from the stakeholder community. Twelve research tracks were identified and defined.

The Road Map Operations Group, a team assembled by ISU's National Concrete Pavement Technology Center (CP Tech Center) under contract to the FHWA, is working with industry and government partners to get the research off the ground. Although some work is ongoing in all 12 tracks, in early 2008 three tracks had framework documents developed and are now "officially" getting underway. These are the Mix Design and Analysis track, the Nondestructive Testing track, and the Surface Characteristics track. While these framework documents cannot be released until reviewed by the Roadmap Executive Committee, the research addressed is briefly described below.

Track 1: Performance-Based Concrete Pavement Mix Design System. The final product of this track will be a practical yet innovative concrete mixture design procedure with new equipment, consensus target values, and common laboratory procedures. Full integration of both structural design and field quality control will define a "lab of the future." This track also lays the groundwork for the concrete paving industry to assume more responsibility for mixture designs as State highway agencies move from method specifications to more advanced acceptance tools.

Track 1 has four subtracks:

- Tests
- Models
- Specifications
- Communication

Track 3: High-Speed Nondestructive Testing and Intelligent Construction Systems. This track will develop high-speed, nondestructive quality control systems to monitor pavement properties continuously during construction. As a result, immediate adjustments can be made to ensure the highest quality finished product that meets given performance specifications. Many problem statements in this track relate to track 1 and others.

Three subtracks have been identified:

- Field Control
- Nondestructive Testing Methods
- Nondestructive Testing and Intelligent Construction Systems Evaluation Implementation

Track 4: Optimized Surface Characteristics for Safe, Quiet, and Smooth Concrete Pavements. This track will improve understanding of concrete pavement surface characteristics. Results will provide tools to help engineers meet or exceed predetermined requirements for friction/safety, tire–pavement noise, smoothness, splash and spray, wheel path wear (hydroplaning), light reflection, rolling resistance, and durability. While each of these functional elements of a pavement is critical, the challenge is to improve one characteristic without compromising another, especially when safety is an issue.

Track 4 has the following subtracks:

- Innovative and Improved Concrete Pavement Surfaces
- Tire–Pavement Noise
- Concrete Pavement Texture and Friction
- Safety and Other Concrete Pavement Surface Characteristics
- Concrete Pavement Profile Smoothness
- Synthesis and Integration of Concrete Pavement Surface Characteristics
- Technology Transfer and Implementation of Concrete Pavement Surface Characteristics Research

Article prepared by Ken McGhee, CPTP Implementation Team (kvmcghee3@aol.com).

FHWA Cooperative Agreements to Advance Concrete Pavement Technologies

The Federal Highway Administration (FHWA) recently entered into three cooperative agreements with outside agencies to advance concrete pavement design and construction technologies in support of the Concrete Pavement Technology Program (CPTP). Each agreement provides an initial year of funding with up to four additional years at FHWA's discretion and available funding.

Technology Transfer of Best Practices for Concrete and Concrete Pavements (American Concrete Institute). This agreement will provide education and training activities for FHWA's customers and partners. Deliverables include—

- Training materials for seminars, workshops, and conferences, including a training syllabus and related materials needed by State departments of transportation (DOTs) to prepare their inspectors for the ACI Transportation Inspector Certification test; and updates of materials used in FHWA-sponsored seminars.

- Seminars and related training activities for State DOTs, FHWA's field offices, and members of the concrete industry.

- Conferences for the highway community, including the coordination of a process for FHWA's participation in and funding of up to three conferences per year.

Advancement of Continuously Reinforced Concrete Pavement (CRCP) Through Technology Transfer and Delivery of Industry Guidance for Design and Engineering (Concrete Reinforcing Steel Institute). Products of the ongoing FHWA CPTP include technical guides addressing the design, construction, and repair and rehabilitation of CRCP that are available for review and implementation by highway agencies. FHWA found that new efforts are needed to create, across the entire pavement community, a national awareness of and a willingness to accept CRCP as a proven pavement technology. The agreement provides for an advisory Expert Task Group (ETG) of pavement community representatives. Deliverables include—

- A strategy for technology transfer of CRCP guidance, including a national communications plan to develop a shared sense of purpose among all of FHWA's partners and customers in the States, the concrete pavement industry, and related supplier groups; and conferences, seminars, and workshops for those stakeholders.

- A strategy for assisting State DOTs in accepting and implementing industry guidance for design and engineering of CRCP. Working groups will meet with State DOTs and prepare lists of action items and a timeline for review and modification, as needed, of FHWA's guidance documents.

Advancement of the Precast-Prestressed Concrete Pavement (PPCP) System Through Technology Transfer and Development of Industry Guidance for Design and Engineering (Precast-Prestressed Concrete Institute). This agreement will encourage timely acceptance and technically sound implementation of PPCP as a proven alternative pavement system. A strategic, national communications plan will be structured to gain the support of decision makers across all elements of the pavement community. Deliverables include—

- A strategy for technology transfer for the PPCP system in the agency/owner and industry communities, including preparation and distribution of informational flyers, videos, and technical reports. An ETG comprised of representatives from industry, State DOTs, and FHWA will advise the contractor.

- A strategy for industry guidance of design and engineering of the PPCP system, including organization of program activities among agency/owner and industry communities through technical committee meetings and related activities. FHWA anticipates that the Institute's recently established Pavement Committee will play a key role in developing and disseminating PPCP guidance.

To support the Concrete Pavement Technology Program's technology transfer mission, FHWA is partnering closely with technical societies:

- American Concrete Institute
- Concrete Reinforcing Steel Institute
- Precast-Prestressed Concrete Institute

Products of the new partnership agreements will include training materials, workshops and conferences, and refinement and dissemination of technical guidance.

Article prepared by Ken McGhee, CPTP Implementation Team (kvmcghee3@aol.com).

Sponsors

Federal Highway Administration • American Association of State Highway and Transportation Officials • American Concrete Pavement Association • Concrete Reinforcing Steel Institute • Missouri Department of Transportation • National Concrete Pavement Technology Center • Portland Cement Association • Transportation Research Board

Preconference Workshop on April 21 — Concrete Pavement Preservation

Conference Forums

- The Decisionmaking Process: Where, When, Why?
- State DOT Practices and Directions
- Alternate Delivery Methods

National Conference on Preservation, Repair, and Rehabilitation of Concrete Pavements

April 22–24, 2009 — St. Louis, Missouri

In today's environment, where highway agency budgets cannot fully meet pavement management needs, it is important that the limited funds available be expended in an optimum manner—to extend the useful life of pavements at the least life cycle cost.

Over the past two decades, there has been much progress in developing effective preservation, repair, and rehabilitation (PRR) techniques. However, many gaps remain, and many practices are not implemented consistently from one region to another. This conference will gather information from around the country to close those gaps.

The conference program will present best practices in the use and timing of various PRR treatments to extend the structural capacity and functional characteristics of concrete pavements. In peer-reviewed papers and invited presentations, the program will address evaluation of concrete pavement condition and new advances in PRR technology. Case studies from highway agencies and industry will be highlighted.

The conference will explore closely related topics such as sustainability, accelerated construction, alternative contracting methods, remaining service life and economic tradeoffs, forensic investigations, mitigation of materials-related distresses, and the latest equipment, materials, and testing methods.

Details on paper submission, program, and venue are posted at http://www.fhwa.dot.gov/Pavement/2009conf.cfm and updated periodically.

For more information, contact Shiraz Tayabji (stayabji@aol.com; 410-997-9020).

For more details, visit http://www.fhwa.dot.gov/Pavement/concrete/2009cptpconf.cfm
Register at http://registeruo.niu.edu/iebms/coe/coe_p2_details.aspx?eventid=9087&OC=40&cc=OTHER

U.S. Department of Transportation
Federal Highway Administration
1200 New Jersey Avenue, SW
Washington, DC 20590

Official Business
Penalty for Private Use $300

FIRST CLASS MAIL
POSTAGE AND FEES PAID
FEDERAL HIGHWAY ADMINISTRATION
PERMIT NO. G-66

August 2008
FHWA-IF-08-015

APPENDIX

III

Foundry Sand Facts for Civil Engineers

OUTLINE

Foundry Sand Facts for Civil Engineers

Technical Report Documentation Page

1. Report No.
FHWA-IF-04-004

2. Government Ascension No.

3. Recipient's Catalog No.

4. Title and Subtitle
Foundry Sand Facts for Civil Engineers

5. Report Date
May 2004

6. Performing Organization Code

7. Author(s)
Foundry Industry Recycling Starts Today (FIRST)

8. Performing Organization Report No.

9. Performing Organization Name and Address
TDC Partners Ltd.
417 S. St. Asaph St.
Alexandria, VA 22314

First
PO Box 333
Fall River, MA 01244

10. Work Unit No. (TRAIS)

11. Contract or Grant No.

12. Sponsoring Agency Name and Address
Federal Highway Administration
Environmental Protection Agency
Washington, DC

13. Type of Report and Period Covered
September 01 - September 03

14. Sponsoring Agency Code

15. Supplementary Notes

16. Abstract

Metal foundries use large amounts of sand as part of the metal casting process. Foundries successfully recycle and reuse the sand many times in a foundry. When the sand can no longer be reused in the foundry, it is removed from the foundry and is termed "foundry sand." Foundry sand production is nearly 6 to 10 million tons annually. Like many waste products, foundry sand has beneficial applications to other industries.

The purpose of this document is to provide technical information about the potential civil engineering applications of foundry sand. This will provide a means of advancing the uses of foundry sand that are technically sound, commercially competitive and environmentally safe.

17. Key Words
Foundry Sand, Materials, Highway Construction, Asphalt, Concrete, Flowable fills, Embankment

18. Distribution Statement
No restrictions. This document is available through the National Technical Information Service, Springfield VA 22161

19. Security Classif. (of this report)
Unclassified

20. Security Classif. (of this page)
Unclassified

21. No. of Pages
80

22. Price

Form DOT F 1700.7 (8-72) Reproduction of complete page authorized

This form was electronically produced by Elite Federal Forms, Inc.

SI* (MODERN METRIC) CONVERSION FACTORS

APPROXIMATE CONVERSIONS TO SI UNITS

Symbol	When You Know	Multiply By	To Find	Symbol
		LENGTH		
in	inches	25.4	millimeters	mm
ft	feet	0.305	meters	m
yd	yards	0.914	meters	m
mi	miles	1.61	kilometers	km
		AREA		
in	square inches	645.2	square millimeters	mm^2
ft^2	square feet	0.093	square meters	m^2
yd^2	square yards	0.836	square meters	m^2
ac	acres	0.405	hectares	ha
mi^2	square miles	2.59	square kilometers	km^2
		VOLUME		
fl oz	fluid ounces	29.57	milliliters	mL
gal	gallons	3.785	liters	L
ft^3	cubic feet	0.028	cubic meters	m^3
yd^3	cubic yards	0.765	cubic meters	m^3
NOTE: Volumes greater than 1000l shall be shown in m^3				
		MASS		
oz	ounces	28.35	grams	g
lb	pounds	0.454	kilograms	kg
T	short tons (2000 lb)	0.907	megagrams (or "metric ton")	Mg (or "t")
		TEMPERATURE		
°F	Fahrenheit temperature	5(F-32)/9 or (F-32)/.18	Celsius temperature	°C
		ILLUMINATION		
fc	foot-candles	10.76	lux	lx
fl	foot-Lamberts	3.426	candela/m^2	cd/m^2
		FORCE and PRESSURE or STRESS		
lbf	poundforce	4.45	newtons	N
lbf/in^2	poundforce per square inch	6.89	kilopascals	kPa

APPROXIMATE CONVERSION FROM SI UNITS

Symbol	When You Know	Multiply By	To Find	Symbol
		LENGTH		
mm	millimeters	0.039	inches	in
m	meters	3.28	feet	ft
m	meters	1.09	yards	yd
km	kilometers	0.621	miles	mi
		AREA		
mm^2	square millimeters	0.0016	square inches	in
m^2	square meters	10.764	square feet	ft^2
m^2	square meters	1.195	square yards	yd^2
ha	hectares	2.47	acres	ac
km^2	square kilometers	0.386	square miles	mi^2
		VOLUME		
mL	milliliters	0.034	fluid ounces	fl oz
L	liters	0.264	gallons	gal
m^3	cubic meters	35.71	cubic feet	ft^3
m^3	cubic meters	1.307	cubic yards	yd^3
NOTE: Volumes greater than 1000l shall be shown in m^3				
		MASS		
g	grams	0.035	ounces	oz
kg	kilograms	2.202	pounds	lb
Mg (or "t")	megagrams (or "metric ton")	1.103	short tons (2000 lb)	T
		TEMPERATURE		
°C	Celsius temperature	1.8C = 32	Fahrenheit temperature	°F
		ILLUMINATION		
lx	lux	0.0929	foot-candles	fc
cd/m^2	candela/m^2	0.2919	foot-Lamberts	fl
		FORCE and PRESSURE or STRESS		
N	newtons	0.225	poundforce	lbf
kPa	kilopascals	0.145	poundforce per square inch	lbf/in^2

* SI is the symbol for the International System of Units.
Appropriate rounding should be made to comply with Section 4 of ASTM E380.
(Revised September 1993)

Forward

Metal foundries use large amounts of sand as part of the metal casting process. Foundries successfully recycle and reuse the sand many times in a foundry. When the sand can no longer be reused in the foundry, it is removed from the foundry and is termed "foundry sand." Foundry sand production is nearly 6 to 10 million tons annually. Like many waste products, foundry sand has beneficial applications to other industries.

The purpose of this document is to provide technical information about the potential civil engineering applications of foundry sand. This will provide a means of advancing the uses of foundry sand that are technically sound, commercially competitive, and environmentally safe.

This document was developed by Foundry Industry Recycling Starts Today (FIRST), in cooperation with the U. S. Department of Transportation and the U.S. Environmental Protection Agency. The United States Government assumes no liability for its contents or use. Neither FIRST nor the Government endorses specific products or manufacturers. This publication does not constitute a standard, specification or regulation.

Acknowledgements

This Document was prepared based on the technical references and examples provided by a variety of sources, including the foundry industry, state departments of transportation, university researchers, and contractors. U.S. Environmental Protection Agency provided partial financial support of the preparation of the document.

U.S. Department of Transportation, through the Federal Highway Administration, supports the expanded use of recycled materials in highway construction and provided overall technical review of the document. The authors appreciate the many agency and industry representatives who provided information and technical comments necessary to compile this document.

Foundry Industry Recycling Starts Today (FIRST) is a non-profit 501 (C) (3) research and education organization whose website www.foundryrecycling.org provides access to the technical references used in the preparation of this document. The American Foundry Society (AFS) is a metalcasting industry association which has sponsored research on foundry sand recycling options.

Table Of Contents

Figures

Tables

Chapter 1:

An Introduction to Foundry Sand

Background Information

What is a foundry? A foundry is a manufacturing facility that produces metal castings by pouring molten metal into a preformed mold to yield the resulting hardened cast. The primary metals cast include iron and steel from the ferrous family and aluminum, copper, brass and bronze from the nonferrous family. There are approximately 3,000 foundries in the U.S.

What is foundry sand? Foundry sand is high quality silica sand that is a byproduct from the production of both ferrous and nonferrous metal castings. The physical and chemical characteristics of foundry sand will depend in great part on the type of casting process and the industry sector from which it originates.

Where does it come from? Foundries purchase high quality size-specific silica sands for use in their molding and casting operations. The raw sand is normally of a higher quality than the typical bank run or natural sands used in fill construction sites.

The sands form the outer shape of the mold cavity. These sands normally rely upon a small amount of bentonite clay to act as the binder material. Chemical binders are also used to create sand "cores." Depending upon the geometry of the casting, sands cores are inserted into the mold cavity to form internal passages for the molten metal (Figure 1). Once the metal has solidified, the casting is separated from the molding and core sands in the shakeout process.

In the casting process, molding sands are recycled and reused multiple times. Eventually, however, the recycled sand degrades to the point that it can no longer be reused in the casting process. At that point, the old sand is displaced from the cycle as byproduct, new sand is introduced, and the cycle begins again. A schematic of the flow of sands through a typical foundry can be found in Figure 2.

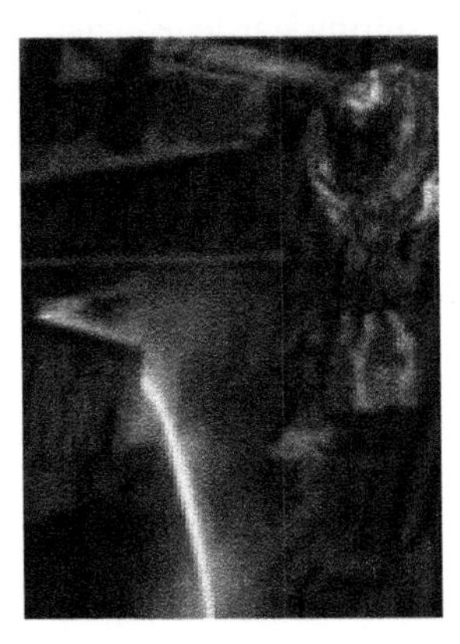

Figure 1. Metal casting in a foundry

How is it produced? Foundry sand is produced by five different foundry classes. The ferrous foundries (gray iron, ductile iron and steel) produce the most sand. Aluminum, copper, brass and bronze produce the rest. The 3,000 foundries in the United States generate 6 million to 10 million tons of foundry sand per year. While the sand is typically used multiple times within the foundry before it becomes a byproduct, only 10 percent of the foundry sand was reused elsewhere outside of the foundry industry in 2001. The sands from the brass, bronze and copper foundries are generally not reused. While exact numbers are not available, the best estimate is that approximately 10 million tons of foundry sand can beneficially be used annually.

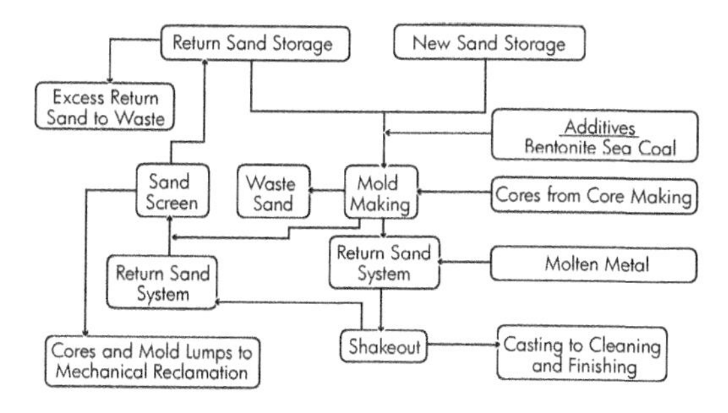

Figure 2. How sand is reused and becomes foundry sand

Foundry Sand Uses and Availability

What makes it useful? Foundry sand is basically fine aggregate. It can be used in many of the same ways as natural or manufactured sands. This includes many civil engineering applications such as embankments, flowable fill, hot mix asphalt (HMA) and portland cement concrete (PCC). Foundry sands have also been used extensively agriculturally as topsoil.

What is being done with it? Currently, approximately 500,000 to 700,000 tons of foundry sand are used annually in engineering applications. The largest volume of foundry sand is used in geotechnical applications, such as embankments, site development fills and road bases.

Where is it available? Foundries are located throughout the United States in all 50 states. However, they tend to be concentrated in the Great Lakes region, with strong foundry presence also found in Texas and Alabama (Figure 3).

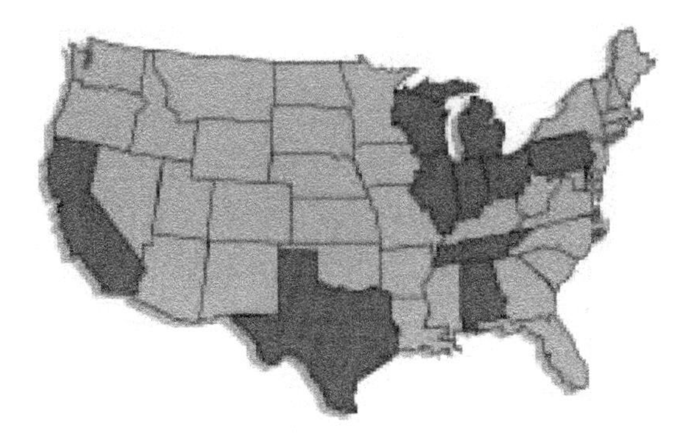

Figure 3. Top ten foundry production states in the U.S.

How does the foundry sand industry operate? Historically, individual foundries have typically developed their own customer base. But over time, foundries have joined together to create regional foundry consortia to pool resources and to develop the recycled foundry sand industry. FIRST (Foundry Industry Recycling Starts Today) is a national coalition of member foundries. FIRST focuses on market development of sustainable options for beneficial reuse of foundry industry byproducts.

Types of Foundry Sand

How many types of foundry sand are there? There are two basic types of foundry sand available, green sand (often referred to as molding sand) that uses clay as the binder material, and chemically bonded sand that uses polymers to bind the sand grains together.

Green sand consists of 85-95% silica, 0-12% clay, 2-10% carbonaceous additives, such as seacoal, and 2-5% water. Green sand is the most commonly used molding media by foundries. The silica sand is the bulk medium that resists high temperatures while the coating of clay binds the sand together. The water adds plasticity. The carbonaceous additives prevent the "burn-on" or fusing of sand onto the casting surface. Green sands also contain trace chemicals such as MgO, K_2O, and TiO_2.

Chemically bonded sand consists of 93-99% silica and 1-3% chemical binder. Silica sand is thoroughly mixed with the chemicals; a catalyst initiates the reaction that cures and hardens the mass. There are various chemical binder systems used in the foundry industry. The most common chemical binder systems used are phenolic-urethanes, epoxy-resins, furfyl alcohol, and sodium silicates.

Foundry Sand Physical Characteristics

What is the typical particle size and shape?
Foundry sand is typically subangular to rounded in shape. After being used in the foundry process, a significant number of sand agglomerations form (Figure 4). When these are broken down, the shape of the individual sand grains is apparent.

Figure 4. Unprocessed foundry sand (Courtesy Lifco Industries)

What are some of the physical properties?
Foundry sand has many of the same properties as natural sands. While one foundry sand will differ statistically from another, recently published properties from Pennsylvania provide fairly typical values. Pennsylvania foundry sands are classified in two categories:

- Foundry sand with clay (5%) – FS #1
- Foundry sand without clay – FS #2

Table 1 shows the results for bulk density, moisture content, specific gravity, dry density, optimum moisture content and permeability measured using the applicable ASTM standard.

Property	ASTM Standard	Foundry Sand with Clay (5%) FS#1	Foundry Sand without Clay FS#2
Bulk density (pcf)	C29	60-70	80-90
Moisture content (%)	D2216	3-5	0.5-2%
Specific gravity	D854	2.5-2.7	2.6-2.8
Dry density (pcf)	D698 Standard Proctor	110-115	100-110
Optimum moisture content (%)	D69	8-12	8-10
Permeability coefficient (cm/s)	D2434	10^{-3}-10^{-7}	10^{2}-10^{-6}

Table 1. Typical physical properties of foundry sand

Figure 5 compares the gradations of these materials to the ASTM C33 upper and lower limits (See Chapter 6). Foundry sand is commonly found to be a uniform fine sand, with 0 to 12% bentonite or minor additives. The quantity of bentonite or minor additives depends on how the green sand has been processed.

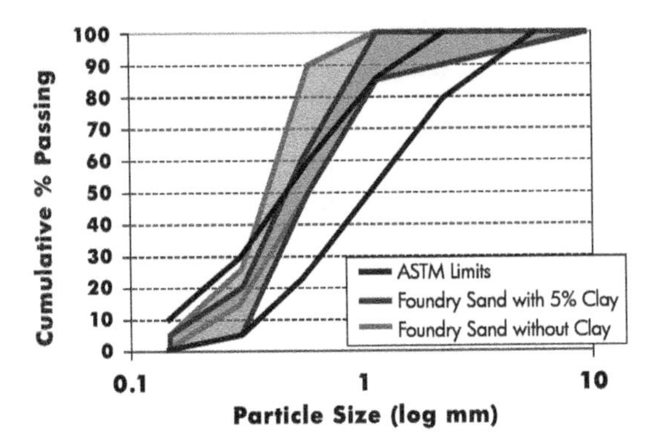

Figure 5. Foundry sand gradation, as compared to ASTM C33

What color is foundry sand? Green sands are typically black, or gray, not green! (Figure 6). Chemically bonded sand is typically a medium tan or off-white color.

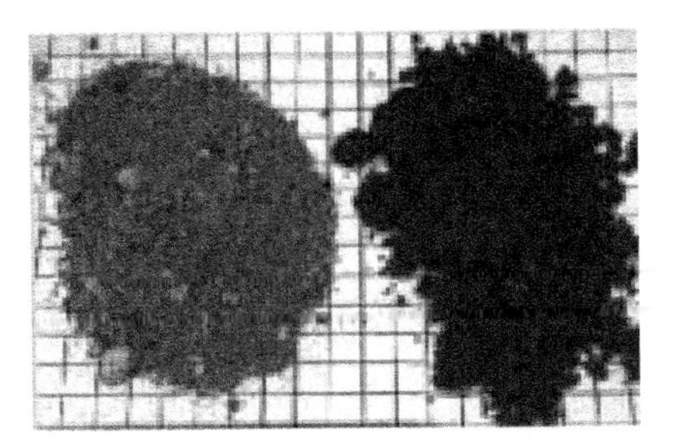

Figure 6. Green sands from a gray iron foundry (after processing and screening)

Foundry Sand Quality

What determines foundry sand quality?
The quality of foundry sand can be quantified by its durability and soundness, chemical composition, and variability. These three characteristics are influenced by various aspects of foundry sand production.

Durability/soundness of foundry sand is important to ensure the long-term performance of civil engineering applications. Durability of the foundry sand depends on how the sand was used at the foundry. Successive molding can cause the foundry sand to weaken due to temperature shock. At later stages of mold use, this can lead to the accelerated deterioration of the original sand particles. However, in civil engineering uses, the foundry sand will not normally be subjected to such severe conditions. In geotechnical applications, foundry sand often demonstrates high durability.

Chemical composition of the foundry sand relates directly to the metal molded at the foundry. This determines the binder that was used, as well as the combustible additives. Typically, there is some variation in the foundry sand chemical composition from foundry to foundry. Sands produced by a single foundry, however, will not likely show significant variation over time. Moreover, blended sands produced by consortia of foundries often produce consistent sands. The chemical composition of the foundry sand can impact its performance.

Variability. Reducing the variability of the foundry sand is critical if consistently good engineering products are to be produced. Foundry sand suppliers should understand and control foundry sand variability so that they can provide customers with a consistent product.

How can I know I'm getting good quality?
Methods to ensure that foundry sands conform to specifications vary from state to state and source to source. Some states require testing and approval before use. Others maintain lists of approved sources and accept project suppliers' certifications of foundry sand quality. More and more, foundry sand generators are determining the engineering properties of their sands.

The degree of quality control necessary depends on experience with the specific foundry sand and its history of variability. Many purchasers require source testing and a certification document to accompany the shipment.

How should foundry sand be handled?
Foundry sand is most often collected and stockpiled outside of the foundries, exposed to the environment. Prior to use in an engineering application, the majority of foundry sand is:

- Collected in closed trucks and transported to a central collection facility;
- Processed, screened, and sometimes crushed to reduce the size of residual core sand pieces. Other objectionable material, such as metals, are removed.

Foundry Sand Economics

The success of using foundry sand depends upon economics. The bottom line issues are cost, availability of the foundry sand and availability of similar natural aggregates in the region. If these issues can be successfully resolved, the competitiveness of using foundry sand will increase for the foundries and for the end users of the sand. This is true of any recycled material.

Foundry Sand Engineering Characteristics

What are the key engineering properties of foundry sand? Since foundry sand has nearly all the properties of natural or manufactured sands, it can normally be used as a sand replacement. It can be used directly as a fill material in embankments. It can be used as a sand replacement in hot mix asphalt, flowable fills, and portland cement concrete. It can also be blended with either coarse or fine aggregates and used as a road base or subbase material. Table 2 shows the relative ranking of foundry sand uses by volume.

Ranking	Application
1	Embankments/Structural Fills
2	Road base/Subbase
3	Hot Mix Asphalt (HMA)
4	Flowable Fills
5	Soil/Horticultural
6	Cement and Concrete Products
7	Traction Control
8	Other Applications

Table 2. Foundry sand applications by volume

According to a recent survey of 10 states, foundry sand has been approved for use by various agencies within the state in the following engineered applications (listed in Table 3). It is expected that more states will be supplementing these uses as they become more familiar with the material.

	IA	IL	IN	MI	MN	NJ	NY	OH	PA	WI
Landfill daily cover				X			X	X		X
Highway embankment				X				X		X
Roadway subbase	X		X	X	X	X		X		X
Parking lot subbase				X	X	X		X		X
Concrete and asphalt		X	X	X	X		X	X	X	X
Foundation subgrade fill	X							X		X
Flowable fill							X	X		X
Generate fill		X	X	X				X		
Other		X	X					X		X

Table 3. Engineered uses of foundry sand

Foundry Sand Environmental Characteristics

What about trace elements in foundry sand? Trace element concentrations present in most clay-bonded iron and aluminum foundry sands are similar to those found in naturally occurring soils. The leachate from these sands may contain trace element concentrations that exceed water quality standards; but the concentrations are no different than those from other construction materials such as native soils or fly ashes. Environmental regulatory agencies will guide both the foundry sand supplier and the user through applicable test procedures and water quality standards. If additional protection from leachate is desired, mechanical methods such as compacting and grading can prevent and further minimize leachate development.

In summary, foundry sand suppliers will work with all potential users to ensure that the product meets environmental requirements for the engineering application under consideration.

Closing

Do I need to know more about this technology to use it confidently? Foundry sand can be used to produce a quality product at a competitive cost under normal circumstances. The remaining chapters of this publication provide a general overview of foundry sand use in various civil engineering applications. It will familiarize highway engineers and inspectors with this technology. This publication is also designed to assist those individuals who have little or no previous experience using foundry sand or no experience in a particular application of foundry sand.

Chapter 2:

Foundry Sand in Structural Fills and Embankments

Engineering Properties For Embankments

How are embankment materials generally classified? Embankment materials used in construction (Figure 7) are generally classified on the basis of soil type, grain size distribution, Atterberg limits, shear strength (friction angle), compactability, specific gravity, permeability and frost susceptibility.

Figure 7. Embankment with foundry sand subbase (Ohio Turnpike, sand from Ford Motor Company supplied by Kurtz Bros. Inc.)

Soil Classification. Foundry sand would normally be classified under the Unified Soil Classification System (USCS) as SP, SM or SP-SM and under AASHTO as A-3, A-2, or A-2-4. It is a nonplastic or low plasticity sand with little or no fines. Some foundries or foundry sand suppliers will process the sand to remove the majority of silts or clays that may be present. The silt or clay content can range from 0 to 12%.

Grain Size Distribution. Foundry sand consists of a uniform sand, with a coarse appearance. Typical gradations are provided in Chapter 1.

Atterberg Limits (Liquid and Plastic). Typically foundry sand without fines is nonplastic. The plastic behavior can depend on the clay content. For foundry sand with 6 to 10% clay, a liquid limit LL greater than 20 and a plastic index PI greater than 2 are typical.

Shear Strength (Friction Angle). Foundry sands have good shear strength. For foundry sands without clay, the direct shear test is used to measure its friction angle. It ranges from 30^0-36^0, which is comparable to conventional sands. Its shear strength is superior to silts, clays or dirty sands, showing that foundry sand is acceptable for use as an embankment material. The triaxial shear strength test can be used to measure the drained shear strength, friction angle and cohesion of foundry sands that contain clay. A typical value of the friction angle and cohesion for these sands is 28^0 and 3700 psf, respectively. But these properties can vary. Foundry sand used on the Ohio Turnpike had a friction angle of 35^0 and cohesion of 6100 psf.

Compaction. Compaction of foundry sand is needed to increase its density during embankment construction. Moisture-density relationships have been developed for green sands with 0 to 5% fines, green sands with 5 to 12% fines and chemically bonded sands. They show the optimal moisture content for maximum dry density for a specified level of compaction. In Figure 8, there is a definite peak in the moisture-density curve for green sands with fines between 5 and 12%. The green sands with few fines and the chemically bonded sands produce a flatter curve. The influence of water is not as significant for them. However, both curves are relatively flat, when compared to plastic soils.

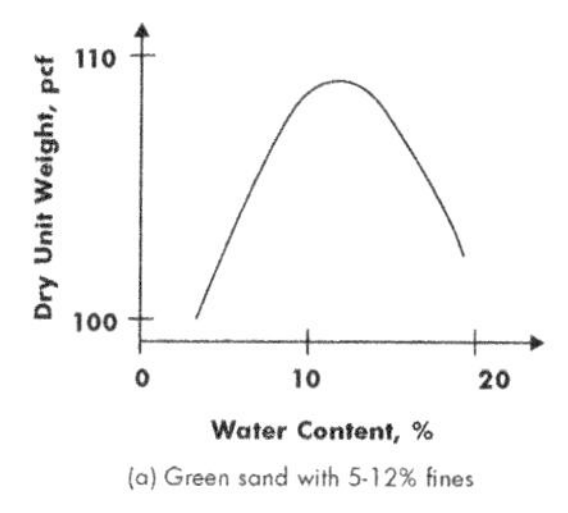

(a) Green sand with 5-12% fines

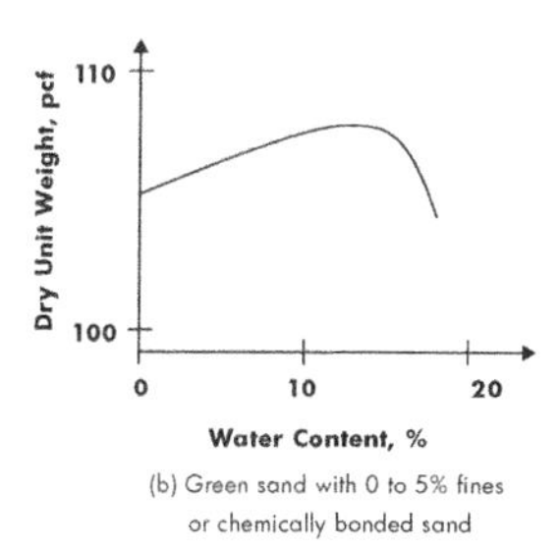

(b) Green sand with 0 to 5% fines or chemically bonded sand

Figure 8. Moisture density relationships for green sand and chemically bonded sand

Specific Gravity. Foundry sands will normally have a specific gravity of 2.50 to 2.80.

Permeability. Green sands with fines less than 6% and chemically bonded sands have permeability values in the range of $6 \ 10^{-4}$ to 5×10^{-3} cm/sec. However, when fines such as bentonite clay are present and greater than 6%, permeability can be lower, between 1×10^{-7} and 3×10^{-6} cm/sec.

Frost Susceptibility. Soils that are not susceptible to frost and that do not produce heave are gravel and clean sands. Fine-grained soils are generally classified as frost susceptible. The fine content of the foundry sand determines its frost susceptibility. Foundry sand without fines can have low to negligible frost susceptibility.

CBR (California Bearing Ratio). In foundry sand, a CBR between 11 and 30 is typical. The resistance to penetration of a 3 in^2 piston in a compacted sample of foundry sand is compared to its resistance in a standard sample of compacted crushed rock. CBR is high when the water content is dry of optimum, and then drops after the optimum water content is reached. The CBR for foundry sand with fines is generally higher than it is for granular sands.

Construction Practices

General. Many contractors have found that working with foundry sand is similar to working with conventional construction materials. Foundry sand has been used effectively in normal embankment construction with and without permeability and leachate control. Foundry sands have also been used in conjunction with geogrid systems and with reinforced earth retaining walls that use straps or grids as horizontal tiebacks.

Standard construction procedures can be adjusted to account for using foundry sand. Many procedures have been developed as the result of the experience gained using foundry sand in trial embankment and construction projects.

Stockpiling. Foundry sand can be stockpiled in a climate-controlled environment or exposed to the elements. The foundry sand stored under controlled climatic conditions can be delivered to meet narrow limitations on moisture content. Conversely, the moisture content of the foundry sand stored outside will vary, depending on its location within the stockpile. It is recommended that foundry sand stockpiled outside be tested at various locations within the stockpile.

Site Preparation. The site should be prepared for foundry sand placement in the same way it is prepared for similar soil fill materials. It should be cleared and grubbed, and the topsoil should be retained for final cover. The normal precautions for draining the site to prevent seeps, pools or springs from contacting the foundry sand should be followed. Also, environmental restrictions may require that the foundry sand be encapsulated in layers of clay.

Delivery and On-Site Storage. As with any fill, foundry sand is hauled to the site in covered dump trucks (Figure 9). The water content of the foundry sand is adjusted to prevent dusting and to enhance compaction. Foundry sand can be stockpiled on-site if the sand is kept moist and if the sand is covered.

Figure 9. Foundry sand delivery (sand supplied by Kurtz Bros., Inc., construction by Trumbull Corp.)

Spreading. Foundry sand is spread using normal construction equipment, such as dozers. Lifts are usually 6 to 12 inches thick. Many contractors then track the dozer for initial compaction. Ideally, the sand is at or near optimal moisture when placed; if not, water should be added.

Compaction. Compaction should begin as soon as the material has been spread (Figure 10) and is at the proper moisture content. Ohio experience has shown it to be preferable to place the foundry sand as close to the optimum moisture content as possible, within 1-2%. Too dry of optimum requires significantly more compactive effort than when the sand is properly moisture-conditioned. However, the required compaction can eventually be achieved. Foundry sands are not normally sensitive to over-rolling and can tolerate a wider range of moisture contents than natural sands.

Figure 10. Spreading foundry sand (Ohio Turnpike, sand supplied by Kurtz Bros. Inc.)

Vibratory smooth drum rollers, pneumatic-tired rollers and vibrating plates have all been used successfully. It is important to properly screen the foundry sand and to remove residual core pieces. In most cases cores are not a problem if they are less than 3 to 4" long. The compaction process will slow down if they are larger.

The lift thickness, the weight and speed of the compaction equipment and the number of passes should be determined for optimal compaction (Figure 11). Many contractors run test strips and relate construction practices to the foundry sand's degree of compaction. When vibratory compaction equipment is used, lifts of 12 inches are acceptable. In fact, thicker lifts may be preferred. They provide greater confinement. If the lift is too thin, the sand may dry out too fast. Also, it will not offer enough confinement for proper compaction.

Figure 11. Grading foundry sand

In a recent project, dynamic compaction was used successfully to compact foundry sand. The moisture-conditioned foundry sand was placed in 10 to 15 foot lifts, and then dynamically compacted by a 12 ton weight dropped from 40 to 60 feet. This height can be adjusted for the required level of compaction and thickness of the sand layer.

In some foundry sand embankment construction, a foundation of coarser material such as rock or shale is placed. The foundry sand is placed on top in uniform horizontal lifts not more than 8 inches deep. It is then compacted according to normal density specifications.

Moisture Control. As with any fill material, controlling its moisture is an important consideration in compaction. Be sure to compare hauling foundry sand that has been moistened to the desired water content at the plant to adding water at the site. Hauling moist foundry sand translates to higher transportation costs, while adding water on-site sacrifices productivity in field placement.

Erosion and Dust Control. To prevent wind and water erosion of the surface of the foundry sand embankment, contractors use the same sediment and erosion control techniques commonly used on other earthwork operations. On a project in Ohio, the contractor installed organic filter socks or berms around the construction area.

Dusting may occur when compacted foundry sand is placed in dry or windy weather, or due to traffic disturbance. During construction, the soil should be kept moist and covered. Clay layers have also been used to cover the face of the embankment to prevent erosion of the foundry sand in a heavy rain. The completed embankment should be covered with topsoil and vegetation.

Three Key Construction Steps. To ensure successful construction of an embankment (Figure 12), with foundry sand and its long-term performance it is important to:

1. ***Assess availability.*** Contact the local foundry sand supplier and determine whether an adequate supply of foundry sand can be provided in the time frame required.
2. ***Investigate site conditions.*** As with any embankment project, use standard geotechnical techniques to evaluate subsurface soil and groundwater conditions. The two most important subsurface characteristics affecting embankment construction and performance are shear strength and compressibility of the foundation soils.
3. ***Evaluate the physical, engineering and chemical properties of the foundry sand.*** The physical and engineering properties that will determine the behavior of a foundry sand embankment (or any embankment) are grain-size distribution, shear strength, compressibility, permeability and frost susceptibility. Laboratory tests designed for testing soil properties apply equally well to testing foundry sands. Most foundry sand distributors can provide information on the physical, engineering and chemical composition of the foundry sand and can provide details on any possible leachate that must be considered during design and construction.

Figure 12. Stepped embankment (Ohio turnpike, sand supplied by Kurtz Bros. Inc., construction by Trumbull Corp.)

Environmental Impacts. The trace element concentrations in most clay-bonded and aluminum foundry sands are similar to those found in naturally occurring soils. The vast majority of foundry sands meet water quality standards for leachate. Additionally, state environmental regulatory agencies can guide you through applicable test procedures and water quality standards. The amount of leachate produced can be controlled by assuring adequate compaction, grading to promote surface runoff, and daily proof-rolling of the foundry sand layer to impede infiltration. When construction is finished, a properly seeded soil cover will reduce infiltration. For highway embankments, the pavement itself can be an effective barrier to infiltration.

Chapter 3:

Foundry Sand in Road Bases

Background Information

What is a road base? A road base is a foundation layer underlying a flexible or rigid pavement and overlying a subgrade of natural soil or embankment fill material. It can be composed of crushed stone, crushed slag or some other stabilized material. It protects the underlying soil from the detrimental effects of environment and from the stresses and strains induced by traffic loads.

Flexible pavement road base. For flexible pavements, there are typically two bases underneath the pavement that comprise the road base, a stabilized base and an untreated or granular subbase (Figure 13). The two different base materials are usually used for economy. Local or cheaper materials are used in the subbase, and the more expensive materials are used in the base.

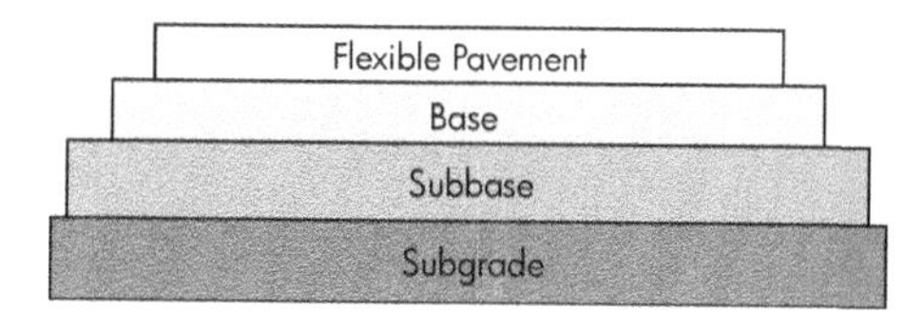

Figure 13. Schematic of a flexible pavement structure

Rigid pavement road base. In contrast to flexible pavements, rigid pavements are typically placed on a single layer of granular or stabilized road base material (Figure 14).

Since there is only one layer under the rigid pavement and above the subgrade, it can be called either a base or a subbase.

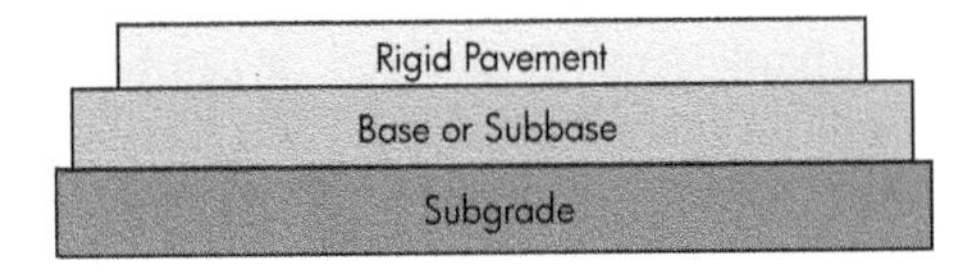

Figure 14. Schematic of a rigid pavement structure

Granular or stabilized road base. The selection of a stabilized base course or a granular base course depends on the traffic loads. Pavements that are subjected to a large number of very heavy wheel loads typically use cement-treated, asphalt-treated, or a pozzolanic stabilized mixture (PSM) base. Granular materials may erode when the heavy traffic induces pumping.

Purpose of the Road Base

Five of the most important reasons for constructing a road base are to:

- Control pumping,
- Control frost action,
- Improve drainage,
- Control shrinkage and swelling of the subgrade, and
- Expedite construction.

Control pumping. For pumping to occur, three conditions must exist simultaneously. The material under the concrete slab must be saturated with free water, the material must be erodible, and frequent heavy wheel loads must pass over the pavement. These loads create large hydrodynamic pressures that transport untreated granular materials and even some weakly cemented materials to the surface. This loss of fines is termed pumping.

Control frost action. Frost action is the combination of frost heave and frost melt. Frost heave causes the pavement to lift up, while frost melt causes the subgrade to soften and the pavement to depress. Both lead to the break up of a pavement. Three factors produce frost action:

1. The soil must be frost susceptible in the depth of frost penetration. These soils generally have more than 3% fines or are uniform sands with more than 10% fines.
2. Water must be available.
3. Temperatures must remain below freezing for a sufficient period of time for water to flow from the water table to where the ice lenses form in the road base.

Improve drainage. A road base can raise the pavement to a desired elevation above the water table, acting as an internal drainage system.

Control shrinkage and swelling of the subgrade. If the subgrade shrinks and expands, the road base can provide a surcharge load to reduce its movement. Dense graded or stabilized base courses reduce the water entering the subgrade, and act as a waterproofing layer. Open-graded base courses serve as a drainage layer.

Expedite construction. A road base can serve as a working platform for heavy construction equipment.

Mix Design Evaluation

The road base material should be made of a mixture of crushed rock and enough fine material to hold the rock in place and to provide good compaction. Foundry sand can be used as the fine material in a road base. Engineering properties that characterize foundry sand as a subbase material are plasticity, shear strength, compaction (moisture-density relationship), drainage and durability.

Plasticity (Shrinkage or Swelling). Green foundry sands without fines and chemically bonded sands are typically non plastic. However, the presence of bentonite clay increases the foundry sand's plasticity. The plasticity index is commonly used to indicate a soil's tendency to undergo volume change (shrinkage or swelling). The plasticity index is typically less than 2 for green sands with no or few fines and chemically bonded sands, and greater than 2 when the clay content increases beyond 6%.

Shear Strength (Friction Angle). A soil's shear strength is its ability to resist deformation. This property is critical when determining a soil's ultimate bearing capacity, which is the largest load that the road base material can support. Shear strength depends on several material properties, such as soil cohesiveness, and the interlocking ability and packing of the particles.

The friction angle of green sands with 6 to 12% clay is higher than it is for chemically bonded foundry sands and green sands without clay. The friction angle ϕ in Table 4 represents the peak strength for dense samples and the ultimate strength for loose ones. The higher friction angle for the green sand with clay is attributed to its fines. Similarly, the cohesive strength of green sands with clay is higher than it is for the chemically bonded sands (Table 5). Green sands without clay are nonplastic. Either ASTM D5311 or ASTM D3080 can be used to measure shear strength.

	ϕ	
	Loose	**Dense**
Green sand with clay (6-12%)	32°-34°	37°-41°
Clean green sand without clay/ Chemically bonded sand	30°	35°
Natural sand	29°-30°	36°-41°

Table 4. Friction angle of foundry sands

	Cohesion (psi)	
	Loose	**Dense**
Green sand with clay (6-12%)	0.60-0.75	1.44-1.82
Chemically bonded sand	0.06	1.04

Table 5. Cohesion of foundry sands

Compaction. The compaction characteristics of a granular base or subbase material depend on the soil's moisture-density relationship. Most specifications require that the granular base be compacted to a specified density that is at least 95% of the Standard Proctor maximum dry density, with the water content near optimum. The Standard Proctor test (ASTM D698 or AASHTO T99) is commonly used to measure compaction. The Modified Proctor test (AASHTO T180) can also be used when the base must have a high shear strength or be dense.

To ensure base stability, angular particles with rough surfaces are preferred over round, smooth particles. Foundry sands are round to subangular in shape, with both smooth and rough surface textures.

Drainage. Since a road base should provide drainage as part of a pavement's structure, the base material's permeability is very important. Materials that are free-draining typically have a permeability between 10^{-2} and 10^{-3} cm/sec.

The permeability of green and chemically bonded sands are given in the table below (Table 6). The presence of clay reduces the permeability of the green sands. Higher permeabilities are associated with foundry sands that have fewer fines, such as green sands that have been processed to remove the clay and chemically bonded sands. Permeability of foundry sands can be measured using ASTM D2434, AASHTO T215 or ASTM D5084.

	Permeability k (cm/sec)
Green sand with clay	2.8×10^{-5} to 2.6×10^{-6}
Green sand without clay	3×10^{-3} to 5×10^{-3}
Chemically bonded sand	4.5×10^{-3} to 5.9×10^{-4}
Natural Sands	10^{-3} to 10^{-4}

Table 6. Permeability of foundry sands

Durability. Foundry sand has sufficient strength to resist excess breakdown when placed in road bases (Figure 15). It has good particle strength. Many states will specify minimum requirements for LA abrasion (ASTM C131 or AASHTO T96) and sodium sulfate soundness (ASTM C88 or AASHTO T104). Foundry sand without clay is normally not susceptible to frost, and this should be assessed using AASHTO T103.

Figure 15. Supplying foundry sand for a road base

Control of Materials

Handling. No deleterious materials (plastic fines, organic matter or extraneous debris) should be in the foundry sand. This will reduce its load carrying capacity and ultimately, the expected performance of the road base. Foundry sand should be screened prior to its use in engineering projects.

Aggregate. To meet state specifications for road bases, blending the foundry sand with another aggregate may be necessary. The gradation of the road base materials influences base stability, drainage and frost susceptibility. Likewise, the aggregate must be sound and able to resist environmental deterioration.

Construction Practices

Blending of Materials. Aggregate used in the construction of road bases should be mixed and processed to produce a uniform blend of material prior to final placement.

Construction Plants. Construction plants should collect and store foundry sand until use. Because of the importance of the fine aggregate moisture content, the foundry sand should have consistent moisture content.

Hauling. Blended mixtures can be hauled to the site in open or covered trucks. The mix in an open truck can dry and dust when hauled long distances.

Spreading. The placement of road base material shall conform to local grading ordinances and agency specifications. The typical road base is a uniform layer of base material that is 6-10 inches thick, without any segregation. The final thickness after compaction is 4-6 inches.

Compaction. Road base material should be rolled to achieve the desired compaction and specified density. It is important that all waste materials be removed from the foundry sand, because it can become entangled in the compaction equipment and delay construction.

Finishing. The final layer of road base shall be finished with equipment capable of shaping and grading the final surface within the tolerances specified by the agency.

What if I can't afford to buy specialized construction equipment? There is no need to! Most plants can be readily adapted to add the foundry sand to the road base mix. For spreading, it can be placed with a jersey spreader.

What advice do you have for a first time user?
Using foundry sand can produce strong durable road bases, but attention should be given to the following precautions:

- **Mix design evaluation.** Proposed mix designs should be evaluated for performance prior to construction. Good quality constituents do not always produce a mix that will perform as desired.
- **Moisture content.** Moisture must be maintained in the mix to ensure optimal compaction. The moisture may be added on site or at the plant.

Marketing Fill

The project specific nature of the fill materials market makes it difficult to quantify the total amount of foundry sand that will be needed on a regular basis. Currently, the rates of foundry sand generation are sufficient to supply construction companies and other related industries with fill material. The marketability of the sand depends on the availability of other fill material at or near the construction site. Transportation costs may quickly offset the low initial cost advantage of foundry sand. Foundry sand use is more advantageous when it is stockpiled close to the construction site.

Chapter 4:

Foundry Sand in Hot Mix Asphalt

Introduction

Asphalt concrete is the most popular paving material used on our highways and roadways in the United States. Over 94% of all pavements in the U.S. are covered with asphalt. This translates to over 2,030,000 miles (Figure 16).

Figure 16. Asphalt pavement (courtesy Asphalt Pavement Alliance)

The most prevalent type of asphalt paving material is hot mixed asphalt (HMA). This consists of a combination of plant-dried coarse and fine aggregates. They are coated with hot asphalt cement, which acts as a binder.

Foundry sand has been used successfully to replace a portion of the fine aggregate used in HMA. Studies have shown that foundry sand can be used to replace between 8 and 25% of the fine aggregate content. When mixes are properly designed using Superpave, Marshall, or Hveem techniques, foundry sand can be an effective sand alternative.

Hot Mix Asphalt Aggregate Requirements

Hot mix asphalt production requires that all constituent products:

- Have inherently good quality characteristics,
- Come from consistent, reputable supply sources,
- Meet all environmental requirements, and
- Are economically competitive with similar materials.

Foundry sand has the potential to be a very high quality material in hot mix applications (Figure 17). However, it is important that the foundry sand be cleaned of clay, dust and other deleterious materials. Additionally, metals present in the sands should be removed either manually or magnetically. Then, it may be blended with other sands, at 8-25% replacement, to provide equal or possibly better results than normal sands.

Gradation. Fine aggregates in hot mixes generally are required to meet the specifications of AASHTO M29. This specification limits materials passing the No. 200 sieve to between 5 and 10%. Many foundry sands have a higher percentage, requiring screening prior to blending or a limit on the maximum amount of foundry sand that can be added to a mix.

Particle Cleanliness. Hot mix asphalt is generally tested by the sand equivalent test (ASTM D-2419) or by the nonplastic index test (AASHTO T-90). These tests detect clay portions, which are very detrimental to aggregate-binder adhesion. It is important that when qualifying foundry sand, the clay content and organic-based additive be quantified and limited in producing an asphalt mix. For many foundry sands, the sand equivalent test is not applicable. According to research done at the University of Wisconsin at Madison, the methylene blue test (AFS 2211-00-S) is a better method for the clay content. The loss on ignition test (AASHTO T 267-86) is a good method for detecting the organic based additives.

Soundness. Nearly all foundry sands meet the loss of soundness specification, AASHTO T104.

Particle Shape and Texture. Many hot mix asphalt specifications now require a fine aggregate angularity test, using AASHTO TP33. Foundry sands typically fall within the specified 40-45% range.

Absorption and Stripping. Foundry sand is generally nonplastic and has low absorption. However, it is primarily silica, which in the past has been linked to stripping. As with all silica-based hot mix asphalt mixtures, a foundry sand mix should be tested using standard stripping tests. The University of Wisconsin at Madison has tested foundry sand mixes for moisture damage and, depending on the clay content and the extent of organic-based additives used, foundry sands can have positive or negative effects on resistance to moisture damage. The University of Wisconsin at Madison is currently developing better methods to quantify clay and organic-based additives to predict how foundry sands influence moisture damage.

Figure 17. Construction of HMA pavement

Case Studies

Pennsylvania, Michigan and Tennessee Departments of Transportation allow the use of recycled foundry sand in HMA. Pennsylvania DOT allows the use of 8 to 10% of the total aggregate portion to be recycled foundry sand in the asphalt wearing course. One hot mix producer in Michigan consistently supplies HMA with 10-20% recycled foundry sand to replace the conventional aggregate, and it meets Michigan DOT specifications. Another hot mix supplier in Tennessee claims that hot mix with foundry sand replacing 10% of the fine aggregate compacts better and outperforms the HMA containing washed river sand. In addition, a hot mix producer in Ontario, Canada, has used foundry sand as a fine aggregate substitute for the past 10 years in both foundation and surface HMA layers.

In Pennsylvania, 10 million tons of asphalt pavement are produced each year. Two million tons of fine aggregate are needed. Although foundry sand cannot be used to replace the total fine aggregate quantity, it can be used to replace up to 15%. This would allow a significant amount of foundry sand to be used each year in Pennsylvania.

Concerns of the Hot Mix Industry

To be used by the hot mix industry, foundry sand has to be a consistent product with adequate supply. The engineering characteristics of the foundry sand have to be relatively similar from batch to batch, especially in gradation, so that the resulting hot mix asphalt is also consistent. Once a proper hot mix asphalt design has been developed and calibrated with the foundry sand, it is not cost-effective to change the mix design. This will result in additional costs. If the foundry sand supply changes during the construction season, the hot mix supplier will be responsible for any out-of-specification material, and will incur any consequent penalties.

Economics

Use of foundry sands can be cost-effective for both the foundries and the HMA industry. Highway agencies and contractors could switch to the recycled material when it is geographically and economically competitive.

Chapter 5:

Foundry Sand in Flowable Fills

Introduction

Flowable fill has several names, but each is essentially the same material:

- Controlled density fill (CDF),
- Controlled low strength material (CLSM),
- Fly ash slurry,
- Lean mix backfill,
- Unshrinkable fill, and
- Soil cement.

Flowable mixtures make up a class of engineering materials having characteristics and uses that overlap those of a broad range of traditional materials including compacted soil, soil-cement, and concrete. Flowable mixtures consist of sand, water, cement and sometimes fly ash. The mixtures are proportioned, mixed and delivered in a very fluid consistency to facilitate placement; they provide an in-place product that is equivalent to a high-quality compacted soil but without the expensive compaction equipment and related labor. ACI defines flowable fill as a cementitious material that is in a flowable state at the time of placement and has a specified compressive strength of 1,200 psi or less at 28 days.

Flowable fills have been used as backfill for bridge structures including abutments, culverts, and trenches. It has been used for embankments, bases, and subbases. It is commonly used as bedding for slabs and pipes. It has also been used to economically fill caissons and piles, abandoned storage tanks, sink holes, shafts and tunnels.

Flowable fill materials usually offer an economic advantage over the cost of placing and compacting earthen backfill materials. Depending on the job conditions and costs involved, significant savings are possible. The closer the project location to the source of the flowable fill, the greater the potential cost savings.

Most foundry sands can be used in flowable fill mixtures. The foundry sand does not have to meet ASTM C33 gradation specification requirements as a concrete fine aggregate to be suitable for use in flowable fill mixes. ACI 229R reports that foundry sand with up to 20% fines produced successful flowable fill mixtures. Because low strength development is desirable in flowable fill, even foundry sand with organic binders may be suitable. Foundry sand for flowable fill can be used in a dry or moisture conditioned form.

Mix Design and Specification Requirements

Flowable fills typically contain portland cement, fly ash, sand and water. Foundry sand can be the major ingredient in flowable fills. The flowable character derives from its distribution of spherical and irregular particle shapes and sizes. When mixed with enough water, the fly ash and sand surfaces are lubricated, so that it flows.

Water requirements for mixture fluidity will depend on the surface characteristics of all solids in the mixture. A range of 50 to 200 gallons per cubic yard would satisfy most materials combinations. As with most flowable fill applications, the wetter it is the better. The water acts as a means of conveyance for the solid particles in the mixture. Portland cement is added, typically in quantities from 50 to 200 pounds per cubic yard, to provide a minimal (weak) cementitious matrix. Table 7 shows mix proportions recommended in ACI 229R. However, contractors familiar with flowable fill and foundry sand have reported using lower quantities of foundry sand, approximately 1,500 pounds per cubic yard.

Component	Typical Mix Design (lb/yd^3)	Range (lb/yd^3)
Fine Aggregate/ Foundry Sand	2850	1850-2910
Cement	100	50-200
Fly Ash	250	0-300
Water	500	325-580

Table 7. Foundry sand mixes (adapted from ACI 229R)

According to the ACI definition, flowable fill should have an upper compressive strength limit of 1,200 psi, but strengths can be designated as low as 50 psi. Most flowable fill mixes are designed to achieve a 28-day maximum unconfined compressive strength of 100 to 200 psi. The goal is to have the flowable fill support early loads without settling, and yet still be readily excavated at a later date. Flowable mixtures can be designed to allow for hand excavation as well.

It is important to remember that flowable fill mixtures with a low ultimate strength in the 50 to 70 psi range have at least two to three times the bearing capacity of well-compacted earthen backfill material.

Mixture Proportioning Concepts For Flowable Fills With Foundry Sand

The following are the most important physical characteristics of flowable fill mixtures:

- Compressive strength development,
- Flowability,
- Time of set, and
- Bleeding and shrinkage.

Strength development in flowable fill mixtures is directly related to its water-to-cement ratio. Water is added to achieve a desired flowability or slump. Just like normal concrete mixes with a given cement content, increasing the water content will usually result in a decrease in compressive strength. The coarser the sand, whether natural or foundry, the higher the bearing capacity is of the hardened flowable fill.

Flowability is primarily a function of the water content and aggregate gradation. The higher the water content and the more uniform and spherical the sand, the more flowable the mixture. It is usually desirable to make the mix as flowable as possible in order to take advantage of the self-compacting qualities of the flowable fill.

Time of set is directly related to the cementitious materials content and type, sand content, water content and weather conditions at the time of placement. Within 24 hours, construction equipment is usually expected to move across the surface of the flowable fill without any apparent damage. The time of set has been found to depend on the type of foundry sand incorporated into the flowable fill. Green foundry sands with low clay content and chemically bonded foundry sands normally require less water in the mixture. The flowable fill also takes less time to harden.

Bleeding and settlement are possible in high water content flowable fill mixtures, since evaporation of the bleed water often results in settlement. As with any cementitious material, plastic shrinkage cracks on the surface of the fill can occur in high water content mixtures as well. The main concern with plastic shrinkage cracking is that water can infiltrate at a later date. Flowable fill mixtures should be checked for settlement and plastic shrinkage.

Structural design procedures with flowable fill materials containing foundry sand are no different than standard geotechnical design procedures for conventional earth backfill materials. The design procedure uses the unit weight and shear strength of the flowable fill to calculate bearing capacity and lateral pressure of the material for given site conditions.

Closing Comments

Although most foundry sands will meet ASTM C33 gradation requirements, it is not a controlling factor in the flowable fill mix design. If the flowable fill meets the general ASTM test requirements shown in Table 8, the flowable fill should be fine for most applications.

ASTM	Title
C39	Standard test method for compressive strength of cylindrical concrete specimens
C88	Standard test method for soundness of agregates by use of sodium sulfate or magnesium sulfate
C138	Standard test method for density (unit weight), yield, and air content (gravimetric) of concrete
C232	Standard test methods for bleeding of concrete
C403	Time of setting of concrete mixtures by penetration resistance
C827	Change in height at early ages of cylindrical specimens from cementitious mixtures
D2166	Unconfined compressive strength of cohesive soils, used where minimal or no cement is added to the blend
D4219	Highly fluid grout-like mixes
D4832	Standard test method for preparation and testing of controlled low strength material (CLSM) test cylinders
D5971	Standard practice for sampling freshly mixed CLSM
D6023	Standard test method for unit weight, yield, cement content and air content (gravimetric) of CLSM
D60624	Standard test method for ball drop on CLSM to determine suitability for load application
D6103	Standard test method for flow consistency of CLSM

Table 8. Recommended test methods for flowable fills

Chapter 6:

Foundry Sand in Portland Cement Concrete

Introduction

Portland cement concrete (PCC) is a mixture of approximately 25% fine aggregate, 45% coarse aggregate, 20% cement and 10% water. Foundry sand can be used beneficially in concrete production as a fine aggregate replacement.

Mixture Design and Specification Requirements

Aggregates are classified based on particle size. Fine aggregates consist of natural sand or crushed stone with particle diameters smaller than 3/8 inch. Coarse aggregates are gravel or crushed stone with particle diameters ranging between 3/8 inch and 2 inches.

The selection of aggregate used in concrete is of great importance. Aggregate properties strongly influence the concrete's freshly mixed and hardened properties. Aggregate must be:

- Clean and free of objectionable materials, including organic material, clay and deleterious contaminants, which can affect bonding of the cement paste to the aggregate,
- Strong, hard and durable, and
- Uniformly graded.

Gradation

Specifications governing the selection and use of aggregates in concrete mixtures generally relate to the particle size distribution or gradation of the aggregates. The recommended gradation for fine aggregate is given in Table 9. This specification is geared toward plain concrete and concrete structures, pavements, sidewalks and precast products.

Sieve No.	Sieve Size (mm)	Percent Passing by Weight
4	4.75	95-100
8	2.36	80-10
16	1.18	50-85
30	0.600	25-60
50	0.600	10-30
100	0.150	2-10

Table 9. Fine aggregate gradation (from ASTM C33)

Generally foundry sand is too fine to permit full substitution. The percentage of materials passing the No. 30, 50 and 100 sieves is too high. To meet the specification, it is necessary to remove the fines or to blend the spent foundry sand with coarser sands. The foundry sands will then comply with the specification. In some areas, natural sands lack finer material. Foundry sand can be blended with them as a partial replacement to satisfy the specification.

Effect of Material Characteristics on Foundry Sand Concrete Quality

Various characteristics of foundry sand can significantly affect the quality of concrete produced. The material characteristics of greatest importance and their effects on the product are discussed below. Foundry sand properties vary in samples taken from one foundry, and there is increased variation from foundry to foundry. This necessitates testing the sand every time prior to use to ascertain its quality.

Particle size distribution. The fine aggregate particle size distribution can affect cement and water requirements, as well as concrete workability, economy, porosity, shrinkage and durability. Too many fine particles can lower the concrete strength and adversely affect durability. ASTM C33 requires that the fine aggregate used in concrete has a fineness modulus, an index of aggregate fineness, in the range of 2.3 to 3.1. The fineness modulus of foundry sand typically ranges from 0.9 to 1.6. The sand has to be blended with a coarser material to meet this specification.

Dust content. ASTM C33 allows a maximum of 5% fine aggregate particles to pass the No. 200 sieve. These particles include clay and other dusts. A large dust content can interfere with the bonding of cement to the aggregate surface, and it can also increase water demand. These factors reduce the durability of hardened concrete. This is a concern when using foundry sands.

Density. Density must be a minimum of 75-110 lbs/ft^3 (1.20 to 1.76 g/cm^3), according to ASTM C330 for general fine aggregate. A higher density aggregate is required when the concrete will be subjected to high compressive loads.

Organics content/deleterious materials content. According to ASTM C33, the maximum amount of clay lumps and friable particles allowed is 3%. Organic content is restricted because it interferes with hydration of the cement and its subsequent strength. The organic content of aggregate can be measured by a color test.

Grain shape. Round particles need less water and cement to coat their surface, and they produce a mixture that is more workable. Angularity increases water demand and cement content to maintain a workable mix. Foundry sand particles are typically angular to rounded.

Specific gravity. Although specific gravity does not directly relate to concrete quality, it can be used as a quality control indicator. The specific gravity of foundry sand varies from 2.5 to 2.8, depending on the source. This compares very favorably to natural sands.

Other Constituents

Coarse aggregate. Coarse aggregates used in the concrete mixture should be appropriately sampled and tested to ensure good quality. Some aggregates of marginal quality have been observed to adversely affect the matrix of hardened concrete.

Cement. Foundry sand can be used in combination with all types of cementitious materials.

Chemical admixtures. In general, foundry sand can be used with any concrete containing chemical admixtures. Retarders and water reducers are compatible with most foundry sands. As with natural sands, any organic material in the foundry sand may affect the dosage and effectiveness of air entraining agents. Trial mixtures should always be examined for any potential compatibility problems.

Construction Practices

General considerations. Foundry sand must be processed prior to reuse, i.e., screened, crushed and magnetic particles should be separated. This will remove waste and deleterious materials, such as tramp metal and core pieces, preventing technical problems at the mix plant.

Plant operations. A separate bin should be reserved for the foundry sand at the plant, as is done for fly ash. Foundry sand can be handled in typical aggregate holding bins. It is important to keep the bins clean and the foundry sand dry to help eliminate any bulking problems at the gate opening.

Case Study

Laboratory use of foundry sand in concrete. An American Foundry Society (AFS) study in Illinois investigated foundry sand as a substitute for fine aggregate in concrete. When foundry sands without fines replaced a portion of the fine aggregate, the concrete produced had compressive strengths, tensile strengths and modulus of elasticity values comparable to mixtures composed of natural sand.

On the other hand, when green foundry sands that had not been processed replaced 33% of the fine aggregate, the resulting concrete compressive strengths at 28 days were between 2,600 psi and 4,000 psi. These low strength concretes can be used in applications not requiring structural grade concrete, such as buried applications like sewer pipe or below grade concrete. The decrease in concrete compressive strength, as well as in tensile strength and modulus of elasticity, were attributed to too many fines and organic materials (clay and dust) in the foundry sand.

Likewise, foundry sand has been used to make paving blocks and bricks. For these applications, foundry sand replaced 35% of the fine aggregate. If the fineness modulus was not exceeded, the concrete product was acceptable provided it met the ASTM specifications for minimum compressive strength, absorption, and bulk density. It is recommended that proper testing be performed to establish the appropriate limits on foundry sand addition prior to using it in commercially produced products.

Applications

Concrete can be used for cast-in-place or precast products such as pipes, ornamental concrete units, load bearing structural units (i.e., beams, girders, etc.), utility structures and concrete blocks. The ultimate use, shape and size of the product will govern the type and gradation of the aggregate required in the concrete mixture. For example, the final dimensions of the precast block will determine the maximum aggregate size. For this reason, when marketing spent foundry sand to precast producers or to a ready mix plant, the particle size distribution requirements should be requested ahead of time.

When the required concrete compressive strength is between 50 psi and 2,500 psi, a 50% fine aggregate substitution with foundry sand has been successful. As with regular concrete, trial mixtures should always be tested prior to production for any potential compatibility problems.

Limitations

Foundry sand is black. In some concretes, this may cause the finished concrete to have a grayish/black tint, which may not be desirable. A 15% fine aggregate replacement with foundry sand produces a minimal color change. Also, the foundry must be able to meet the quantity requirements of the precast manufacturer.

Chapter 7:

Foundry Sand in Other Engineering Applications

Introduction

Other engineering applications of foundry sand will be presented in this chapter. They include using foundry sand in:

- Portland cement manufacturing,
- Mortars,
- Agricultural / soil amendments,
- Traction material on snow and ice,
- Vitrification of hazardous materials,
- Smelting,
- Rock wool manufacturing,
- Fiberglass manufacturing, and
- Landfill cover or hydraulic barriers.

Foundry Sand in Portland Cement Manufacturing

Portland cement reacts chemically with water when hydrating, causing it to set and to harden. When mixed with fine and coarse aggregate, concrete is formed. There are several specifications for portland cement, as designated by ASTM C150 and ASTM C1157.

Production of Portland Cement. Portland cement is manufactured using materials with the appropriate proportions of calcium oxide, silica, alumina, and iron oxide. These ingredients are found in natural rock, like shale, dolomite and limestone. It is the chemistry of the foundry sand as a silica source that is more important in cement production than is its grain size or shape. The requirements that must be met for foundry sand to be used in portland cement production are:

- Its silica content equals or exceeds 80%,
- It is a low alkali material,
- A large quantity of sand is available, and
- It has uniform particle sizes.

Foundry sand may be one of the highest quality sources of silica available to the cement industry. The major chemical constituents of raw portland cement available in foundry sand include silica and alumina and iron oxides. By using foundry sands to replace virgin sands, the quantity of mined virgin sands can be reduced.

Blended Cements. Blended hydraulic cements are of particular interest in the beneficial use of foundry sand, since these cements are produced by blending together two or more types of fine materials. Historically, blended cements have included portions of blast furnace slag or fly ash.

Foundry Sand Acceptance by Portland Cement Manufacturers. The cement manufacturer has to evaluate the foundry sand to confirm its compatibility with other raw materials. In addition, cement producers need a chemical oxide analysis, TCLP results, annual volume and a sample. The chemical oxide analysis shows the amount of silica contained in the sand, while the TCLP shows whether or not the sand is hazardous. Also, because limestone, silica and clay are all common materials, the cement manufacturer has to be willing to use foundry sand.

Limitations. Factors that may limit foundry sand use in portland cement manufacturing involve limits on the quantity of foundry sand available and cost issues. Cement manufacturers require significant quantities of silica, 10,000–40,000 tons annually for a plant. It is unlikely that a single foundry can provide that much sand. The sand from several foundries should be pooled at a community storage site to meet the demands of a single cement plant.

Cement manufacturers will pay nominal fees to foundries for the use of their spent sand, but will consider it waste disposal. Additional charges may be levied for handling fees and shipping. However, there is potential for cost savings.

Case Studies

Laboratory Study of Foundry Sand in Cement. The American Foundry Society (AFS) in Illinois studied using green sand from a gray iron foundry in portland cement manufacture. First, a chemical analysis of the sand was performed to see if it met AASHTO specifications. Based on the chemical content, the foundry sand appeared to be an attractive alternative to raw material for cement kiln feed. Four mixtures were designed using 0%, 4.45%, 8.9% and 13.36% of foundry sand. The chemical characteristics of the resulting clinkers showed little difference between those made with and without foundry sand. The cement produced with the foundry sand met all the relevant chemical specifications. The properties of the cement, namely set time and compressive strength, were not affected by the presence of foundry sand. There was even a slight increase in compressive strength.

Commercial Study by Frazer & Jones. In January 1994, a green sand manufacturer (Frazer & Jones) in upstate New York shipped 15,000 tons of foundry sand to a cement manufacturer in Ontario, Canada. It was used successfully as a replacement for excavated silica materials in the manufacture of low-alkali portland cement. The finished product was a high quality portland cement.

Foundry Sand in Grouts and Mortars

Mortars primarily consist of sand, cement and other additives, and are used in masonry construction. Its primary uses are to joint and seal concrete masonry units, to strengthen masonry structures by bonding with steel reinforcing and to provide architectural quality. Approximately one cubic foot of sand is used to make one cubic foot of mortar. The cement paste occupies the space between the sand particles and makes it workable.

Sand used in masonry mortar mixes is generally specified based on grading. ASTM C144 recommends the following gradation (Table 10). Sands that are deficient in fines make a coarse mix and ones that have too many fines produce a weak mortar.

Sieve No.	Sieve Size (mm)	Percent Passing by Weight
4	4.75	100
8	2.36	95-100
16	1.18	70-85
30	0.600	40-75
50	0.300	20-40
100	0.150	10-25
200	0.075	0-10

Table 10. ASTM C144 sand gradation for mortars

A well-suited aggregate will fall into the middle of this range. Adequate gradation reduces mortar segregation and bleeding and improves mortar water retention and workability. Foundry sand is generally finer than the ASTM gradation requirements, but it can be blended with coarser sands to meet the specification.

In addition to aggregate gradation, the aggregate should be clean and free from wastes, and also have a consistent moisture content. Likewise, the color of the foundry sand can impart a dark tint to the mortar, so it should be deemed acceptable prior to its use in architectural projects.

Foundry Sand in Agricultural/Soil Amendments

Foundry sand can be used as an additive in topsoil and compost materials. It is ideal for topsoil manufacture because of its uniformity, consistency and dark color. Since a high sand content is required in topsoil, it is an essential ingredient. In composting, foundry sand reduces the formation of clumps and prevents the mix from compacting. This allows air to circulate through the material and to stimulate decomposition. Ohio nurseries have been blending foundry sands with soils and compost for use on ornamentals. Kurtz Bros. of Ohio has also used it successfully as golf green topdressing. For these applications, the presence of clay is beneficial, since it promotes the retention of nutrients.

Regulatory issues for foundry sand use in agricultural applications vary from state to state on a case-by-case basis.

Foundry Sand to Vitrify Hazardous Materials

Because of foundry sand's high silica content, it is an ideal candidate to encapsulate, vitrify or neutralize hazardous materials. Preliminary tests show that this is a feasible option.

Foundry Sand as a Traction Material on Snow and Ice

Another possible use of foundry sand is as an anti-skid material on roads covered with snow and ice. Particularly its angular particles improve traction on highways in the winter. Likewise, its black color (i.e., green sand) will hold the heat longer and will melt the ice faster. To be used as an anti-skid material, the foundry sand must meet each state's requirements. Typically, foundry sand is too fine to comply with anti-skid regulations, but when mixed with a coarser material, it does comply. Trial mixes should be formulated and evaluated prior to use. In addition, the foundry sand should be free from glass, metals or other substances that could be harmful to cars and vehicles.

Foundry Sand for Smelting

Another potential use for foundry sand is as a raw material in zinc and copper smelting. Foundry sand can be used in place of virgin sand. Guidelines for using silica sands in zinc and copper smelting are that the sand should be relatively pure silica (minimum 99.0%), have a maximum particle size of 2 mm and a bulk phenol content of less than 2 mg/kg (ppm).

The foundry should demonstrate to the smelting plant that the foundry sand meets these criteria and will produce quality zinc and copper.

Case Study. A smelter in California has been using recycled foundry sand for some time. All of the foundry sands used in area brass and steel foundries were deemed acceptable due to their high silica content.

Foundry Sand in Rock Wool Manufacturing

Rock wool fibers are commonly used to reinforce other materials, such as building material insulation, and are similar to fiberglass. Foundry sand can serve as a source of silica in the rock wool production process. It is produced by combining blast furnace slag with silica or alumina in a cupola furnace and then fiberizing the molten material. To be used in production, the foundry sand has to be pre-treated and formed into briquettes.

Foundry Sand in Fiberglass Manufacturing

Foundry sand can be used in the manufacture of fiberglass. Fiberglass is produced by melting silica sand and straining it through a platinum sieve with microscopic holes, thereby forming the desired glass fibers. Manufacturers of fiberglass, such as Owens-Corning and CertainTeed Corporation, have specifications for silica content and particle size distribution, so to be used in this application, the foundry sand has to have the needed properties.

Foundry Sand for Landfill Cover or Hydraulic Barriers

Foundry sand has been used for some time as a daily cover soil on municipal waste landfills and as a landfill liner. Foundry sand containing clay (>6%) and the following Atterberg limits (liquid limit greater than 20 and plastic index greater than 3) has a low permeability. It is an ideal material for final or top cover. It is also expected to be resistant to permeation by brines and leachates in the short term.

Each state may have different requirements placed on landfill materials and foundry sand usage should be decided on a case-by-case basis.

Notes:

Notes:

Notes:

Notes:

Notes:

Notes:

APPENDIX

IV

Increasing the Quality of Concrete and Concrete Related Products

OUTLINE

INCREASING THE QUALITY OF CONCRETE AND CONCRETE RELATED PRODUCTS

KOSOVO CLUSTER AND BUSINESS SUPPORT PROJECT

April 19, 2007

This publication was produced for review by the United States Agency for International Development. It was prepared by the KCBS project team of Chemonics International Inc. based on a Final Report prepared by Short Term Technical Advisor, Jeff Groom.

INCREASING THE QUALITY OF CONCRETE AND CONCRETE RELATED PRODUCTS

THIS REPORT ADDRESSES AN ASSEMENT OF THE STATE OF PORTLAND CEMENT CONCRETE AND PORTLAND CEMENT CONCRETE PRODUCTION THROUGHOUT KOSOVO. THE REPORT PRESENTS FIELD REPORTS AND RECOMMENDATIONS TO INCREASE THE DURABILITY OF PORTLAND CEMENT CONCRETE AND INCREASE THE EFFICIENCY IN PRODUCTION OF PORTLAND CEMENT CONCRETE IN AN ATTEMPT TO INCREASE THE PROPER USE OF PORTLAND CEMENT CONCRETE.

Kosovo Cluster and Business Support project – "Increasing the Quality of Concrete and Concrete Related Products"

Contract No. AFP-I-00-03-00030-00, TO #800

This report submitted by Chemonics International Inc. / April 19, 2007

CONTENTS

PURPOSE OF ASSIGNMENT

The purpose of the assignment was to provide technical assistance to ready mixed concrete producers regarding quality control (QC) testing based on European standards, developing new portland cement concrete mixes to meet the European standards, and introduce the use of chemical additives, specifically water-reducing and air-entraining admixtures. The goal is to improve the quality of concrete products, increase production, and decrease operation costs in order to expand the local market by completing the following tasks:

- Review current principles and practices of proportioning portland cement concrete mixes.
- Determine the current level of use for chemical admixtures in concrete mixes, specifically the use of water-reducing admixtures (used to increase strength by reducing the amount of water needed to achieve a desired consistency), air-entraining admixtures (a relatively inexpensive admixture used to increase freeze/thaw durability), and superplasticizers (used to make flowable concrete).
- Use of statistical analysis to evaluate test results for fresh and hardened concrete properties and to make adjustments, based on the analysis, to concrete mix proportions to meet European standards.
- Review the needed calculations to determine concrete mix proportions to include aggregate moisture content, specific gravity of constituents, and aggregate gradations.
- Demonstrations of proper concrete testing techniques based on European standards. The demonstrations will include testing of fresh and hardened concrete.
- Virtual Concrete Laboratory demonstrations and videos on concrete sampling and test.
- Develop mix design for usage of recycled concrete and usage of wasted material.

BACKGROUND

Many of the concrete producers in Kosovo use mix designs handed down from past employees or experience for ready mix concrete and production of concrete elements such as curbs, concrete pipes, blocks, pavement blocks, and other concrete products. The mix proportions are based on antiquated methods that result in less durable and extremely conservative concrete.

Only three concrete producers have established laboratories for the purpose of quality control/quality assurance (QC/QA) of their products. In addition there are two other independent private laboratories that provide concrete mix proportioning and concrete testing, as well as the laboratory of Pristina University. The companies that currently have working concrete laboratories are Renelual Tahir and Vellezerit e Bashkuar from Prizren and Papenburg and Adriani from Ferizaj.

In Kosovo concrete is the main construction material used to construct all types of structures, whereas precast concrete products such as pipes, blocks, pavement concrete stones, and curbs are used in the road construction industry. One of the main constraints in the road industry is the production of low quality concrete curbs and other elements, which do not last more than a few years. In addition, the concrete producers are facing high operation costs due to high cement contents or use of improper aggregates/river stones. It is believed that in Kosovo there are more than 50 concrete plants in use and over 90% of those are second-hand plants older than 20 years in age, some of which are former SOE concrete plants that were installed 25 years ago.

In Kosovo, the primary material used in concrete is river stones and only a few producers use manufactured (crushed) aggregates. The aim of the government of Kosovo is to protect rivers and encourage the private sector to start using crushed aggregates for concrete production. A small number of concrete producers use chemical additives but the majority of producers are not aware of the benefits obtained by using chemical additives. There is a cement factory in Kosovo that produces over 350,000 tons annually; a further 350,000 to 500,000 tons are imported. The water used for production is mainly drinking water, which depletes the limited supply of drinking water in Kosovo.

Market Position:

In Kosovo over 800,000 tons of cement is used in an estimated 2.0 million cubic meters. The cost of concrete is estimated to be 140 to 160 million euros. The main competition to ready mixed concrete are handmade concrete products, which are often used for housing and road construction. Road construction accounts for over 30% of the total market. Individual investors compensate with higher prices for ready mix concrete.

Production:

Raw materials used by concrete producers are produced locally: river stone, cement from Sharrcem (south of Kosovo), and water; additives are imported from Slovenia. Concrete plants are very old and manufacturing quality varies. Taking into consideration the total capacity of these concrete plants, Kosovo currently operates at 25% or less of production capacity reflecting a glut of concrete plants in Kosovo. There are several concrete plants that are in poor working condition and therefore application of standards will force them to improve. The Kosovo Standardization Agency publicly announced the adoption of EU concrete standards, which could be approved in March 2007.

In Kosovo there are over 35 producers of precast concrete elements including concrete stones for pavement, blocks, curbs, concrete pipes, concrete pillars for electricity, manholes, decorative stones from concrete, etc.

EXECUTIVE SUMMARY

Interviews, laboratory tours, and production facility (batch plant) tours were performed with KCBS personnel to assess the current local practices for proportioning portland cement concrete and to determine the methods used to proportion concrete mixes. Based on the information obtained, demonstrations and presentations were prepared and delivered as a means to transfer knowledge regarding European standards and current concrete technology with respect to chemical additives and the production of durable concrete.

Laboratory test results of fresh and hardened concrete are used to assess the consistency and quality of concrete. In addition, results can also be used to measure the effects of adjusting concrete mix proportions. Current test methods in Kosovo are based on past Yugoslav test methods. Although the tests are relatively easy to perform, the methods and equipment used are not standardized making it extremely difficult for the industry to track changes in materials and trends. The industry could benefit greatly by the translation and distribution of a few basic European standards.

The concrete production facilities observed are more than adequate to produce concrete meeting the European standards. All but one batch plant observed were computerized.

The quality control personnel are using concrete mix proportions, which have been handed down over time. At each company interviewed, mix proportions were based on different methods of proportioning, none of which were very accurate. The current mindset is that greater strength equates to better concrete. The concept of designing for durability seemed foreign to most participants. Very few, if any, of the concrete mixes contain chemical additives. Most of the concrete suppliers have used chemical additives in the past but only when requested by the client.

Methods currently in use to determine concrete mix proportions result in concrete with relatively high strength. Most concrete is proportioned to meet a compressive strength of 30 MPa (4,350 psi). Most of the hardened concrete tested achieved a compressive strength of over 45 MPa (6,530 psi). Compressive strength is inversely proportional to the ratio of the mass of water compared to the mass of cement (know as the water/cement ratio). Greater strength is achieved by lowering the water/cement ratio. The mixes observed have a water/cement ratio of 0.60 or higher. The concrete producers would greatly benefit by reducing the amount of cement in concrete mixes to reach a more reasonable compressive strength. Methods to achieve this were discussed with all participants and example-mixes were proportioned incorporating methods to achieve the reduced cement content. In addition, the use of water-reducing and air-entraining additives could also play an important role in reducing the cement content while maintaining quality, gaining durability, and lowering the material cost of the concrete.

FIELD ACTIVITIES TO ACHIEVE PURPOSES

Numerous field trips were taken for the purpose of observing laboratory practices, concrete batching methods, and materials used in the production of portland cement concrete. Presentations were given and discussions held with individuals involved in the concrete industry to present European standards, to explain the methods used to establish concrete mix proportions, and to describe methods to adjust concrete mix proportions based on test results. Reports of the field trips are presented in Annex I.

TASK FINDINGS AND RECOMMENDATIONS

Findings:

- The concrete industry is capable of producing very sophisticated concrete if specified. All participants were aware of newer concrete technology but have not been able to apply the technology because project specifications and standards are nonexistent.
- The concrete producers observed have some knowledge of quality control (QC) testing and are performing QC testing at a relatively basic level. Test results are recorded and stored but there was no evidence that the results were being tracked so that large variations in materials could be identified.
- The primary function of the testing laboratories observed was to provide in-house quality control for the concrete producer where the lab was housed.
- Methods for tracking and storing test results using electronic spreadsheets were presented and discussed. Statistical methods of data analysis and European standards for determining required compressive strength (using average compressive strength and standard deviation of results) were presented. Since data have not been organized, performing the statistical analysis on actual data was impossible.
- The QC test methods being used were consistent from company to company. Each company performed slump testing and made compressive strength cubes. Other tests for determining the air content and fresh unit weight (density) of concrete were not being utilized. Although the methods were somewhat consistent, the test equipment used for determining slump was different. There is a need to distribute the interpreted European standards for basic QC test methods.
- The concrete production facilities observed are modern and in good working condition. It is assumed the batch plants are capable of dispensing admixtures into the concrete during batching and mixing. If not, dispensing units are inexpensive and in the U.S. are supplied and installed free of charge by the chemical admixture companies.
- There exists a mindset within the industry that greater strength means better concrete. The goal of several of the QC departments was to see how strong they could make their concrete. There was significant pride shown by individuals when concrete tested at a compressive strength of 60 MPa (8,700 psi).
- Durability in the U.S. concrete industry is defined as the concrete's ability to resist the negative impacts of the environment where it is placed. As an example concrete containing no entrained air will deteriorate in an environment where freezing and thawing conditions exist. Designing concrete for durability seemed like a new concept for the participants primarily because compressive strength is the only property specified.

Currently durability in concrete is a function of the high compressive strength. It is true that concrete strength is directly proportional to durability. Increased strength is achieved by increasing the cement content. Since cement is the most expensive constituent in concrete, increasing the cement content increases the cost of the materials.

- The industry could benefit by designing concrete to achieve durability. The use of chemical additives (admixtures) would be the most cost-effective way to achieve this.
- The individuals interviewed had a good working knowledge of the Yugoslav test methods. They did however struggle with methods used to determine concrete mix proportions that differ from what they are using. The current method relies on an empirical formula used to determine the water/cement ratio. Since concrete is one of the only materials sold by volume but batched by mass, newer methods use the specific gravity (the mass of a constant volume of material as compared to the mass of the same volume of water) of each material to determine the volume/mass relationship. Although the manual used by most QC managers addresses this relationship, mix proportions are still determined based on the empirical formula.
- It is going to be difficult to convince the producers that purposely entraining air into concrete will increase the freeze/thaw durability. One producer actually puts an admixture in the concrete to remove air. There are several technical journal articles available that prove entrained air in concrete increases freeze/thaw durability. These types of materials would be a benefit if they could be translated. In addition testing could also help in convincing producers of the benefits of air-entraining admixtures.
- The benefits of using water-reducing admixtures will be easier for the producer to accept since one can see the benefits immediately. The main benefit is reducing the amount of water needed, which will increase the compressive strength for a given mass of cement. The supplier can then reduce the amount of cement to achieve the same strength gained prior to the use of the water reducer.
- Discussions were held with a local architect to determine how concrete is specified for projects. Currently, concrete compressive strength is the only property specified. Concrete construction in general could also benefit by educating the design professionals on new concrete technologies.
- The cooperation within the industry is impressive. Numerous times competing QC managers would share information help each other by explaining their understanding of the concepts presented.
- There exists a concrete technical committee within the Road Construction Association of Kosovo. The committee is in its infancy but once mature, i would be a good clearing-house for new concrete technologies.
- A meeting was attended on the final day that included representatives of industry, private laboratories, and the government. The purpose of the meeting was to discuss laboratory accreditation. It was encouraging to see industry and government working toward a common goal. Based on the meeting there seems to be some confusion between the meanings of accreditation (evidence that a laboratory is capable of testing materials to a set of standards) and authorization (being allowed to provide testing on government projects). After the meeting there were numerous discussions regarding producer's laboratories being allowed to test competitor's materials on project sites, basically providing commercial testing services. It was suggested that the producers' laboratories be used to provide internal quality control testing and let the independent laboratories provide testing commercially.

Recommendations:

The following are recommendations, in order of importance, for the implementation of the presented tasks.

- Distribute the translated European standards and test methods for concrete-related materials. A list of the most important standards is presented in Annex II. (These standards have been made available to the Kosovo Standardization Agency, but the Agency has not promulgated them. KCBS will distribute to the four lab-operating companies immediately). Once the standards are distributed and the producers have used the standards and test methods, reinforce the use by having an expert, familiar with concrete materials testing, go into each laboratory to address questions and verify that the standards and test methods are being applied properly.
- In the U.S., the chemical admixture and the cement manufacturing companies are developing new concrete technology. The new technology is introduced to the industry by technical sales staff working directly with the concrete producers. Since the admixture companies have far more to gain by increasing sales in this area, the Kosovo Association of Concrete Producers (KACP) should contact the admixture companies and work with them to establish technical sales representatives in Kosovo to work with the individual concrete producers. Initially one visit to Kosovo per month would be adequate until the individual establishes working relationships with the suppliers. KCBS could facilitate by making the initial introductions between the admixture personnel and the QC managers of the concrete suppliers. The QC manager for one concrete company has attended a class given by one of the admixture companies; the class was given in Slovenia. It would be a benefit to have the admixture companies come to Kosovo.
- Expand the current education process to include architects engineers. Educate the design community on the European standards and teach them how to specify concrete for strength and durability and how to interpret test results to determine conformance to the specifications. In addition, the industry is in dire need of construction documents that better specify construction materials including concrete.
- Encourage monthly meetings between designers, contractors, materials suppliers, and government representatives to facilitate the exchange of new ideas and technologies. Have guest speakers attend and speak on the new technologies. As an example, ask a technical sales representative from an admixture company to speak on air-entrained concrete at one of the meetings. Make it a social event. Suggested topics are:
  - Durability of concrete
  - Accurately measuring strength of field placed concrete
  - Specifying concrete based on European standards
  - Proper concrete placement techniques
- Find individuals within the industry to champion the cause of more durable concrete through the use of chemical admixtures. Place the individuals on the concrete technical committee of the KACP. These individuals should be well trained. We know that one chemical admixture company has a training seminar that is held outside of Kosovo. Sending individuals to the class would give them an opportunity to visit a commercial laboratory to see how laboratories conduct the business side of the industry.

CONCLUSIONS AND RECOMMENDATIONS FOR FUTURE ACTIVITY

Portland cement concrete is probably the predominant construction material used in Kosovo. Concrete is used in structural elements in residential and commercial construction as well as precast concrete pavers, curbs, and pipe. There are numerous untapped markets for portland cement concrete in Kosovo. If the durability of concrete can be increased then concrete can be used in others types of construction such as paving, parking lots, sidewalks, or other types of flatwork. Concrete could easily be marketed as an inexpensive alternative to the concrete pavers currently being used in flatwork applications.

Making durable concrete and concrete products can be achieved by the use of inexpensive chemical admixtures. Utilizing the admixtures can result in using less cement, which is the most expensive constituent, per cubic meter of concrete. By lowering the cost of concrete the producer can increase profits while supplying a better product.

Introducing the concrete industry to newer technology will benefit the industry by lowering the material costs per cubic meter and increasing the durability of the concrete. Initially the introduction of air-entraining and water-reducing admixtures will greatly increase the consistency and durability of concrete and concrete-related products. In addition to materials, increasing the understanding of proportioning concrete mixes and how to adjust concrete mix proportions will benefit the individual producer by increasing the efficiency of each concrete mix.

At this time, the producers should focus on working with water-reducing and air-entraining admixtures. The introduction of mineral admixtures (fly ash and silica fume) would be somewhat overwhelming. As a side note, someone should investigate the feasibility of converting the local coal-burning power plant into a source of fly ash (fly ash is the byproduct of coal-burning power plants collected in the smoke stacks). There are many uses for fly ash including cement replacement in concrete (up to 25%), treating industrial waste in landfills, and increasing strength in construction subgrades. In addition, collecting the particulate emissions from the power plant would reduce air pollution.

There exists a need in the industry to establish an individual, or individuals, who can constantly promote European standards, proper testing techniques, and the use of new technologies in concrete production. This would result in constant reinforcement of the new concepts in hopes that the new methods become the accepted standard of care. If the admixture companies choose not to participate, KCBS should consider funding a one-year position to promote new technologies in concrete along with proper testing techniques for concrete and aggregates. In addition, the individual would be responsible to work with the government to establish construction standards and project specifications.

It is rumored that a commercial testing laboratory from Germany is considering moving to Kosovo. This may be a benefit to the concrete industry in that the German laboratory may force the local industry to raise its current standards to meet European laboratory standards. The impact on the concrete industry and acceptance of the German laboratory into the local market will be interesting to observe.

ANNEXES

ANNEX I—FIELD TRIP REPORTS

FIELD TRIP SUMMARY REPORT

Date of Visit: April 3, 2007
Site Visited: Renelual Tahiri, Prizren, Kosovo; Vellezerit e Bashkuar, Prizren, Kosovo
Attending: Valdet Osmani, Eljesi Surdulli, Jeff Groom, Hetem Muharremi, Feriz Shabani, and Driton Kryeziu

The purpose of the field trip was to observe the laboratories of Renelual Tahiri, and Vellezerite Bashkuar and discuss portland cement concrete mix proportioning and basic quality control (QC) test procedures and ascertain the current level of QC within the organization. Our tour included observation of equipment and discussions regarding the testing procedures and within laboratory QC procedures. A summary of our findings is presented below.

Renelual Tahiri

- The laboratory is well equipped for routine in-house QC testing. The laboratory manager Hetem is knowledgeable in most QC test procedures (Yugoslav) and has the computer capability to track QC test results. Hetem uses an equation based on the strength of cement to determine the compressive strength of his concrete. The equation is as follows:

  $$F'c = k * C((1/ w/c)-0.5) \text{ where}$$

  k = constant between 0.55 and 0.65
  C = strength of cement (MPa)
  w/c = water cement ratio

  From the equation Hetem determines the desired water/cement ratio and then determines the water content for a given cement content.

- The company does not use water-reducing or air-entraining admixtures, although they do use an air-reducing admixture.

Vellezerit e Bashkuar

- We made a short visit to Vellezerit e Bashkuar to make contact with Driton to plan the following day's activities. During the visit we discussed the use of air-entraining admixtures and water reducers, both of which Driton would like to use in concrete production. In addition, we discussed Driton's method of determining mix proportions. His method is based on each constituent being a specified percentage of the total mix mass.

CONCLUSION

Both laboratories are currently using Yugoslav test results and both have some understanding of European Norms. Each laboratory could benefit from EU procedures and adjusting their mix proportion techniques to include the specific gravity of all constituents.

FIELD TRIP SUMMARY REPORT

Date of Visit: April 4, 2007
Site Visited: Renelual Tahiri, Prizren, Kosovo
Attending: Valdet Osmani, Jeff Groom, Hetem Muharremi, Feriz Shabani, Driton Kryeziu, Bujar Hyseni, and Mondi Sako

The purpose of the field trip was to demonstrate the EU procedures for determining basic concrete QC testing of fresh and hardened concrete. The demonstrations were performed using the laboratory and equipment of Renelual Tahiri. Specifically we demonstrated the following on a small sample of fresh concrete:

- Calibration of the unit weight bucket and air content gauge. (EN 12350-6:2000 Density and EN 12350-7:2000 Air Content (Pressure Method))
- Slump test EN 12350-2:2000
- Determination of density EN 12350-6:2000
- How to determine the total air content of fresh concrete using the pressure method EN 12350-7:2000
- Casting and curing of strength specimens in accordance with EN 12390-2:2000
- Proper testing technique to determine the compressive strength of hardened concrete in accordance with EN 12390-3:2000

In addition, we briefly discussed tracking the test results and general testing procedures.

CONCLUSION

As anticipated the individuals were somewhat familiar with the basic testing procedures such as casting tests samples in three layers and compacting each layer with 25 stokes of a tamping rod, etc. There were certain very specific parts of each procedure that were new, such as the target time for raising the slump cone off of the test sample. Each client would benefit greatly by obtaining copies of the EU norms.

FIELD TRIP SUMMARY REPORT

Date of Visit: April 5, 2007
Site Visited: Renelual Tahiri office, Prizren, Kosovo
Attending: Valdet Osmani, Jeff Groom, Hetem Muharremi, Feriz Shabani, Driton Kryeziu, and Bujar Hyseni

The purpose of the field trip was to present a rigorous PowerPoint presentation on quality control testing and analyzing test results to adjust portland cement concrete mix proportions. The presentation included the following information:

- A review of European Norm testing methods for fresh and hardened concrete.
- Storing test results to easily conduct statistical analysis of the results.
- Statistical analysis techniques utilized by the European Norms.
- A new method of determining concrete mix proportions using the specific gravity of each constituent to easily convert mass to volume and volume to mass. This method is more accurate than the methods currently used in Prizren area for determining concrete mix proportions.
- Adjustment of concrete mix proportions based on the statistical analysis and European Norm requirements for average compressive strength.

There was a long discussion on statistics and how to apply the statistics to testing data. In addition there was a discussion of different methods to test for compressive strength including the differences between the use of cylinder and cube-shaped specimens.

CONCLUSION

The EU standards use statistics to determine what the average compressive strength of a given class of concrete should be. The analysis uses the average compressive strength as well as the standard deviation of the data to determine an appropriate over design for compressive strength. I believe this was a new concept to some of the attendees. Follow-up visits to each attendee may benefit the transfer of knowledge for the concepts presented.

FIELD TRIP SUMMARY REPORT

Date of Visit: April 6, 2007
Site Visited: Vellezerit & Bashkuar Laboratory, Prizren, Kosovo
Attending: Valdet Osmani, Jeff Groom, Hetem Muharremi, Feriz Shabani, Driton Kryeziu, and Bujar Hyseni

The purpose of the field trip was to present EN procedures for determining the class of concrete recommended for different corrosion-inducing exposures. The presentation was based on information contained in EN 206-1:2000 Concrete – *Specification, performance, production and conformity.* In addition, proportioned concrete mixes utilizing superplasticizing and air-entraining admixtures. The following information was presented and discussed:

- Different types of concrete needed for different corrosion inducing environments.
- A list of job duties for QC laboratory technicians and QC Manager position.
- Several concrete mixes were batched utilizing superplasticizing and air-entraining admixtures. The mixes were being placed on forms for concrete curbs. The mixes were proportioned to have a w/c ratio of 0.33. The first mix was very stiff (dry) and hard to place. The second mix was made with white cement and had a slightly higher slump. The third mix had a very high slump.
- Testing was performed by QC technicians that were employed by Vellezerit and Bashkuar. Testing methods were observed and recommendations were made based on the procedures observed.
- Placement procedures were also observed. We observed the concrete being over consolidated (vibrated), which results in driving the air (lowering the durability) from the mix. The concrete was being over consolidated to achieve a very smooth surface once the curbs were removed from the forms. We discussed changing the consolidation procedures. The change in procedures could result in a less smooth surface but the curbs would be more durable. Procedures were changed to accommodate the recommendations. We were notified later in the day that the surface of the curbs was acceptable; therefore, I hope the change in procedures will be permanent.

There was further discussion regarding different methods of consolidation such as vibrating tables versus stinger type vibrators.

CONCLUSION

The attendees were somewhat familiar with the nomenclature adopted by the EN standard. The concrete mixes made showed the trial and error approach needed when using the types of admixtures being utilized. The testing methods, in general, conform with EN standards except for a few minor changes that will result in more accurate testing. (The laboratory technicians were not familiar with the proper use of the air meter. The proper procedures were reviewed.) The revised placement procedures should result in a more durable product and ultimately will save time during production. The primary change was eliminating the over vibration of each curb and utilizing a method were the production crew "stings" each curb five times across the length of the curb.

FIELD TRIP SUMMARY REPORT

Date of Visit: April 7, 2007
Site Visited: Office of United Consultants Group, Laborator per Arkitekture, Pristina, Kosovo
Attending: Valdet Osmani, Jeff Groom, and Ilir Murseli, Director

The purpose of the field trip was to gain an understanding of how the different parties interact on a construction project. We discussed various phases of construction and in particular one project that was being designed by the United Consultants group. The following information was discussed:

- Communication between consultants, contractors, and material suppliers on construction projects.
- How concrete is specified for construction projects.
- A design change for a project the United Consultants Group is currently designing. The current design calls for cast-in-place architectural concrete. We discuss the use of precast panels in lieu of the cast-in-place concrete. The architect was not aware that precast panels could be cast in Kosovo. We discussed the benefit of better quality control in preparing precast panels. In addition, this design change could benefit a client of KCBS by allowing them to work directly with the design architect and providing a more technical concrete mix to achieve the appropriate esthetics the architect is seeking.

CONCLUSION

Ilir Murseli has worked throughout Europe and understands the importance of project specifications and standardization of procedures. It is my opinion he could be an excellent resource to KCBS in their ongoing commitment to introduce standards to the construction industry in that he has seen the benefits of such standardization as compared to the current method of construction in Kosovo. Based on our visit, I also believe the architectural and engineering communities could benefit from exposure to EN standards.

FIELD TRIP SUMMARY REPORT

Date of Visit: April 9, 2007
Site Visited: Granit, Istog, Kosovo
Attending: Valdet Osmani, Jeff Groom, and Ismet Loshaj, Company Director

The purpose of the field trip was to discuss the potential for Granit to develop a testing laboratory for its own use. In addition we hoped to discuss the use of additives in the concrete produced by Granit. The following information was discussed:

- Granit's primary focus at this time is the production of asphalt for road construction. Ismet said he primarily produces concrete for use on his projects as a secondary material.
- Ismet said he has focused on asphalt but he intends to focus more on concrete and the technological advances made in concrete production.
- Istog is a beautiful region surrounded by high mountains. We discussed the causes of concrete deterioration in the Istog area (freeze/thaw) and proper materials to be used for durable concrete. In addition, we discussed concrete mix proportions needed for durable concrete. Specifically we discussed how the use of water-reducing additives can help lower water/cement ratios. A lower water/cement ratio is necessary for durable concrete.
- There was a good discussion on the mechanisms involved in concrete deteriorated by freezing and thawing and how salt accelerates the deterioration.
- Ismet is interested in manufacturing lightweight concrete block. Lightweight block is manufactured with lightweight aggregates that are predominately volcanic tuft or pumice. Without a local source, the cost involved in shipping the material my make the blocks too costly to sell.

CONCLUSION

Ismet Loshaj is in a prime position to benefit from the efforts of KCBS. He is showing some interest in establishing a laboratory and is currently erecting a new steel building that could easily house the laboratory. In addition, when Ismet investigates the newer technology (chemical additives) KCBS is in a good position to encourage Ismet to investigate the benefits of using chemical additives.

FIELD TRIP SUMMARY REPORT

Date of Visit: April 10, 2007
Site Visited: Papenburg and Adiani, Farizaj, Kosovo
Attending: Valdet Osmani, Jeff Groom, Hetem Muharremi, Feriz Shabani, Driton Kryeziu, and Bujar Hyseni

The purpose of the field trip was to review mix design procedures using the specific gravity method of determining mix proportions. In addition, two concrete mixes were made and testing procedures were reviewed. A water-reducing admixture was used in the second mix to demonstrate the benefit of using water-reducing admixtures. The following was noted:

- The participants continue to struggle with the use of specific gravity to determine concrete mix proportions. The manual being used by most participants presents this method but boils the procedure down to an equation that can be used by individuals without truly understanding the procedure.
- The testing procedures and equipment used at Papenburg are vastly different than the methods and equipment used at Tahiri during a previous field trip. This is an indication that the procedures and equipment are in dire need of standardization.
- A concrete mix was made with and without water-reducing additives. The 0.5 M^3 mix was initially batched without the additive. The participants noted how dry the mix appeared as observed at the back of the mixer truck. We then poured the water-reducing admixture into the concrete and mixed for approximately minutes. We then placed the concrete into a wheelbarrow and the participants observed that the concrete flowed and was more workable. The mix contained roughly 10% less water then the concrete mixes currently in use. Unfortunately the cube molds utilized by Papenburg where being used and a compressive strength specimen could not be cast.

CONCLUSION

Basic quality control test methods need to be translated and distributed. Once the test methods and standards are distributed and being used, the principals discussed during these field trips should be reinforced with follow-up visits from concrete experts.

The benefit of using water-reducing additives was demonstrated. Based on the observed differences between the concrete with and without the additive, the benefit of the additive was obvious.

FIELD TRIP SUMMARY REPORT

Date of Visit: April 12, 2007
Site Visited: Road Construction Project Albania (Border to Morine-Kukes)
Attending: Valdet Osmani, Jeff Groom, Feriz Shabani, Driton Kryeziu, and Eljesi Surdulli

The purpose of the field trip was to observe the concrete work associated with the road construction. We observed several retaining walls and box culverts in addition to one of the concrete plants and testing laboratories being used for the project. The following was observed:

- Several box culverts
- Several relatively tall retaining walls
- Aggregate crushing facilities, concrete batch plant, and testing laboratory.
- We met with the project engineers in Morine-Kukes and looked at the project specifications.

OBSERVATIONS / CONCLUSION

The project is very impressive; a huge undertaking by any standards. Some of the concrete is being supplied by Vellezerit e Bashkuar out of their Prizren plant. Other concrete is being supplied by a concrete plant located about halfway through the project The on-site operation produces its own aggregates and all of the operations look to be relatively new.

We observed several concrete cubes being tested for strength. The testing on the project is specified to be in accordance with British standards. The testing observed did not meet the British standards because the loading of the specimens was too fast.

The project specifications were well written and easy to follow. The specified properties are easily attainable based on the materials and batching facilities. The project specifications allowed the use of air-entraining admixtures but the use was not specifically required. This amazes me since some of the concrete will be used for bridge structures exposed to freezing and thawing and deicing chemicals.

ANNEX II —STANDARDS AND TEST METHODS NEEDING TRANSLATION AND DISTRIBUTION

STANDARDS AND TESTS FOR FRESH AND HARDENED CONCRETE

EN 206-1:2000 CONCRETE – Specification, performance, production, and conformity

EN 12350-1:2000 Sampling

EN 12350-2:2000 Slump

EN 12350-6:2000 Unit Weight (Density)

EN 12350-7:2000 Air Content (Pressure)

EN 12390-2:2000 Making and Curing test Specimens

EN 12390-3:2000 Compressive Strength of Test Specimens

STANDARDS AND TESTS FOR AGGREGATES

EN 12620:2002 Aggregates for Concrete

EN 932-1:1997 Sampling

EN 932-2:1997 Reducing Sample Size

EN 932-5:2000 Equipment Calibration

EN 933-1:1997 Particle Size Distribution (Sieving)

EN 1097-6:2000 Particle Density and Absorption

ANNEX III—RECOMMENDED QUALITY CONTROL TESTING AND FREQUENCIES

AGGREGATE TESTING

PROPERTY	TEST METHOD	FREQUENCY
Grading (Sieve Analysis)	EN 9330-1	1 per week for each aggregate
Aggregate density (specific gravity) and water absorption	EN 1097-6	1 per month for each aggregate

FRESH CONCRETE*

PROPERTY	TEST METHOD	FREQUENCY	NOTES
Slump	EN 12350-2	1 per day	
Density	EN 12350-6	1 per day	Compare actual density to density of mix based on proportions. Adjust mass of sand as necessary for yield.
Air Content	EN 12350-7	1 per day	
Cast compressive strength specimens	EN 12390-2	2 per week	Make 4 cubes for each test. Test 2 cubes at 7 days and test 2 cubes at 28 days

* These tests should be performed every time concrete is sampled for testing. Tests should be conducted daily for each class of concrete batched.

HARDENED CONCRETE

PROPERTY	TEST METHOD	FREQUENCY	NOTES
Compressive Strength	EN 12390-3	2 cubes at 7 days age 2 cubes at 28 days age	Cast 4 cubes per set

ANNEX IV—FIELD VISIT PHOTOGRAPHS

Demonstrating air content test

Demonstrating positioning of the slump cone

Learning to read air meter

Performing unit weight test

Determining the slump of superplasticized concrete

Production crew vibrating concrete curbs. Note the line made by the vibrator down the middle of each curb. This is an indication of over vibration.

Discussing EU standards for classification of concrete

Participants in presentation

Performing slump test at Papenburg and Adriani. Note the nonconfoming method of consolidation.

Air content test at Papenburg and Adriani

Testing compressive strength cubes

Adding water-reducing admixture to concrete

APPENDIX

V

Concrete and Masonry Construction

OSHA 3106 1998 (Revised)

OUTLINE

This informational booklet is intended to provide a generic, nonexhaustive overview of a particular standards-related topic. This publication does not alter or determine compliance responsibilities set forth in OSHA standards and the Occupational Safety and Health Act. Moreover, because interpretations and enforcement policy may change over time, for additional guidance on OSHA compliance requirements, the reader should consult current administrative interpretations and decisions by the Occupational Safety and Health Review Commission and the courts.

CONTENTS

WHAT DOES OSHA'S CONCRETE AND MASONRY STANDARD COVER?

The Occupational Safety and Health Administration's standard for concrete and masonry construction — *Subpart Q, Concrete and Masonry Construction, Title 29 of the Code of Federal Regulations* (CFR), Part 1926.700 through 706 — sets forth requirements with which construction employers must comply to protect construction workers from accidents and injuries resulting from the premature removal of formwork, the failure to brace masonry walls, the failure to support precast panels, the inadvertent operation of equipment, and the failure to guard reinforcing steel.

Subpart Q prescribes performance-oriented requirements designed to help protect all construction workers from the hazards associated with concrete and masonry construction operations at construction, demolition, alteration, or repair worksites. Other relevant provisions in both general industry and construction standards (29 CFR Parts 1910 and 1926) also apply to these operations.

WHAT ARE THE KEY, NEW CHANGES TO THE STANDARD?

OSHA's concrete and masonry standard includes the following important changes:

- Expands and toughens protection against masonry wall collapses by requiring bracing and a limited access zone prior to the construction of a wall
- Permits employers to use several more recently developed methods of testing concrete instead of just the one currently recognized method
- Sets and clarifies requirements for both cast-in-place concrete and pre-cast concrete during construction

WHAT ARE THE COMPONENTS OF THE NEW STANDARD?

Subpart Q is divided into the following major groups each of which is discussed in more detail in the following paragraphs:

- Scope, application, and definitions (29 CFR 1926.700);
- General requirements (29 CFR 1926.702);
- Equipment and tools (29 CFR 1926.702);
- Cast-in-place concrete (20 CFR 1926.703);
- Pre-cast concrete (29 CFR 1926.704);
- Lift-slab construction (29 CFR 1926.705); and
- Masonry construction (29 CFR 1926.706).

WHAT ARE THE GENERAL REQUIREMENTS OF THE STANDARD?

Construction Loads

Employers must not place construction loads on a concrete structure or portion of a concrete structure unless the employer determines, based on information received from a person who is qualified in structural design, that the structure or portion of the structure is capable of supporting the intended loads.

Reinforcing Steel

All protruding reinforcing steel, onto and into which employees could fall, must be guarded to eliminate the hazard of impalement.

Post-Tensioning Operations

Employees (except those essential to the post-tensioning operations) must not be permitted to be behind the jack during tensioning operations.

Signs and barriers must be erected to limit employee access to the post-tensioning area during tensioning operations.

Concrete Buckets

Employees must not be permitted to ride concrete buckets.

Working Under Loads

Employees must not be permitted to work under concrete buckets while the buckets are being elevated or lowered into position.

To the extent practicable, elevated concrete buckets must be routed so that no employee or the fewest employees possible are exposed to the hazards associated with falling concrete buckets.

Personal Protective Equipment

Employees must not be permitted to apply a cement, sand, and water mixture through a pneumatic hose unless they are wearing protective head and face equipment.

Equipment and Tools

The standard also includes requirements for the following equipment and operations:

- Bulk cement storage
- Concrete mixers
- Power concrete trowels
- Concrete buggies
- Concrete pumping systems
- Concrete buckets
- Tremies
- Bull floats
- Masonry saws
- Lockout/tagout procedures

WHAT ARE THE REQUIREMENTS FOR CAST-IN-PLACE CONCRETE?

General Requirements for Formwork

Formwork must be designed, fabricated, erected, supported, braced, and maintained so that it will be capable of supporting without failure all vertical and lateral loads that might be applied to the formwork. As indicated in the Appendix to the standard, formwork that is designed, fabricated, erected, supported, braced, and maintained in conformance with Sections 6 and 7 of the *American National Standard for Construction and Demolition Operations — Concrete and Masonry Work* (ANSI) A10.9-1983 also meets the requirements of this paragraph.

Drawings or Plans

Drawings and plans, including all revisions for the jack layout, formwork (including shoring equipment), working decks, and scaffolds, must be available at the jobsite.

Shoring and Reshoring

All shoring equipment (including equipment used in reshoring operations) must be inspected prior to erection to determine that the equipment meets the requirements specified in the formwork drawings.

Damaged shoring equipment must not be used for shoring. Erected shoring equipment must be inspected immediately prior to, during, and immediately after concrete placement. Shoring equipment that is found to be damaged or weakened after erection must be immediately reinforced.

The sills for shoring must be sound, rigid, and capable of carrying the maximum intended load. All base plates, shore heads, extension devices, and adjustment screws must be in firm contact and secured, when necessary, with the foundation and the form.

Eccentric loads on shore heads must be prohibited unless these members have been designed for such loading.

If single-post shores are used one on top of another (tiered), then additional shoring requirements must be met. The shores must be as follows:

- Designed by a qualified designer and the erected shoring must be inspected by an engineer qualified in structural design.
- Vertically aligned.
- Spliced to prevent misalignment.
- Adequately braced in two mutually perpendicular directions at the splice level. Each tier also must be diagonally braced in the same two directions.

Adjustment of single-post shores to raise formwork must not be made after the placement of concrete.

Reshoring must be erected, as the original forms and shores are removed, whenever the concrete is required to support loads in excess of its capacity.

Vertical Slip Forms

The steel rods or pipes on which jacks climb or by which the forms are lifted must be (1) specifically designed for that purpose and (2) adequately braced where not encased in concrete. Forms must be designed to prevent excessive distortion of the structure during the jacking operation. Jacks and vertical supports must be positioned in such a manner that the loads do not exceed the rated capacity of the jacks.

The jacks or other lifting devices must be provided with mechanical dogs or other automatic holding devices to support the slip forms whenever failure of the power supply or lifting mechanisms occurs.

The form structure must be maintained within all design tolerances specified for plumbness during the jacking operation.

The predetermined safe rate of lift must not be exceeded.

All vertical slip forms must be provided with scaffolds or work platforms where employees are required to work or pass.

Reinforcing Steel

Reinforcing steel for walls, piers, columns, and similar vertical structures must be adequately supported to prevent overturning and collapse.

Employers must take measures to prevent unrolled wire mesh from recoiling. Such measures may include, but are not limited to, securing each end of the roll or turning over the roll.

Removal of Formwork

Forms and shores (except those that are used for slabs on grade and slip forms) must not be removed until the employer determines that the concrete has gained sufficient strength to support its weight and superimposed loads. Such determination must be based on compliance with one of the following:

- The plans and specifications stipulate conditions for removal of forms and shores, and such conditions have been followed, or
- The concrete has been properly tested with an appropriate American Society for Testing and Materials (ASTM) standard test method designed to indicate the concrete compressive strength and the test results indicate that the concrete has gained sufficient strength to support its weight and superimposed loads.

Reshoring must not be removed until the concrete being supported has attained adequate strength to support its weight and all loads placed upon it.

Pre-Cast Concrete

Pre-cast concrete wall units, structural framing, and tilt-up wall panels must be adequately supported to prevent overturning and to prevent collapse until permanent connections are completed.

Lifting inserts that are embedded or otherwise attached to tilt-up wall panels must be capable of supporting at least two times the maximum intended load applied or transmitted to them; lifting inserts for other pre-cast members must be capable of supporting four times the load. Lifting hardware shall be capable of supporting at least five times the maximum intended load applied or transmitted to the lifting hardware.

Only essential employees are permitted under pre-cast concrete that is being lifted or tilted into position.

Lift-Slab Operations

- Lift-slab operations must be designed and planned by a registered professional engineer who has experience in lift-slab construction. Such plans and designs must be implemented by the employer and must include detailed instructions and sketches indicating the prescribed method of erection. The plans and designs must also include provisions for ensuring lateral stability of the building/structure during construction.
- Jacking equipment must be marked with the manufacturer's rated capacity and must be capable of supporting at least two and one-half times the load being lifted during jacking operations and the equipment must not be overloaded. For the purpose of this provision, jacking equipment includes any load bearing component that is used to carry out the lifting operation(s). Such equipment includes, but is not limited to, the following: threaded rods, lifting attachments, lifting nuts, hook-up collars, T-caps, shearheads, columns, and footings.
- Jacks/lifting units must be designed and installed so that they will neither lift nor continue to lift when loaded in excess of their rated capacity; and jacks/lifting units must have a safety device which will cause the jacks/lifting units to support the load at any position in the event of their malfunction or loss of ability to continue to lift.
- No employee, except those essential to the jacking operation, shall be permitted in the building/structure while any jacking operation is taking place unless the building/structure has been reinforced sufficiently to ensure its integrity during erection. The phrase "reinforced sufficiently to ensure its integrity" as used in this paragraph means that a registered professional engineer, independent of the engineer who designed and planned the lifting operation, has determined from the plans that if there is a loss of support at any jack location, that loss will be confined to that location and the structure as a whole will remain stable.
- Under no circumstances shall any employee who is not essential to the jacking operation be permitted immediately beneath a slab while it is being lifted.

Masonry Construction

Whenever a masonry wall is being constructed, employers must establish a limited access zone prior to the start of construction. The limited access zone must be as follows:

- Equal to the height of the wall to be constructed plus 4 feet (1.2 m), and shall run the entire length of the wall.
- On the side of the wall that will be unscaffolded.
- Restricted to entry only by employees actively engaged in constructing the wall.
- Kept in place until the wall is adequately supported to prevent overturning and collapse unless the height of the wall is more than 8 feet (2.4 m) and unsupported, in

which case it must be braced. The bracing must remain in place until permanent supporting elements of the structure are in place.

WHAT OTHER HELP CAN OSHA PROVIDE?

Safety and Health Program Management Guidelines

Effective management of worker safety and health protection is a decisive factor in reducing the extent and severity of work-related injuries and illnesses and their related costs. To assist employers and employees in developing effective safety and health programs, OSHA published recommended Safety and Health Program Management Guidelines (Federal Register 54(18):3908-3916, January 26, 1989). These voluntary guidelines apply to all places of employment covered by OSHA.

The guidelines identify four general elements that are critical to the development of a successful safety and health management program:

- Management commitment and employee involvement
- Worksite analysis
- Hazard prevention and control
- Safety and health training

The guidelines recommend specific actions under each of these general elements to achieve an effective safety and health program. A single free copy of the guidelines can be obtained from the U.S. Department of Labor, OSHA/OICA Publications, P.O. Box 37535, Washington, DC 20013-7535, by sending a self-addressed mailing label with your request.

State Programs

The Occupational Safety and Health Act of 1970 encourages states to develop and operate their own job safety and health plans. States administering occupational safety and health programs through plans approved under Section 18(b) of the Act must adopt standards and enforce requirements that are "at least as effective" as federal requirements. There are currently 25 state plan states: 23 cover the private and public sector (state and local governments) and 2 cover the public sector only. For more information on state plans, see the list of states with approved plans at the end of this publication.

Free On-Site Consultation

Consultation assistance is available on request to employers who want help in establishing and maintaining a safe and healthful workplace. Largely funded by OSHA, the service is provided at no cost to the employer. Primarily developed for smaller employers with more hazardous operations, the consultation service is delivered by state government agencies or universities employing professional safety consultants and health consultants. Comprehensive assistance includes an appraisal of all work practices and environmental hazards of the workplace and all aspects of the employer's present job safety and health program.

The program is separate from OSHA's inspection efforts. No penalties are proposed or citations issued for any safety or health problems identified by the consultant. The service is confidential.

For more information concerning consultation assistance, see the list of consultation projects at the end of this publication.

Voluntary Protection Programs

Voluntary Protection Programs (VPPs) and on-site consultation services, when coupled with an effective enforcement program, expand worker protection to help meet the goals of the Act. The three VPPs — Star, Merit, and Demonstration — are designed to recognize outstanding achievement by companies that have successfully incorporated comprehensive safety and health programs into their total management system. They motivate others to achieve excellent safety and health results in the same outstanding way as they establish a cooperative relationship among employers, employees, and OSHA.

For additional information on VPPs and how to apply, contact the OSHA area or regional offices listed at the end of this publication.

Training and Education

OSHA area offices offer a variety of information services, such as publications, audiovisual aids, technical advice, and speakers for special engagements. The OSHA Training Institute in Des Plaines, Illinois, provides basic and advanced courses in safety and health for federal and state compliance officers; state consultants; federal agency personnel; and private sector employers, employees, and their representatives.

OSHA also provides funds to nonprofit organizations through grants to conduct workplace training and education in subjects where OSHA believes there is a lack of workplace training. Grants are awarded annually and grant recipients are expected to contribute 20% of the total grant cost.

For more information on grants, training, and education, contact the OSHA Training Institute, Office of Training and Education, 1555 Times Drive, Des Plaines, IL 60018; telephone: (847) 297-4810.

For further information on any OSHA program, contact your nearest OSHA area or regional office listed at the end of this publication.

Electronic Information

Internet—OSHA standards, interpretations, directives, technical advisors, compliance assistance, and additional information are now on the World Wide Web at http://www.osha.gov.

CD-ROM—A wide variety of OSHA materials, including standards, interpretations, directives, and more, can be purchased on CD-ROM from the U.S. Government Printing Office. To order, write to the Superintendent of Documents, P.O. Box 371954, Pittsburgh, PA 15250-7954 or telephone (202) 512-1800. Specify OSHA Regulations, Documents, and Technical

Information on CD-ROM (ORDT), GPO Order No. S/N 729-013-00000-5. The price is $38 per year ($47.50 foreign), $15 per single copy ($18.75 foreign).

Emergencies

For life-threatening situations, call (800) 321-OSHA. Complaints will go immediately to the nearest OSHA area or state office for help.

For further information on any OSHA program, contact your nearest OSHA area or regional office listed at the end of this publication.

GLOSSARY

Bull Float A tool used to spread out and smooth concrete.

Formwork The total system of support for freshly placed or partially cured concrete, including the mold or sheeting (form) that is in contact with the concrete as well as all supporting members including shores, reshores, hardware, braces, and related hardware.

Jacking Operation Lifting vertically a slab (or group of slabs) from one location to another — for example, from the casting location to a temporary (parked) location, or from a temporary location to another temporary location, or to the final location in the structure — during a lift-slab construction operation.

Lift Slab A method of concrete construction in which floor and roof slabs are cast on or at ground level and, using jacks, are lifted into position.

Limited Access Zone An area alongside a masonry wall, that is under construction, and that is clearly demarcated to limit access by employees.

Pre-Cast Concrete Concrete members (such as walls, panels, slabs, columns, and beams) that have been formed, cast, and cured prior to final placement in a structure.

Reshoring The construction operation in which shoring equipment (also called reshores or reshoring equipment) is placed, as the original forms and shores are removed in order to support partially cured concrete and construction loads.

Shore A supporting member that resists a compressive force imposed by a load.

Tremie A pipe through which concrete may be deposited under water.

Vertical Slip Forms Forms that are jacked vertically during the placement of concrete.

OSHA-RELATED PUBLICATIONS

Single free copies of the following publication(s) can be obtained from the U.S. Department of Labor, OSHA/OICA Publications, P.O. Box 37535, Washington, DC 20013-7535. Send a self-addressed mailing label with your request.

Asbestos Standard for Construction Industry — OSHA 3096

The following publications are available from the Superintendent of Documents, U.S. Government Printing Office, Washington, DC 20402; telephone (202) 512-1800. Include GPO Order No. And make checks payable to the Superintendent of Documents.

Construction Industry Digest — OSHA 2202

Order No. 029-016-00151-4. Cost $2.25.

Excavations — OSHA 2226.

Order No. 029-016-00176-1. Cost $1.25.

Title 29 CFR Part 1926 — (Construction)

Order No. 029-016-00122-1. Cost $30.00.

STATES WITH APPROVED PLANS

Commissioner

Alaska Department of Labor
1111 West 8th Street
Room 306
Juneau, AK 99801
(907) 465-2700

Director

Industrial Commission of Arizona
800 W. Washington
Phoenix, AZ 85007
(602) 542-5795

Director

California Department of Industrial Relations
45 Fremont Street
San Francisco, CA 94105
(415) 972-8835

Commissioner

Connecticut Department of Labor
200 Folly Brook Boulevard
Wethersfield, CT 06109
(203) 566-5123

Director

Hawaii Department of Labor and Industrial Relations
830 Punchbowl Street
Honolulu, HI 96813
(808) 586-8844

Commissioner

Indiana Department of Labor State Office Building
402 West Washington Street
Room W195
Indianapolis, IN 46204
(317) 232-2378

Commissioner

Iowa Division of Labor Services
1000 E. Grand Avenue
Des Moines, IA 50319
(515) 281-3447

Secretary

Kentucky Labor Cabinet
1047 U.S. Highway South, Suite 2
Frankfort, KY 40601
(502) 564-3070

Commissioner

Maryland Division of Labor and Industry
Department of Labor
Licensing and Regulation
1100 N. Eutaw St., Rm. 613
Baltimore, MD 21202-2206
(410) 767-2215

Director

Michigan Department of Consumer and Industry Services
4th Floor, Law Building
P.O. Box 30004
Lansing, MI 48909
(517) 373-7230

Commissioner

Minnesota Department of Labor and Industry
443 Lafayette Road
St. Paul, MN 55155
(612) 296-2342

Administrator

Nevada Division of Industrial Relations
400 West King Street
Carson City, NV 89710
(702) 687-3032

Secretary

New Mexico Environment Department
1190 St. Francis Drive
P.O. Box 26110
Santa Fe, NM 87502
(505) 827-2850

Commissioner

New York Department of Labor
W. Averell Harriman State
Office Building - 12,
Room 500
Albany, NY 12240
(518) 457-2741

Commissioner

North Carolina Department of Labor
319 Chapanoke Road
Raleigh, NC 27603
(919) 662-4585

Administrator

Department of Consumer
and Business Services
Occupational Safety and Health Division
(OR-OSHA)
350 Winter Street, N.E.
Room 430
Salem, OR 97310-0220
(503) 378-3272

Secretary

Puerto Rico Department of Labor
and Human Resources
Prudencio Rivera Martinez Building
505 Munoz Rivera Avenue
Hato Rey, PR 00918
(809) 754-2119

Director

South Carolina Department of Labor
Licensing and Regulation
Koger Office Park, Kingstree Bldg.
P.O. Box 11329
Columbia, SC 29210
(803) 896-4300

Commissioner

Tennessee Department of Labor
Attention: Robert Taylor
710 James Robertson Parkway
Nashville, TN 37243-0659
(615) 741-2582

Commissioner

Industrial Commission of Utah
160 East 300 South, 3rd Floor
P.O. Box 146650
Salt Lake City, UT 84114-6650
(801) 530-6898

Commissioner

Vermont Department of Labor
and Industry
National Life Building – Drawer 20
120 State Street
Montpelier, VT 05620
(802) 828-2288

Commissioner

Virgin Islands Department of Labor
2131 Hospital Street, Box 890
Christiansted
St. Croix, VI 00820-4666
(809) 773-1994

Commissioner

Virginia Department of Labor
and Industry
Powers-Taylor Building
13 South 13th Street
Richmond, VA 23219
(804) 786-2377

Director

Washington Department of Labor
and Industries
General Administration Building
P.O. Box 44001
Olympia, WA 98504-4001
(360) 902-4200

Administrator

Worker's Safety and Compensation
Division (WSC)
Wyoming Department of Employment
Herschler Building, 2nd Floor East
122 West 25th Street
Cheyenne, WY 82002
(307) 777-7786

OSHA CONSULTATION PROJECT DIRECTORY

State	Telephone
Alabama	(205) 348-7136
Alaska	(907) 269-4957
Arizona	(602) 542-5795
Arkansas	(501) 682-4522
California	(415) 982-8515
Colorado	(970) 491-6151
Connecticut	(860) 566-4550
Delaware	(302) 761-8219
District of Columbia	(202) 576-6339
Florida	(904) 488-3044
Georgia	(404) 894-2643
Guam	011 (671) 475-0136
Hawaii	(808) 586-9100
Idaho	(208) 385-3283
Illinois	(312) 814-2337
Indiana	(317) 232-2688

State	Telephone
Iowa	(515) 965-7162
Kansas	(913) 296-7476
Kentucky	(502) 564-6895
Louisiana	(504) 342-9601
Maine	(207) 624-6460
Maryland	(410) 880-4970
Massachusetts	(617) 727-3982
Michigan	(517) 332-1817(H) (517) 322-1809(S)
Minnesota	(612) 297-2393
Mississippi	(601) 987-3981
Missouri	(573) 751-3403
Montana	(406) 444-6418
Nebraska	(402) 471-4717
Nevada	(702) 486-5016
New Hampshire	(603) 271-2024
New Jersey	(609) 292-2424
New Mexico	(505) 827-4230
New York	(518) 457-2481
North Carolina	(919) 662-4644
North Dakota	(701) 328-5188
Ohio	(614) 644-2246
Oklahoma	(405) 528-1500
Oregon	(503) 378-3272
Pennsylvania	(412) 357-2561
Puerto Rico	(787) 754-2188
Rhode Island	(401) 277-2438
South Carolina	(803) 896-4300
South Dakota	(605) 688-4101
Tennessee	(615) 741-7036
Texas	(512) 440-3809

(*Continued*)

State	Telephone
Utah	(801) 530-7606
Vermont	(802) 828-2765
Virginia	(804) 786-6359
Virgin Islands	(809) 772-1315
Washington	(360) 902-5638
West Virginia	(304) 558-7890
Wisconsin	(608) 266-8579(H) (414) 521-5063(S)
Wyoming	(307) 777-7786

(H) — Health
(S) — Safety

OSHA AREA OFFICES

Area	Telephone
Albany, NY	(518) 464-4338
Albuquerque, NM	(505) 248-5302
Allentown, PA	(610) 776-0592
Anchorage, AK	(907) 271-5152
Appleton, WI	(414) 734-4521
Austin, TX	(512) 916-5783
Avenel, NJ	(908) 750-3270
Baltimore, MD	(410) 962-2840
Bangor, ME	(207) 941-8177
Baton Rouge, LA	(504) 389-0474
Bayside, NY	(718) 279-9060
Bellevue, WA	(206) 553-7520
Billings, MT	(406) 247-7494
Birmingham, AL	(205) 731-1534
Bismarck, ND	(701) 250-4521
Boise, ID	(208) 334-1867
Bowmansville, NY	(716) 684-3891

Area	Telephone
Braintree, MA	(617) 565-6924
Bridgeport, CT	(203) 579-5581
Calumet City, IL	(708) 891-3800
Carson City, NV	(702) 885-6963
Charleston, WV	(304) 347-5937
Cincinnati, OH	(513) 841-4132
Cleveland, OH	(216) 522-3818
Columbia, SC	(803) 765-5904
Columbus, OH	(614) 469-5582
Concord, NH	(603) 225-1629
Corpus Christi, TX	(512) 888-3420
Dallas, TX	(214) 320-2400
Denver, CO	(303) 844-5285
Des Plaines, IL	(847) 803-4800
Des Moines, IA	(515) 284-4794
Englewood, CO	(303) 843-4500
Erie, PA	(814) 833-5758
Fort Lauderdale, FL	(954) 424-0242
Fort Worth, TX	(817) 428-2470
Frankfort, KY	(502) 227-7024
Harrisburg, PA	(717) 782-3902
Hartford, CT	(860) 240-3152
Hasbrouck Heights, NJ	(201) 288-1700
Guaynabo, PR	(787) 277-1560
Honolulu, HI	(808) 541-2685
Houston, TX (South Area)	(281) 286-0583
Houston, TX (North Area)	(281) 591-2438
Indianapolis, IN	(317) 226-7290
Jackson, MS	(601) 965-4606
Jacksonville, FL	(904) 232-2895
Kansas City, MO	(816) 483-9531

(Continued)

Area	Telephone
Lansing, MI	(517) 377-1892
Little Rock, AR	(501) 324-6291
Lubbock, TX	(806) 743-7681
Madison, WI	(608) 264-5388
Marlton, NJ	(609) 757-5181
Methuen, MA	(617) 565-8110
Milwaukee, WI	(414) 297-3315
Minneapolis, MN	(612) 348-1994
Mobile, AL	(334) 441-6131
Nashville, TN	(615) 781-5423
New York, NY	(212) 466-2482
Norfolk, VA	(804) 441-3820
North Aurora, IL	(630) 896-8700
North Syracuse, NY	(315) 451-0808
Oklahoma City, OK	(405) 231-5351
Omaha, NE	(402) 221-3182
Parsippany, NJ	(201) 263-1003
Peoria, IL	(309) 671-7033
Philadelphia, PA	(215) 597-4955
Phoenix, AZ	(602) 640-2007
Pittsburgh, PA	(412) 644-2903
Portland, OR	(503) 326-2251
Providence, RI	(401) 528-4669
Raleigh, NC	(919) 856-4770
Salt Lake City, UT	(801) 487-0073
Sacramento, CA	(916) 566-7470
San Diego, CA	(619) 557-2909
Savannah, GA	(912) 652-4393
Smyrna, GA	(770) 984-8700
Springfield, MA	(413) 785-0123

Area	Telephone
St. Louis, MO	(314) 425-4249
Tampa, FL	(813) 626-1177
Tarrytown, NY	(914) 524-7510
Toledo, OH	(419) 259-7542
Tucker, GA	(770) 493-6644
Westbury, NY	(516) 334-3344
Wichita, KS	(316) 269-6644
Wilkes-Barre, PA	(717) 826-6538
Wilmington, DE	(302) 573-6115

OSHA REGIONAL OFFICES

Region I

(CT,[1] MA, ME, NH, RI, VT[1])
JKF Federal Building
Room E-340
Boston, MA 02203
Telephone: (617) 565-9860

Region II

(NJ, NY,[1] PR,[1] VI[1])
201 Varick Street
Room 670
New York, NY 10014
Telephone: (212) 337-2378

Region III

(DC, DE, MD,[1] PA, VA,[1] WV)
Gateway Building, Suite 2100
3535 Market Street
Philadelphia, PA 19104
Telephone: (215) 596-1201

Region IV

(AL, FL, GA, KY,[1] MS, NC, SC,[1] TN[1])
Atlanta Federal Center
61 Forsyth Street, SW,
Room 6T50
Atlanta, GA 30303
Telephone: (404) 562-2300

Region V

(IL, IN,[1] MI,[1] MN,[1] OH, WI)
230 South Dearborn Street
Room 3244
Chicago, IL 60604
Telephone: (312) 353-2220

Region VI

(AR, LA, NM,[1] OK, TX)
525 Griffin Street
Room 602
Dallas, TX 75202
Telephone: (214) 767-4731

Region VII

(IA,[1] KS, MO, NE)
City Center Square
1100 Main Street, Suite 800
Kansas City, MO 64105
Telephone: (816) 426-5861

Region VIII

(CO, MT, ND, SD, UT,[1] WY[1])
1999 Broadway, Suite 1690
Denver, CO 80202-5716
Telephone: (303) 844-1600

Region IX

(American Samoa, AZ,[1] CA,[1] Guam, HI,[1] NV,[1] Trust Territories of the Pacific)
71 Stevenson Street
Room 420
San Francisco, CA 94105
Telephone: (415) 975-4310

Region X

(AK,[1] ID, OR,[1] WA[1])
1111 Third Avenue
Suite 715
Seattle, WA 98101-3212
Telephone: (206) 553-5930

[1]These states and territories operate their own OSHA-approved job safety and health programs (Connecticut and New York plans cover public employees only). States with approved programs must have a standard that is identical to, or at least as effective as, the federal standard.

APPENDIX

VI

Glossary of Abbreviations

AAN American Association of Nurserymen
AAR Association of American Railroads
AASHTO American Association of State Highway and Transportation Officials
AITC American Institute of Timber Construction
AC alternating current
ACI American Concrete Institute
AGC Associated General Contractors of America, Inc.
AIA American Institute of Architects
AISC American Institute of Steel Construction
AISI American Iron and Steel Institute
AITC American Institute of Timber Construction
ANSI American National Standards Institute
ARA American Railway Association
AREA American Railway Engineering Association
ASCE American Society of Civil Engineers
ASLA American Society of Landscape Architects
ASME American Society of Mechanical Engineers
ASTM American Society of Testing and Materials
ATR Automatic Traffic Recorder
AWPA American Wood Preservers Association
AWS American Welding Society
AWG American Wire Gauge
AWWA American Water Works Association
CCTV closed circuit television
CMP communications plenum cable; corrugated metal pipe
CMS changeable message sign
COAX radio frequency transmission cable (coaxial cable)
CRSI Concrete Reinforcing Steel Institute
CV compacted volume
DBE Disadvantage Business Enterprise
QA quality assurance
QC quality control
RCS Ramp Control Signal
REA Rural Electrification Association
RF radio frequency
RHW moisture and heat-resistant; cross linked synthetic polymer
RMS root mean square
RSC rigid steel conduit

SAE Society of Automotive Engineers
SI International System of Units (The Modernized Metric System)
SPDT single pole double throw
SPST single pole single throw
SSPC Society for Protective Coatings
SV stockpiled volume
SWPPP Storm Water Pollution Prevention Plan
TH trunk highway
TMC Traffic Management Center
TMS Traffic Management System
TSM Traffic System Management
UL Underwriters Laboratories, Inc.
USD United States Department of Agriculture
UV Ultra violet
VAC volt alternating current (60 Hz)
VDC volt direct current
XHHW moisture and heat-resistant cross linked synthetic polymer

1101.2 Metric Units

The International System of Units (SI; the Modernized Metric System) according to ASTM E 380 are used in these Specifications. ASTM E 380 also provides conversion factors and commentary.

Metric Prefix and Magnitude

M mega (10^6)
k kilo (10^3)
m milli (10^{-3})
μ micro (10^{-6})
n nano (10^{-9})
p pico (10^{-12})

Metric Symbols

A ampere (electric current)
cd candela (luminous intensity)
F farad (electric capacitance)
g gram (mass)
H henry (inductance)
ha hectare (area)
Hz hertz (frequency — cycles or impulses per second)
J joule (energy)
km/h kilometer per hour (velocity)
km^2 square kilometer (area)
L liter (volume)
m/s meters per second (velocity)
m meter (length)
m^2 square meter (area)
m^3 cubic meter (volume)
m^3/s cubic meters per second (flow rate)
N newton (force)

N•m newton meter (torque)
Pa pascal (pressure, stress)
s second (time)
S siemens (electrical conductance)
t metric ton (mass)
V volt (electric potential)
W watt (power)
Ω ohm (electric resistance)
°C degree Celsius (temperature)

APPENDIX

VII

Common Definitions

Unless another intention clearly appears, words and phrases (including technical words and phrases and such others as have acquired a special meaning) shall be construed according to rules of grammar and according to general usage.

Wherever the following terms, or pronouns in place of them, are used in these Specifications, the Plans, or other Contract documents, the intent and meaning shall be interpreted as follows:

Addendum A supplement to the proposal form as originally issued or printed, covering additions, corrections, or changes in the bidding conditions for the advertised work, that is issued by the Contracting Authority to prospective bidders prior to the date set for opening of proposals.

Advertisement for Bids The public announcement, as required by law, inviting bids for the work to be performed or materials to be furnished.

Aggregate Natural materials such as sand, gravel, crushed rock, or taconite tailings, and crushed concrete or salvaged bituminous mixtures, usually with a specified particle size, for use in base course construction, paving mixtures, and other specified applications.

Auxiliary Lane The portion of the roadway adjoining the traveled way for parking, speed-change, or other purposes supplementary to through traffic movement.

Award The acceptance by the Contracting Authority of a bid, subject to execution and approval of the Contract.

Bid Schedule A listing of Contract items in the proposal form showing quantities and units of measurement that provides for the bidder to insert unit bid prices.

Bidder An individual, firm, or corporation submitting a Proposal for the advertised work.

Bridge A structure, including supports, erected over a depression or an obstruction, such as a water course, highway, or railway, and having a track or passageway for carrying traffic or other moving loads. Traffic or other moving loads are carried directly on the upper portion of the superstructure (called the bridge deck).

Brush Shrubs, trees, and other plant life having a diameter of 100 mm (4 inches) or less at a point 600 mm (approx. 24 inches) above ground surface as well as fallen trees and branches.

Calendar Day Every day shown on the calendar.

Carbonate Sedimentary rock composed primarily of carbonate minerals, including dolostone (dolomite, $CaMg(CO_3)_2$), limestone (calcite, $CaCO_3$), and mixtures of dolostone and limestone.

Certificate of Compliance A certification provided by a manufacturer, producer, or supplier of a product that the product, as furnished to the Contractor, complies with the pertinent specification or Contract requirements. The certification shall be signed by a person who is authorized to bind the company supplying the material covered by the certification.

Certified Test Report A test report provided by a manufacturer, producer, or supplier of a product indicating actual results of tests or analyses, covering elements of the specification requirements for the product or workmanship, and including validated certification.

Certified CCTV Technician An individual certified by the Contractor and approved by the Engineer to perform all work associated with a CCTV system.

Change Order A written order issued by the Engineer to the Contractor covering permissible adjustments, minor Plan changes or corrections, and rulings with respect to omissions, discrepancies, and intent of the Plans and Specifications, but not including any Extra Work or other alterations that are required to be covered by Supplemental Agreement.

Orders issued to implement changes made by mutual agreement shall not become effective until signed by the Contractor and returned to the Engineer.

Changed Condition The Contract clause (1402) that provides for adjustment of Contract terms for site conditions that differ from those in the Contract, for suspension of work ordered by the Department, and for significant changes in the character of the work.

City, Village, Township, Town, or Borough A subdivision of the county used to designate or identify the location of the proposed work.

Commissioner The Commissioner of the Minnesota Department of Transportation, or the chief executive of the department or agency constituted for administration of Contract work within its jurisdiction.

Contract The written agreement between the Contracting Authority and the Contractor setting forth their obligations, including, but not limited to, the performance of the work, the furnishing of labor and materials, the basis of payment, and other requirements contained in the Contract documents.

The Contract documents include the advertisement for bids, Proposal, Contract form, Contract bond, these Specifications, supplemental Specifications, Special Provisions, general and detailed Plans, notice to proceed, and orders and agreements that are required to complete the construction of the work in an acceptable manner, including authorized extensions, all of which constitute one instrument.

Contract Bond The approved form of security executed by the Contractor and Surety or Sureties, guaranteeing complete execution of the Contract and all Supplemental Agreements pertaining thereto and the payment of all legal debts pertaining to construction of the Project.

Contract Item (Pay Item) A specifically described unit of work for which a price is provided for in the Contract.

Contract Time The completion date, number of working days, or number of calendar days allowed for completion of the Contract, including authorized time extensions.

Completion date and calendar day Contracts shall be completed on or before the day indicated even where that date is a Saturday, Sunday, or holiday.

Contracting Authority The political subdivision, governmental body, board, department, commission, or officer making the award and execution of Contract as the party of the first part.

Contractor The individual, firm, or corporation contracting for and undertaking prosecution of the prescribed work; the party of the second part to the Contract, acting directly or through a duly authorized representative.

County The county in which the prescribed work is to be done; a subdivision of the State, acting through its duly elected Board of County Commissioners.

Culvert A structure constructed entirely below the elevation of the roadway surface and not a part of the roadway surface, which provides an opening under the roadway for the passage of water or traffic.

Department The Department of Transportation of the State of Minnesota, or the political subdivision, governmental body, board, commission, office, department, division, or agency constituted for administration of the Contract work within its jurisdiction.

Detour A road or system of roads, usually existing, designated as a temporary route by the Contracting Authority to divert through traffic from a section of roadway being improved.

Divided Highway A highway with separated traveled ways for traffic in opposite directions.

Dormant Seeding Seeding allowed in the late fall when the ground temperature is too low to cause seed germination so that the seed remains in a dormant condition until spring.

Dormant Sodding Sodding allowed in the late fall when the ground temperature is too low so that normal rooting does not take place until spring.

Easement A right acquired by public authority to use or control property for a designated highway purpose.

Engineer The duly authorized engineering representative of the Contracting Authority, acting directly or through the designated representatives who have been delegated responsibility for engineering supervision of the construction, each acting within the delegated scope of duties and authority.

Equipment All machinery, tools, and apparatus, together with the necessary supplies for upkeep and maintenance, necessary for the proper construction and acceptable completion of the Contract within its intended scope.

Erosion Control Schedule An oral commitment or written document by the Contractor illustrating construction sequences and proposed methods to control erosion.

Extra Work Any work not required by the Contract as awarded, but which is authorized and performed by Supplemental Agreement, either at negotiated prices or on a Force Account basis as provided elsewhere in these Specifications.

Frontage Road (Or Street) A local road or street auxiliary to and located on the side of a highway for service to abutting property and adjacent areas and for control of access.

Grade Separation A bridge with its approaches that provides for highway or pedestrian traffic to pass without interruption over or under a railway, highway, road, or street.

Gravel Naturally occurring rock or mineral particles produced by glacial and water action. Particle size ranges from 76 mm (3 inches) diameter to the size retained on a 2.0 mm diameter (#10 sieve).

Guaranteed Analysis A guarantee from a manufacturer, producer, or supplier of a product that the product complies with the ingredients or specifications indicated on the product label.

Highway, Street, or Road A general term denoting a public way for purposes of vehicular travel, including the entire area within the right of way.

Holidays The days of each year set aside by legal authority for public commemoration of special events, and on which no public business shall be transacted except as specifically provided in cases of necessity. Unless otherwise noted, holidays shall be as established in MS 645.44.

Incidental Whenever the word "incidental" is used in the plan or special provisions it shall mean no direct compensation will be made.

Industry Standard An acknowledged and acceptable measure of quantitative or qualitative value or an established procedure to be followed for a given operation within the given industry. This will generally be in the form of a written code, standard, or specification by a creditable association.

Inspector The Engineer's authorized representative assigned to make detailed inspections of Contract performance.

Interchange A grade-separated intersection with one or more turning roadways for travel between intersection legs.

Intersection The general area where two or more highways join or cross, within which are included the roadway and roadside facilities for traffic movements in the area.

Limestone See Carbonate.

Loop A one-way turning roadway that curves about 270 degrees to the right, primarily to accommodate a left-turning movement, but which may also include provisions for another turning movement.

Materials Any substances specified for use in the construction of the Project and its appurtenances.

Materials Laboratory The Mn/DOT Central Materials Laboratory and, for those tests so authorized, a Mn/DOT District Materials Laboratory.

Maximum Density The maximum density of a particular soil as determined by the method prescribed in the Mn/DOT Grading and Base Manual.

Metric The International System of Units (SI; the Modernized Metric System) according to ASTM E 380 is used in these Specifications. ASTM E 380 also provides conversion factors and commentary.

Metric Ton (T) A mass of 1,000 kg.

Minor Extra Work Extra Work ordered by the Engineer and required to complete the project as originally intended and authorized in writing, (Work Order/Minor Extra Work), as specified in 1403.

MGal 1,000 gallons.

Nominal The intended, named, or stated value, as opposed to the actual value. The nominal value of something is the value that it is supposed or intended to have, or the value by which it is commonly known. The actual value may differ from these statements by a greater or lesser amount depending on the accuracy and precision of the process used to determine the actual value.

Notice to Proceed Written notice to the Contractor to proceed with the Contract work including, when applicable, the date of beginning of Contract time.

Npdes Permit The general permit issued by the MPCA that authorized the discharge of storm water associated with construction activity under the National Pollutant Discharge Elimination System Program.

Optimum Moisture The moisture content of a particular soil at maximum dry density as determined by the method prescribed in the Mn/DOT Grading and Base Manual.

(P) A designation in the summary of quantities in the Plan meaning that the Plan quantity will be the quantity for payment. Measurement or recomputation will not be made except as provided in 1901.

Pay Item (Contract Item) A specifically described unit of work for which a price is provided for in the Contract.

Pavement Structure The combination of subbase, base course, and surface course placed on a subgrade to support the traffic load and distribute it to the roadbed.

Permanent Erosion Control Measures Soil-erosion control measures such as curbing, culvert aprons, riprap, flumes, sodding, erosion mats, and other means to permanently minimize erosion on the completed Project.

Plan The Plan, profiles, typical cross-sections, and supplemental drawings that show the locations, character, dimensions, and details of the work to be done.

Plan Quantity The quantity listed in the summary of quantities in the Plan. The summary of quantities will usually be titled Statement of Estimated Quantities, Schedule of Quantities for Entire Bridge, or Schedule of Quantities.

Profile Grade The trace of a vertical plane intersecting the top surface of the roadbed or pavement structure, usually along the longitudinal centerline of the traveled way. Profile grade means either elevation or gradient of such trace according to the context.

Project The specific section of the highway, the location, or the type of work together with all appurtenances and construction to be performed under the Contract.

Proposal The offer of a bidder on the prescribed Proposal form to perform the work and furnish the labor and materials at the prices quoted.

Proposal Form The approved form on which the Contracting Authority requires bids to be prepared and submitted for the work.

Proposal Guaranty The security furnished with a bid to guarantee that the bidder will enter into the Contract if the bid is accepted.

Pure Live Seed (Percentage) A percentage determined by the percent of seed germination times the percent of seed purity.

Quality Assurance (QA) The activities performed by the Department that have to do with making sure the quality of a product is what it should be.

Quality Compaction A compaction method as defined in 2105.3F2.

Quality Control (QC) The activities performed by the Contractor that have to do with making the quality of a product what it should be.

Questionnaire The specified forms on which a bidder may be required to furnish information as to ability to perform and finance the work.

Ramp A connecting roadway for travel between intersection legs at or leading to an interchange.

Right of Way A general term denoting land, property, or interest therein, usually in a strip, acquired for or devoted to a highway.

Road A general term denoting a public way for purposes of vehicular travel, including the entire area within the right of way.

Roadbed The graded portion of a highway within top and side slopes, prepared as a foundation for the pavement structure and shoulders.

Roadway The portion of a highway within limits of construction.

Scale A device used to measure the mass or the proportion of a liquid or solid. This definition includes metering devices.

Shoulder The portion of the roadway contiguous with the traveled way for accommodation of stopped vehicles, for emergency use, and for lateral support of the base and surface courses.

Sidewalk That portion of the roadway primarily constructed for the use of pedestrians.

Sieve A woven wire screen meeting the requirements of AASHTO M-92 for the size specified.

Special Provisions Additions and revisions to the standard and supplemental Specifications covering conditions peculiar to an individual Project.

Specifications A general term applied to all directions, provisions, and requirements pertaining to performance of the work.

Specified Completion Date The date on which the Contract work is specified to be completed.

Specimen Tree Historic or otherwise significant tree indicated in the Contract or determined by the Engineer.

State The State of Minnesota acting through its elected officials and their authorized representatives.

Street A general term denoting a public way for purposes of vehicular travel, including the entire area within the right of way.

Structures Bridges, culverts, catch basins, drop inlets, retaining walls, cribbing, manholes, endwalls, buildings, sewers, service pipes, underdrains, foundation drains, and other man-made features.

Subcontractor An individual, firm, or corporation to whom the Contractor sublets part of the Contract.

Subgrade The top surface of a roadbed upon which the pavement structure and shoulders are constructed. Also, a general term denoting the foundation upon which a base course, surface course, or other construction is to be placed, in which case reference to subgrade operations may imply depth as well as top surface.

Substructure The part of a bridge below the bearings of simple and continuous spans, skewbacks, or arches and tops of footings for rigid frames, together with the backwalls, wingwalls, and wing protection railings.

Superintendent The Contractor's authorized representative in responsible charge of the work.

Superstructure The entire bridge except the substructure.

Supplemental Agreement A written agreement between the Contracting Authority and the Contractor, executed on the prescribed form and approved as required by law, covering the performance of Extra Work or other alterations or adjustments as provided for within the general scope of the Contract, but which Extra Work or Change Order constitutes a modification of the Contract as originally executed and approved.

Supplemental Drawings An approved set of drawings consisting of standard plates or plans showing the details of design and construction for various structures and products for which standards have been developed. These standard plates and plans shall govern by reference as identified and supplemented or amended in the general Plans and Specifications.

Supplemental Specifications Additions and revisions to the standard Specifications that are approved subsequent to issuance of the printed book of standard Specifications.

Surety The Corporation, partnership, or individual, other than the Contractor, executing a bond furnished by the Contractor.

Temporary By-Pass A section of roadway, usually within an existing right of way, provided to temporarily carry all traffic around a specific work site.

Temporary Erosion Control Measures Soil-erosion control measures such as bale checks, silt curtains, sediment traps, and other means to temporarily protect the Project from erosion before and during the installation of permanent erosion control measures. Temporary erosion control measures may also be used to supplement the permanent measures.

Traffic Lane The portion of a traveled way for the movement of a single line of vehicles.

Traveled Way The portion of the roadway for the movement of vehicles, exclusive of shoulders and auxiliary lanes.

Turn Lane An auxiliary lane for left or right turning vehicles.

Work The furnishing of all labor, materials, equipment, and other incidentals necessary or convenient to the successful completion of the Project and the carrying out of all the duties and obligations imposed by the Contract upon the Contractor. Also used to indicate the construction required or completed by the Contractor.

Working Day A calendar day, exclusive of Saturdays, Sundays, and State recognized legal holidays, on which weather and other conditions not under the control of the Contractor will permit construction operations to proceed for at least 2 hours, with the normal working force engaged in performing the progress-controlling operations.

Working Drawings Stress sheets, shop drawings, erection plans, falsework plans, framework plans, cofferdam plans, bending diagrams for reinforcing steel, or any other supplementary plans or similar data that the Contractor is required to furnish and submit to the Engineer.

Work Order A written order signed by the Engineer of a contractual status requiring performance or other action by the Contractor without negotiation of any sort.

Work Order/Minor Extra Work A written order signed by the Engineer requiring performance of Minor Extra Work.

APPENDIX

VIII

Industry Resources

American Concrete Institute
William Tolley, Executive Vice President
P.O. Box 9094
Farmington Hills, MI 48333
Tel: (248) 848-3800; Fax: 248/848-3801

American Concrete Pavement Association
Jerry Voigt, President
5420 Old Orchard Road
Skokie, IL 60077
Tel: (847) 966-2272; Fax: 847/966-9970

American Concrete Pipe Association
John J. Duffy, President
222 W. Las Colinas Blvd., Suite 641
Irving, TX 75039-5423
Tel: (972) 506-7216; Fax: 972/506-7682

American Concrete Pressure Pipe Association
David Prosser, President
11800 Sunrise Valley Drive, #309
Reston, VA 20191
Tel: (703) 391-9135; Fax: 703/391-9136

American Concrete Pumping Association
Christi E. Collins
676 Enterprise Dr., Suite B
Lewis Center, OH 43035
Tel: (614) 431-5618; Fax: 614/431-6944

American Segmental Bridge Institute
Clifford L. Freyermuth, Manager
9201 N. 25th Ave., Suite 150B
Phoenix, AZ 85021
Tel: (602) 997-9931; Fax: 602/997-9965

American Shotcrete Association
Thomas H. Adams, Executive Director
38800 Country Club Drive
Farmington Hills, MI 48331
Tel: (248) 848-3780; Fax: 248/848-3740

American Society of Concrete Contractors
Bev Garnant, Executive Director
2025 S. Brentwood Blvd.
St. Louis, MO 63144
Tel: (314) 962-0210; Fax: 314/968-4367

Architectural Precast Association
Fred L. McGee, Executive Director
P.O. Box 08669
16521 San Carlos Blvd., Ste. H
Ft. Myers, FL 33908-0669
Tel: (941) 454-6989; Fax: 941/454-6787

Brick Industry Association
Nelson J. Cooney, President
11490 Commerce Park Drive
Reston, VA 22091
Tel: (703) 620-0010; Fax: 703/620-3928

Cement Association of Canada
Francois Lacroix, President
60 Queen Street, Suite 206
Ottawa, Ontario K1P 5Y7
Tel: (613) 236-9471; Fax: 613/563-4498

Cement Kiln Recycling Coalition
Mike Benoit, President
1225 Eye Street, N.W., Suite 300
Washington, DC 20005
Tel: (202) 789-1948; Fax: 202/408-0877

Concrete Foundations Association
Ed Sauter, Executive Director
P.O. Box 204
Mount Vernon, IA 52314
Tel: (319) 895-6940; Fax: 319/895-8830

Concrete Reinforcing Steel Institute
Robert J. Risser, Jr., President
933 North Plum Grove Road
Schaumburg, IL 60173
Tel: (847) 517-1200; Fax: 847/517-1206

Concrete Sawing and Drilling Association
Patrick O'Brien, Executive Director
11001 Danka Way North, Suite 1
St. Petersburg, FL 33716
Tel: (727) 577-5004; Fax: 727/577-5012
Web Site: www.csda.org
Email: pat@csda.org

Construction Industry Manufacturers Association
Jim Stollenwerk, President
111 E. Wisconsin Avenue, Ste. 940
Milwaukee, WI 53202-4879
Tel: (414) 272-0943; Fax: 414/272-1170

Expanded Shale, Clay and Slate Institute
John P. Ries, Managing Director
2225 E. Murray Holladay Road, Suite 102
Salt Lake City, UT 84117
Tel: (801) 272-7070; Fax 801/272-3377

Federacion Interamericana del Cemento
Callao No. 2970, Of.
316 Las Condes
Santiago, Chile
Tel: (011) 56-2-233-3977

Insulating Concrete Form Association
960 Harlem Avenue, #112
Glenview, IL 60025
Tel: (847) 657-9730; Fax: 847/657-9728

Interlocking Concrete Pavement Institute
Charles A. McGrath
1444 I Street, NW, Suite 700
Washington, DC 20005
Tel: (202) 712-9036; Fax: 202/408-0285

International Concrete Repair Institute
Kelly Page, Executive Director
3166 S. River Road
Suite 132
Des Plaines, IL 60018
Tel: (847) 827-0830; Fax: 847/827-0832

International Grooving & Grinding Association
12573 Route 9W
West Coxsackie, NY 12192
Tel: (518) 731-7450; Fax: 518/731-7490

Mason Contractors Association of America
Michael M. Adelizzi, Executive Director
1910 South Highland Ave., Ste. 101
Lombard, IL 60148
Tel: (630) 705-4200; Fax: 630/705-4209

National Concrete Masonry Association
Mark B. Hogan, President
13750 Sunrise Valley Drive
Herndon, VA 20171-4662
Tel: (703) 713-1900; Fax: 703/713-1910
Web Site: www.ncma.org

National Precast Concrete Association
Ty E. Gable, President/CEO
10333 N. Meridian, Suite 272
Indianapolis, IN 46290
Tel: (317) 571-9500; Fax: 317/571-0041

National Ready Mixed Concrete Association
Robert Garbini, Executive Vice President and Chief Operating Officer
900 Spring St.
Silver Spring, MD 20910
Tel: (301) 587-1400; Fax: 301/585-4219

National Slag Association
Terry Wagaman, President
25 Stevens Avenue
Building A
West Lawn, PA 19609
Tel: (610) 670-0701; Fax: 610/670-0702

National Stone, Sand & Gravel Association
Jennifer Joy Wilson, President
2101 Wilson Blvd., Ste. 100
Arlington, VA 22201
Tel: (703) 525-8788; Fax: 703/525-7782

Post-Tensioning Institute
Theodore L. Neff, Executive Director
1717 W. Northern Avenue, Suite 218
Phoenix, AZ 85021
Tel: (602) 870-7540; Fax: 602/870-7541

Precast/Prestressed Concrete Institute
James Toscas, President
209 W. Jackson Blvd. Suite 500
Chicago, IL 60606-6938
Tel: (312) 786-0300; Fax: 312/786-0353

The Masonry Society
Phil Samblanet, Executive Director
3970 Broadway, Suite 201-D
Boulder, CO 80304
Tel: (303) 939-9700; Fax: 312/786-0353

Tilt-Up Concrete Association
Ed Sauter, Executive Director
P.O. Box 204
Mount Vernon, IA 52314
Tel: (319) 895-6911; Fax: 319/895-8830

TRIP (The Road Information Program)
Lisa Templeton, Director of Development
1726 M Street, NW
Suite 401
Washington, DC 20036
Tel: (202) 466-6706; Fax: 202/785-4722

Wire Reinforcement Institute, Inc.
Roy H. Reiterman, P.E., Technical Director
P.O. Box 450
301 E. Sandusky St.
Findlay, OH 45839-0450
Tel: (419) 425-9473; Fax: 419/425-5741

GLOSSARY

Abrasion resistance Ability of a surface to resist being worn away by rubbing and friction.

Acrylic resin One of a group of thermoplastic resins formed by polymerizing the esters or amides of acrylic acid; used in concrete construction as a bonding agent or surface sealer.

Adhesives The group of materials used to join or bond similar or dissimilar materials; for example, in concrete work, the epoxy resins.

Air-water jet A high velocity jet of air and water mixed at the nozzle; used in cleanup of surfaces of rock or concrete such as horizontal construction joints.

Alkali–aggregate reaction Chemical reaction in mortar or concrete between alkalis (sodium and potassium) from Portland cement or other sources and certain constituents of some aggregates; under certain conditions, deleterious expansion of the concrete or mortar may result.

Alkali-carbonate rock reaction The reaction between alkalis (sodium and potassium) from Portland cement and certain carbonate rocks, particularly calcitic dolomite and dolomitic limestones, present in some aggregates; the products of the reaction may cause abnormal expansion and cracking of concrete in service.

Alkali reactivity (of aggregate) Susceptibility of aggregate to alkali–aggregate reaction.

Alkali-silica reaction The reaction between the alkalis (sodium and potassium) in Portland cement and certain siliceous rocks or minerals, such as opaline chert and acidic volcanic glass, present in some aggregates; the products of the reaction may cause abnormal expansion and cracking of concrete in service.

Autogenous healing A natural process of closing and filling of cracks in concrete or mortar when kept damp.

Bacterial corrosion The destruction of a material by chemical processes brought about by the activity of certain bacteria which may produce substances such as hydrogen sulfide, ammonia, and sulfuric acid.

Blistering The irregular raising of a thin layer at the surface of placed mortar or concrete during or soon after completion of the finishing operation, or in the case of pipe after spinning; also bulging of the finish plaster coat as it separates and draws away from the base coat.

Bug holes Small regular or irregular cavities, usually not exceeding 15 mm in diameter, resulting from entrapment of air bubbles in the surface of formed concrete during placement and compaction.

Butyl stearate A colorless oleaginous, practically odorless material ($C_{17}H_{35}COOC_4H_9$) used as an admixture for concrete to provide dampproofing.

Cavitation damage Pitting of concrete caused by implosion; i.e., the collapse of vapor bubbles in flowing water which form in areas of low pressure and collapse as they enter areas of higher pressure.

Chalking Formation of a loose powder resulting from the disintegration of the surface of concrete or an applied coating such as cement paint.

Checking Development of shallow cracks at closely spaced, but irregular, intervals on the surface of plaster, cement paste, mortar, or concrete.

Cold-joint lines Visible lines on the surface of formed concrete indicating the presence of joints where one layer of concrete had hardened before subsequent concrete was placed.

Concrete, preplaced-aggregate Concrete produced by placing coarse aggregate in a form and later injecting a Portland-cement-sand grout, usually with admixtures, to fill the voids.

Corrosion Destruction of metal by chemical, electrochemical, or electrolytic reaction with its environment.

Cracks, active The cracks for which the mechanism causing the cracking is still at work. Any crack that is still moving.

Cracks, dormant Those cracks not currently moving or which the movement is of such magnitude that the repair material will not be affected.

Craze cracks Fine, random cracks or fissures in a surface of plaster, cement paste, or mortar.

Crazing The development of craze cracks; the pattern of craze cracks existing in a surface. (See also Checking.)

Dampproofing Treatment of concrete or mortar to retard the passage or absorption of water or water vapor, either by application of a suitable coating to exposed surfaces or by use of a suitable admixture, treated cement, or preformed films such as polyethylene sheets under slabs on grade. (See also Vapor barrier.)
D-cracking A series of cracks in concrete near and roughly parallel to joints, edges, and structural cracks.
Delamination A separation along a plane parallel to a surface as in the separation of a coating from a substrate or the layers of a coating from each other, or in the case of a concrete slab, a horizontal splitting, cracking, or separation of a slab in a plane roughly parallel to, and generally near, the upper surface; found most frequently in bridge decks and caused by the corrosion of reinforcing steel or freezing and thawing; similar to spalling, scaling, or peeling except that delamination affects large areas and can often be detected only by tapping.
Deterioration Decomposition of material during testing or exposure to service. (See also Disintegration.)
Diagonal crack In a flexural member, an incline's crack caused by shear stress, usually at about 45 degrees to the neutral axis of a concrete member; a crack in a slab, not parallel to the lateral or longitudinal directions.
Discoloration Departure of color from that which is normal or desired.
Disintegration Reduction into small fragments and subsequently into particles.
Dry-mix shotcrete Shotcrete in which most of the mixing water is added at the nozzle.
Drypacking Placing a zero slump concrete, mortar, or grout by ramming it into a confined space.
Durability The ability of concrete to resist weathering action, chemical attack, abrasion, and other conditions of service.
Dusting The development of a powdered material at the surface of a hardened concrete.
Efflorescence A deposit of salts, usually white, formed on a surface, the substance having emerged in solution from within concrete or masonry and subsequently having been precipitated by evaporation.
Epoxy concrete A mixture of epoxy resin, catalyst, and fine aggregate. (See also Epoxy resin.)
Epoxy resin A class of organic chemical bonding systems used in the preparation of special coatings or adhesives for concrete or as binders in epoxy resin mortars and concretes.
Erosion Progressive disintegration of a solid by the abrasive cavitation action of gases, fluids, or solids in motion. (See also Abrasion resistance and Cavitation damage.)
Evaluation Determining the condition, degree of damage or deterioration, or serviceability and, when appropriate, indicating the need for repair, maintenance, or rehabilitation. (See also Repair, Maintenance, and Rehabilitation.)
Exfoliation Disintegration occurring by peeling off in successive layers; swelling up and opening into leaves or plates like a partly opened book.
Exudation A liquid or viscous gel-like material discharge through a pore, crack, or opening in the surface of concrete.
Feather edge Edge of a concrete or mortar patch or topping that is beveled at an acute angle.
Groove joint A joint created by forming a groove in the surface of a pavement, floor slab, or wall to control random cracking.
Hairline cracks Cracks in an exposed concrete surface having widths so small as to be barely perceptible.
Honeycomb Voids left in concrete due to failure of the mortar to effectively fill the spaces among coarse aggregate materials.
Incrustation A crust or coating, generally hard, formed on the surface of concrete or masonry construction or on aggregate particles.
Joint filler Compressible material used to fill a joint to prevent infiltration of debris and to provide support for sealants.
Joint sealant Compressible material used to exclude water and solid foreign material from joints.
Laitance A layer of weak and nondurable material containing cement and fines from aggregates, brought by bleeding water to the top of overwet concrete, the amount of which is generally increased by overworking or overmanipulating concrete at the surface by improper finishing or by job traffic.
Latex A water emulsion of a high molecular weight polymer used especially in coatings, adhesives, and leveling and patching compounds.
Maintenance Taking periodic actions that will prevent or delay damage or deterioration or both. (See also Repair.)
Map cracking See Crazing.
Microcracks Microscopic cracks within concrete.
Monomer An organic liquid of relatively low molecular weight that creates a solid polymer by reacting with itself or other compounds of low molecular weight or with both.

Overlay A layer of concrete or mortar, seldom thinner than 25 mm (1 in.), placed on and usually bonded onto the worn or cracked surface of a concrete slab to restore or improve the function of the previous surface.

Pattern cracking Intersecting cracks that extend below the surface of hardened concrete; caused by shrinkage of the drying surface which is restrained by concrete at a greater depth where little or no shrinkage occurs; varies in width and depth from fine and barely visible to open and well defined.

Peeling A process in which thin flakes of mortar are broken away from a concrete surface, such as by deterioration or by adherence of surface mortar to forms as forms are removed.

Pitting Development of relatively small cavities in a surface caused by a phenomena such as corrosion or cavitation, or in concrete localized disintegration such as popout.

Plastic cracking Cracking that occurs in the surface of fresh concrete soon after it is placed and while it is still plastic.

Polyester One of a large group of synthetic resins, mainly produced by a reaction of dibasic acids with dihydroxyl alcohols, commonly prepared for application by mixing with a vinyl-group monomer and free-radical catalyst at ambient temperatures and used as bonders for resin mortars and concretes, fiber laminates (mainly glass), adhesives, and the like. (See also Polymer concrete.)

Polyethylene A thermoplastic high molecular weight organic compound used in formulating protective coatings; in sheet form, used as a protective cover for concrete surfaces during the curing period, or to provide a temporary enclosure for construction operations.

Polymer The product of polymerization; more commonly, a rubber or resin consisting of large molecules formed by polymerization.

Polymer concrete Concrete in which an organic polymer serves as the binder; also known as resin concrete; sometimes erroneously employed to designated hydraulic-cement mortars or concretes in which part or all of the mixing water is replaced by an aqueous dispersion of thermoplastic copolymer.

Polymer-cement concrete A mixture of water, hydraulic cement, aggregate, and a monomer or polymer polymerized in place when a monomer is used.

Polymerization The reaction in which two or more molecules of the same substance combine to form a compound containing the same elements in the same proportions, but of higher molecular weight, from which the original substance can be generated, in some cases only with extreme difficulty.

Polystyrene resin Synthetic resins varying in color from colorless to yellow formed by the polymerization of styrene, or heated, with or without catalysts, that may be used in paints for concrete or for making sculpted molds or as insulation.

Polysulfide coating A protective coating system prepared by polymerizing a chlorinated alkypolyether with an inorganic polysulfide.

Polyurethane Reaction product of an isocyanate with any of a wide variety of other compounds containing an active hydrogen group; used to formulate tough, abrasion-resistant coatings.

Polyvinyl acetate Colorless, permanently thermoplastic resin, usually supplied as an emulsion or water-dispersible powder characterized by flexibility, stability toward light, transparency to ultraviolet rays, high dielectric strength, toughness, and hardness; the higher the degree of polymerization, the higher softening temperature; may be used in paints for concrete.

Polyvinyl chloride A synthetic resin prepared by the polymerization of vinyl chloride; used in the manufacture of non-metallic water-stops for concrete.

Popout The breaking away of small portions of concrete surface due to internal pressure, which leaves a shallow, typically conical, depression.

Pot life Time interval after preparation during which a liquid or plastic mixture is usable.

Reactive aggregate Aggregate containing substances capable of reacting chemically with products of solution or hydration of the Portland cement in concrete or mortar under ordinary conditions of exposure, resulting in some cases in harmful expansion, cracking, or staining.

Rebound hammer An apparatus that provides a rapid indication of the mechanical properties of concrete based on the distance of rebound of a spring-driven missile.

Rehabilitation The process of repairing or modifying a structure to a desired useful condition.

Repair Replace or correct deteriorated, damaged, or faulty materials components, or elements of a structure.

Resin A natural or synthetic, solid or semisolid organic material of indefinite and often high molecular weight having a tendency to flow under stress that usually has a softening or melting range and usually fractures conchoidally.

Resin mortar (or concrete) See Polymer concrete.

Restraint (of concrete) Restriction of free movement of fresh or hardened concrete following completion of placement in framework or molds within an otherwise confined space; restraint can be internal or external and may act in one or more directions.

Rock pocket A porous, mortar-deficient portion of hardened concrete consisting primarily of coarse aggregate and open voids, caused by leakage of mortar from form, separation (segregation) during placement, or insufficient consolidation. (See also Honeycomb.)

Sandblasting A system of cutting or abrading a surface such as concrete by a stream of sand ejected from a nozzle at high speed by compressed air; often used for cleanup of horizontal construction joints or for exposure of aggregate in architectural concrete.

Sand streak A streak of exposed fine aggregate in the surface of formed concrete that is caused by bleeding.

Scaling Local flaking or peeling away of the near-surface portion of hardened concrete or mortar; also of a layer from metal. (See also Peeling and Spalling.)

Shotcrete Mortar or concrete pneumatically projected at high velocity onto a surface; also known as air-blown mortar; pneumatically applied mortar or concrete, sprayed mortar, and gunned concrete. (See also Dry-mix shotcrete and Wet-mixing shotcrete.)

Shrinkage Volume decrease caused by drying and chemical changes; a function of time but not temperature or stress caused by external load.

Shrinkage crack Crack due to restraint of shrinkage.

Shrinkage cracking Cracking of a structure or member from failure in tension caused by external or internal restraints as reduction in moisture content develops or as carbonation occurs, or both.

Spall A fragment, usually in the shape of a flake, detached from a larger mass by a blow, section of weather, pressure, or expansion within the larger mass; a small spall involves a roughly circular depression not greater than 20 mm in depth nor 150 mm in any dimension; a large spall may be roughly circular or oval or, in some cases, elongated more than 20 mm in depth and 150 mm in greatest dimension.

Spalling The development of spalls.

Sulfate attack Chemical or physical reaction, or both, between sulfates, usually in soil or in ground water and concrete or mortar, primarily with calcium aluminate hydrates in the cement-paste matrix, often causing deterioration.

Sulfate resistance Ability of concrete or mortar to withstand sulfate attack. (See also Sulfate attack.)

Swiss hammer (See also Rebound hammer.)

Temperature cracking Cracking as a result of tensile failure caused by temperature drop in members subjected to internal restraints.

Thermal shock The subjection of newly hardened concrete to a rapid change in temperature, which may be expected to have a potentially deleterious effect.

Thermoplastic Becoming soft when heated and hard when cooled.

Thermosetting Becoming rigid by chemical reaction and not meltable.

Transverse cracks Cracks that develop at right angles to the long direction of a member.

Tremie A pipe or tube through which concrete is deposited underwater, having at its upper end a hopper for filling and bail for moving the assemblage.

Tremie concrete Subaqueous concrete placed by means of a tremie.

Tremie seal The depth to which the discharge end of the tremie pipe is kept embedded in the fresh concrete placed in a cofferdam for the purpose of preventing the intrusion of water when the cofferdam is dewatered.

Vapor barrier A membrane placed under concrete floor slabs that are placed on grade and intended to retard transmission of water vapor.

Waterstop A thin sheet of metal, rubber, plastic, or other material inserted across a joint to obstruct seepage of water through a joint.

Water void Void along the underside of an aggregate particle or reinforcing steel that formed during the bleeding period and initially filled with bleed water.

Weathering Changes in color, texture, strength, chemical composition, or other properties of a natural or artificial material caused by the action of weather.

Wet-mixing shotcrete Shotcrete in which ingredients, including mixing water, are mixed before introduction into the delivery hose; accelerator, if used, is normally added at the nozzle.

ABBREVIATIONS

ACI	American Concrete Institute
ASTM	American Society for Testing and Materials
AWA	anti-washout mixture
CDMS	Continuous Deformation Monitoring System
CE	Corps of Engineers
CERC	Coastal Engineering Research Center
CEWES-SC	U.S. Army Engineer Waterways Experiment Station, Structures Laboratory, Concrete Technology Division
CMU	concrete masonry units
CRD	Concrete Research Division, Handbook for Concrete and Cement
EM	Engineer Manual
EP	Engineer Pamphlet
ER	Engineer Regulation
FHWA	Federal Highway Administration
GPS	Global Positioning System
HAC	high alumina cement
HMWM	high molecular weight methacrylate
HQUSACE	Headquarters, U.S. Army Corps of Engineers
HRWRA	high-range water-reducing admixture
ICOLD	International Commission on Large Dams
MPC	magnesium phosphate cement
MSA	maximum size aggregate
MSDS	Manufacturer's Safety Data Sheet
NACE	National Association of Corrosion Engineers
NCHRP	National Cooperative Highway Research Program
NDT	nondestructive testing
OCE	Office, Chief of Engineers
PC	polymer concrete
PCA	Portland Cement Association

PIC	polymer-impregnated concrete
PMF	Probable Maximum Flood
PVC	polyvinyl chloride
R-values	rebound readings
RCC	roller compacted concrete
REMR	Repair, Evaluation, Maintenance and Rehabilitation Research Program
ROV	remotely operated vehicle
TM	technical manual
TOA	time of arrival
UPE	ultrasonic pulse-echo
UV	ultraviolet
WES	Waterways Experiment Station
WRA	water-reducing admixture

Index

Page numbers followed by *f* indicates a figure and *t* indicates a table.

D

M

N

O

Printed and bound by CPI Group (UK) Ltd, Croydon, CR0 4YY
08/06/2025
01896872-0012